CRC Series in Chromatography

Editors-in-Chief

Gunter Zweig, Ph.D. and Joseph Sherma, Ph.D.

General Data and Principles
Gunter Zweig, Ph.D. and
Joseph Sherma, Ph.D.

Carbohydrates
Shirley C. Churms, Ph.D.

Drugs
Ram Gupta, Ph.D.

Terpenoids
Carmine J. Coscia, Ph.D.

Steroids
Joseph C. Touchstone, Ph.D.

Pesticides and Related Organic Chemicals
Joseph Sherma, Ph.D. and
Joanne Follweiler, Ph.D.

Lipids
Helmut K. Mangold, Dr. rer. nat.

Hydrocarbons
Walter L. Zielinski, Jr., Ph.D.

Inorganics
M. Qureshi, Ph.D.

Phenols and Organic Acids
Toshihiko Hanai, Ph.D.

Amino Acids and Amines
S. Blackburn, Ph.D.

Polymers
Charles G. Smith,
Norman E. Skelly, Ph.D.,
Carl D. Chow, and Richard A. Solomon

Plant Pigments
Hans-Peter Köst, Ph.D.

Nucleic Acids and Related Compounds
Ante M. Krstulovic, Ph.D.

CRC Handbook of Chromatography: Drugs

Volume IV

Editor
Ram N. Gupta, Ph.D.
Professor
Department of Pathology
McMaster University
and
Head, Toxicology
Department of Laboratory Medicine
St. Joseph's Hospital
Hamilton, Ontario, Canada

Editors-in-Chief

Gunter Zweig, Ph.D.
President
Zweig Associates
Arlington, Virginia
(Deceased)

Joseph P. Sherma, Ph.D.
Professor of Chemistry
Lafayette College
Easton, Pennsylvania

CRC Press
Taylor & Francis Group
Boca Raton London New York

CRC Press is an imprint of the
Taylor & Francis Group, an informa business

First published 1989 by CRC Press
Taylor & Francis Group
6000 Broken Sound Parkway NW, Suite 300
Boca Raton, FL 33487-2742

Reissued 2018 by CRC Press

ISBN 13: 978-1-315-89193-4 (hbk)
ISBN 13: 978-1-351-07103-1 (ebk)

Visit the Taylor & Francis Web site at http://www.taylorandfrancis.com and the
CRC Press Web site at http://www.crcpress.com

CRC SERIES IN CHROMATOGRAPHY

SERIES PREFACE

The fat-soluble photosynthetic pigments present in plants and algae, including chlorophylls, carotenoids, and related pigments, comprise an important class of compounds with an extensive literature. Dr. Köst and his co-authors have done an admirable job in searching out and organizing much of the critical chromatographic data and methodology in the present volume.

Because of the chemical nature of these prenyllipid compounds, liquid chromatography is preferred for their isolation, separation, and determination. The most widely used methods include low pressure column LC, paper chromatography, TLC, and, most recently, HPLC. All of these methods are covered by Dr. Köst.

Chromatography was "invented" in the early 1900s by Michael Tswett, a Russian botanist and plant physiologist who first applied liquid-solid chromatography on a column of chalk to resolution of the complex natural mixture of yellow and green chloroplast pigments in the extracts of leaves he was studying. On a personal note, I was fortunate to work with Dr. Harold Strain for five summers at the Argonne National Laboratory when I first began to teach. Dr. Strain was one of the first important American chromatography experts and used all variations of liquid chromatography extensively in his studies of photosynthetic pigments. My experience with Dr. Strain set the foundation for my lifelong career of research and writing in chromatography.

Readers of this Handbook are asked to contact the Series Editor if they find errors or omissions in coverage as well as with suggestions for future volumes and authors within the Handbook of Chromatography series.

Joseph Sherma

PREFACE

The phenomenal growth in the application of liquid chromatographic (LC) techniques for the determination of drugs in pharmaceutical preparations and in biological fluids has continued in the first half of this decade. In the mid to late 1970s, a large number of papers were published describing gas chromatographic (GC) procedures for the drug groups anticonvulsants and antidepressants. In the last few years, a large number of publications have appeared describing LC procedures for the same drugs.

There have been a number of improvements in the LC instrumentation and column technology. Variable wavelength absorbance detectors are now available which match fixed wavelength detectors in sensitivity. Some of these detectors allow monitoring of absorbance at multiple wavelengths. A number of manufacturers now market photo diode array absorbance detectors which allow instant absorbance scanning over a wide wavelength range of any eluting peak, check peak purity, and complement component separation by mathematical manipulation of absorbance data of incomplete chromatographic separation. There are also improvements in the design of electrochemical detectors. Multielectrode detectors are now available which require little maintenance and allow ultra-high sensitivity. Fluorescence detectors with monochromators and high energy power sources have also become available. However, use of lasers as power sources for fluorescence detectors is not yet common.

A number of manufacturers market on-line sample preparation systems, samplers allowing precolumn derivatization, and efficient postcolumn reactors. Although instrumentation for narrow bore LC is commercially available, this technique has not yet been commonly applied for the determination of drugs. There is improved quality control in the manufacture of columns. For convenience, cartridge-type columns and fittings requiring no tools have become available. Good quality silica-based columns can now be purchased at economical proces from general suppliers. However, polymer-based columns have failed to gain popularity and are relatively more expensive than silica-based columns because of limited sales.

In general, GC is now the preferred technique only when the required sensitivity is not available with an LC procedure for the determination of a particular drug. However, separation of widely different compounds is more efficiently accomplished with temperature programming GC analysis than by solvent programming LC analysis. Thus, considerable GC retention data of drugs with the use of capillary columns have been published for the identification of an unknown drug in a given matrix. However, the use of capillary columns for the determination of drugs has not been as widespread as was anticipated. The nitrogen detector is now the most widely used detector for the GC determination of drugs. Gas chromatography-mass spectrometry remains the ultimate standard to confirm the identification of an unknown drug.

There has been a further decline in the popularity of thin-layer chromatography in the past few years. In the majority of the laboratories, drugs of abuse are now screened by immunoassay for improved sensitivity and convenience.

The purpose of this handbook is to provide a reference source and summaries of different chromatographic techniques published in refereed journals during the past 6 years. When the number of publications of a given drug was numerous, only recent papers were selected, even if they described only the modification of the original key publications. Despite the size of this work, a number of publications or drugs might have been missed as the literature search was carried out manually. In some cases, copies of the required papers could not be obtained. A number of publications could not be included as they were either theoretical or did not provide information compatible with the format of this handbook.

There is a significant difference between the present volumes and Volumes I and II of this handbook. For a number of drugs, e.g., cyclosporine, the chromatographic parameters of a number of publications are identical. However, they differ in the sample preparation techniques.

Therefore, detailed summaries of extraction procedures have now been provided for comparison of the different publications for the determinations of a given drug.

I am grateful to Dr. Gillian Luxton, Head of the Clinical Chemistry Laboratory, St. Joseph's Hospital for her encouragement to accept this project and for providing all the required facilities.

Mrs. S. Rogers and Mrs. J. Maragno of this hospital library made a special effort to get copies of the published papers from different sources.

Mrs. D. Thompson, Director of the hospital pharmacy, arranged to get information from the Drug Information Center in Toronto.

Miss Maelly Lew went to different libraries to get the information in emergency situations when a paper under review would refer to earlier papers.

I thank Miss Elisa Capretta, Mrs. Mary Bruce, Miss Rhita Gilners, and Miss Abha Gupta for preparing this manuscript.

Mrs. Diane Kirshenblat provided moral support when there was a temptation to abandon the project.

I am grateful to Ms. Sandy Pearlman, Director of Editing and Mrs. Amy Skallerup, Senior Editor, CRC Press for their help during the early phases of manuscript preparation.

Mr. J. C. Richardson, Senior Coordinating Editor, had the difficult task of making this manuscript uniform within the constraints of space limitations. I thank him for his courteous response to my various suggestions and changes.

Finally, I thank my family members, who tolerated my absence for more than a year.

Ram N. Gupta
December, 1986

THE EDITORS-IN-CHIEF

Gunter Zweig, Ph.D., received his undergraduate training at the University of Maryland, College Park, where he was awarded the Ph.D. in biochemistry in 1952. Two years following his graduation, Dr. Zweig was affiliated with the late R. J. Block, pioneer in paper chromatography of amino acids. Zweig, Block, and Le Strange wrote one of the first books on paper chromatography, which was published in 1952 by Academic Press and went into three editions, the last one authored by Gunter Zweig and Dr. Joe Sherma, the co-Editor-in-Chief of this series. *Paper Chromatography* (1952) was also translated into Russian.

From 1953 to 1957, Dr. Zweig was research biochemist at the C. F. Kettering Foundation, Antioch College, Yellow Springs, Ohio, where he pursued research on the path of carbon and sulfur in plants, using the then newly developed techniques of autoradiography and paper chromatography. From 1957 to 1965, Dr. Zweig served as lecturer and chemist, University of California, Davis and worked on analytical methods for pesticide residues, mainly by chromatographic techniques. In 1965, Dr. Zweig became Director of Life Sciences, Syracuse University Research Corporation, New York (research on environmental pollution), and in 1973 he became Chief, Environmental Fate Branch, Environmental Protection Agency (EPA) in Washington, D.C. From 1980 to 1984 Dr. Zweig was Visiting Research Chemist in the School of Public Health, University of California, Berkeley, where he was doing research on farmworker safety as related to pesticide exposure.

During his government career, Dr. Zweig continued his scientific writing and editing. Among his works are (many in collaboration with Dr. Sherma) the now 11-volume series on *Analytical Methods for Pesticides and Plant Growth Regulators* (published by Academic Press); the pesticide book series for CRC Press; co-editor of *Journal of Toxicology and Environmental Health;* co-author of basic review on paper and thin-layer chromatography for *Analytical Chemistry* from 1968 to 1980; co-author of applied chromatography review on pesticide analysis for *Analytical Chemistry,* beginning in 1981.

Among the scientific honors awarded to Dr. Zweig during his distinguished career were the Wiley Award in 1977, the Rothschild Fellowship to the Weizmann Institute in 1963/64; and the Bronze Medal by the EPA in 1980.

Dr. Zweig authored or co-authored over 80 scientific papers on diverse subjects in chromatography and biochemistry, besides being the holder of three U.S. patents. In 1985, Dr. Zweig became president of Zweig Associates, Consultants in Arlington, Va.

Following his death on January 27, 1987, the Agrochemicals Section of the American Chemical Society posthumously elected him a Fellow and established the Gunther Zweig Award for Young Chemists in his honor.

Joseph Sherma, Ph.D., received a B.S. in Chemistry from Upsala College, East Orange, N.J., in 1955 and a Ph.D. in Analytical Chemistry from Rutgers University in 1958, carrying on his thesis research in ion exchange chromatography under the direction of the late William Rieman III. Dr. Sherma joined the faculty of Lafayette College in September, 1958, and is presently Charles A. Dana Professor and Head of the Chemistry Department.

Dr. Sherma, independently and with others, has written over 300 research papers, chapters, books, and reviews involving chromatography and other analytical methodology. He is editor for residues and trace elements of the *Journal of the Association of Official Analytical Chemists* and a member of the advisory board of the *Journal of Planar Chromatography.* He is a consultant on analytical methodology for many companies and government agencies.

Dr. Sherma has received two awards for superior teaching at Lafayette College and the 1979 Distinguished Alumnus Award from Upsala College for outstanding achievements as an educator, researcher, author, and editor. He is a member of the ACS, Sigma Xi, Phi Lambda Upsilon, SAS, AIC, and AOAC. Dr. Sherma's current interests are in quantitative TLC, mainly applied to clinical analysis, pesticide residues, and food additives.

THE EDITOR

Ram N. Gupta, Ph.D., is Head of Toxicology in the Department of Laboratory Medicine at St. Joseph's Hospital and Professor in the Department of Pathology at McMaster University in Hamilton, Ontario, Canada.

Dr. Gupta received his M.Sc. degree in 1962 and Ph.D. degree in 1963 in Organic Chemistry from McMaster University. He continued working in the Chemistry Department of McMaster University as a Research Associate until 1971 when he moved to the Department of Pathology at the same university.

Dr. Gupta has been elected as a fellow of the Chemical Institute of Canada. He is a member of the American Chemical Society, American Association of Clinical Chemists, Canadian Society of Clinical Chemists, and the Association of Clinical Biochemists (U.K.). He is the author of more than 40 scientific publications.

His present research interests are the development of chromatographic procedures for the assay of drugs and other biochemicals in biological fluids.

TABLE OF CONTENTS

ORGANIZATION OF TABLES AND EXPLANATION OF ABBREVIATIONS

Gas Chromatography (GC)

Specimen: Cerebrospinal fluid (CSF); not available (NA). The number in parenthesis refers to milliliters of plasma or serum used for the preparation of sample extract unless stated otherwise. There is no indication when volumes of other specimens are different from that of plasma or serum.
Extraction: In this column, the extraction procedure is given a number and the corresponding procedure is described at the end of the table for the extraction of plasma or serum unless indicated otherwise. Any difference in the extraction procedure of another type of specimen is not indicated.
Column: Columns are made of glass or fused silica unless noted otherwise. Length is given in meters and inner diameter in millimeters.
Packing: The number in the parenthesis shows the mesh size of the support. The film thickness of the capillary columns is given in μm and indicated by a footnote.
Gas: Gas flow, if given in units other than milliliters per minute, has been indicated by a footnote.
DET: Detector. Flame ionization detector (FID); nitrogen phosphorous detector (NPD); also, alkali flame ionization detector; thermionic sensitive detector; or nitrogen specific detector; electron capture detector (ECD); electron-impact mass spectrometer (MS-EI); chemical ionization mass spectrometer (MS-CI); negative ion chemical ionization mass spectrometer (MS-NCI). Any other detector used and the reagent gas used for chemical ionization, if different from the carrier gas, are indicated by footnotes.
RT min: Retention time in minutes of the title drug. It may be the retention time of the parent drug or its derivative. A dash "—" indicates that the title drug is not determined in the procedure under review, whereas NA indicates that the retention time is not available.
Internal Standard: The names of the compounds used as internal standards are given in full. Any abbreviation used to describe the internal standard is explained by a footnote. A dash "—" indicates that no internal standard was used in the procedure. The retention time in minutes is given in parenthesis as it appears in the chromatogram. It may be of the parent compound or its derivative. The retention time when the internal standard is an isotropically labelled drug is considered the same as of the drug itself.
Deriv: Derivative. This column indicates the type of derivative formed at some stage of the sample preparation. The details of derivatization reagent and procedure are included in the corresponding extraction procedure. A dash "—" indicates that no derivative was prepared.
Other Compounds: Metabolites of the parent drug or other similar or unrelated drugs when determined simultaneously with the title drug are listed in this column. Their retention times are given in parenthesis.
Ref: Reference.

Liquid Chromatography (LC)

This includes column liquid chromatography, high pressure liquid chromatography, and high performance liquid chromatography (see under GC for the explanation of common columns).
Column: Columns are made of steel unless noted otherwise. Length is in centimeters and inner diameter in millimeters.
Packing: Packing is described by the trade names as used by the authors. Footnotes indicate if a precolumn, a guard column, or a temperature other than ambient were used.
Elution: The eluting solvent is given a number and the corresponding solvent is described at the end of the table. The procedure is isocratic unless indicated as gradient. The conditions for gradient elution are described with the description of the elution solvent.

Flow Rate: Flow rate given in other units has been changed to milliliter per minute; a footnote indicates that only the pump pressure is given. Detector (DET); absorbance (ABS). Wavelength (nm) for absorbance detection is given. Two numbers are given when the absorbance is monitored simultaneously at two different wavelengths. A footnote indicates a programmed change of absorbance wavelength. Fluorescence (FL). The first number in the parenthesis is the excitation wavelength (nm), and the second, the emission wavelength. Other detectors are described without the use of abbreviations. Potentials for electrochemical detectors and procedures involving post-column reactors are indicated by footnotes.

Thin-Layer Chromatography (TLC)

See under GC and LC for the explanation of common columns.
Plate: Unless otherwise noted, plates are made of glass. Laboratory indicates that the plates have been coated by the authors in their laboratory.
Layer: High performance thin-layer chromatography (HPTLC).
Solvent: Developing solvent is given a number which is described at the end of the table.
Post-Separation Treatment: (sp) The plate is spayed with the described reagent. (D) The plate is dipped in the described reagent. (E) The plate is exposed to the vapors of the described reagent.
Det: Detection. Qualitative detection is indicated as visual. Wavelength (nm) for short or long wave UV lamp is given when fluorescence or quenching of fluorescence is observed under UV light. When the plate is scanned with the densitometer for quantitative determination of drug concentration, the mode of scanning is indicated as reflectance, transmission or reflectance/transmission for simultaneous mode. Wavelength (nm) for scanning and for fluorescence scanning, the excitation (first) and emission (second) are given.

CADRALAZINE

Liquid Chromatography

Specimen (mℓ)	Extraction	Column (cm × mm)	Packing (μm)	Elution	Flow (mℓ/min)	Det. (nm)	RT (min)	Internal Standard (RT)	Other compounds (RT)	Ref.
Plasma (1)	I-1	25 × 4.6	LiChrosorb RP-8 (10)[a]	E-1	2.7	ABS (254)	5.5	CGP 24751 (13)	—	1
Plasma, urine (2, 1)	I-2	25 × 4	LiChrosorb RP-8 (10)[a,b]	E-2	2.7	ABS (254)	6	c (9.2)	d	2

[a] Protected by a precolumn (50 × 3.2 nm) packed with Perisorb RP-8 (30 to 40 μm).
[b] Column temp = 30°C.
[c] N-ethyl-N-2-hydroxyethyl-(3-4,5-trimethyoxy) benzamide.
[d] Chromatographic separation of various metabolites is shown.

Extraction — I-1. The sample was spiked with 0.5 mℓ of the internal standard solution 1 μg/mℓ in 0.005 *M* sulfuric acid , diluted with 1 mℓ of water, and extracted with 6 mℓ of chloroform-ethanol (95:5, v/v). The organic layer was back extracted into 0.5 mℓ of 0.09 *M* KCl/HCl buffer, and 100 μℓ of the aqueous phase were injected.
I-2. The sample was spiked with 1 mℓ of the internal standard solution (621 ng/mℓ is phosphate buffer, pH 7.4), the mixture diluted to 5 mℓ with pH 7.4 phosphate buffer and extracted twice with 6 mℓ portions of chloroform. The combined organic layers were evaporated to dryness under a stream of nitrogen. The residue was reconstituted with 2 mℓ of methanol and the solution again evaporated under nitrogen. The residue was dissolved in 100 μℓ of methanol, 50 μℓ of 0.05 *M* sulfuric acid added and 50 μℓ of this solution were injectied.

Elution — E-1. Acetonitrile-0.1 *M* phosphate buffer, pH 6 (15:85).
E-2. Acetonitrile-0.1 *M* phosphate buffer, pH 6 (15:85).

REFERENCES

1. **Hauffe, S. A. and Dubois, J. P.,** Determination of cadralazine in human plasma and urine by high-performance liquid chromatography, *J. Chromatogr.*, 290, 223, 1984.
2. **Crolla, T., Santini, F., Visconti, M., and Pifferi, G.,** High-performance liquid chromatographic separation of cadralazine from its potential metabolites and degradation products. Quantitation of the drug in human plasma and urine, *J. Chromatogr.*, 310, 139, 1984.

CAFFEINE

Gas Chromatography

Specimen (mℓ)	Extraction	Column (m × mm)	Packing (mesh)	Oven temp (°C)	Gas (mℓ/min)	Det.	RT (min)	Internal standard (RT)	Deriv.	Other compounds (RT)	Ref.
Plasma (0.05)	I-1	1.5 × 2	3% SP-1000 Chromosorb W (100/120)	210	N_2 (20)	FID	9	7-Ethyltheophylline (7.2)	—	—	1
Plasma, urine (0.02, 1)	I-2	1.8 × 2	1.8% OV-17 Chromosorb W (100/120)	240	He (30)	NPD	1.3	N-Methylphenothiazine (2.1)	—	—	2
Urine (2)	I-3	2 × 2.5	3% OV-17 Chromosorb W (NA)	195	N_2 (25)	NPD	4.4	Mepivacaine (8.7)	—	—	3

Liquid Chromatography

Specimen (mℓ)	Extraction	Column (cm × mm)	Packing (μm)	Elution	Flow (mℓ/min)	Det. (nm)	RT (min)	Internal standard (RT)	Other compounds (RT)	Ref.
Plasma, serum (0.1)	I-4	25 × 4.6	Ultrasphere-ODS (5)	E-1	1.0	ABS (254)	8.8	β-Hydroxyethyltheo-phylline (5.7)	Theophylline (5.2) Theobromine (4)	4
Blood, plasma, urine (0.1)	I-5	15 × 4.6	Perkin-Elmer-C_{18} (5)	E-2	NA	ABS (276)	6.8	8-Chlorotheophylline (8.8)	Dimethyluric acid (1.9) 3-Methylxanthine (2.1) 1,3-Dimethyluric acid (2.4) 3,7-Dimethylxanthine (3.8)	5
Plasma (1)	I-6	30 × 4	μ-Bondapak-C_{18} (10)[a]	E-3	1.0	ABS (273)	9	—	—	6

Plasma (0.5)	I-7	25 × 4	LiChrosorb Si 60 (5)	E-4	2.0	ABS (280)	3.5	7-Ethyltheophylline (3.0) 1,3,7-Trimethyl uric acid (12)	Theobromine (5) Theophylline (6.5) Paraxanthine (8)	7
Beans	I-8	NA	Fine Pak C_{18} (5)	E-5	1.2	ABS (272)	7.5	—	—	8
Plasma (0.1)	I-9	10 × 8	Radial-Pak C_{18} (4)[b]	E-6	2.7	ABS (273)	8	β-Hydroxyethyltheophylline (5.3)	Theophylline (4.6) Paraxanthine (4.1) Theobaromine (2.8)	9
Urine (0.05)	I-10	30 × 7.5	BioGel TSK-20 (10)	E-7	0.8	ABS (254)	—	Benzyloxyurea (41)	5-Acetylamino-6-amino-3-methyluracil (16.2) 1-Methylxanthine (32)	10
Urine (0.2)	I-11	25 × 4.6	Hypersil-ODS (5)	E-8 Gradient	1.5	ABS (280)	16.7	Proxyphylline (18)	Uric acid (3.6) 7-Methyl uric acid (5.3) 7-Methylxanthine (6) 1-Methyl uric acid (6.8) 3-Methylxanthine (7.2) 1-Methylxanthine (8.6) 1,3-Dimethyluric acid (9.5) Theobromine (9.7) 1,7-Dimethyluric acid (12) 1,7-Dimethylxanthine (12.7) Theophylline (13.5)	11, 12

CAFFEINE (continued)

Thin-Layer Chromatography

Specimen (mℓ)	Extraction	Plate (Manufacturer)	Layer (mm)	Solvent	Post-separation treatment	Det (nm)	Rf	Internal standard (Rf)	Other compounds (Rf)	Ref.
Plasma (1)	I-12	10 × 20 cm (Merck)	Silica gel PF_{254} (0.25)	S-1	—	Visual (254)	0.67	—	Theophylline (0.57) 1,7-Dimethylxanthine (0.48)	13

[a] The column was protected by a precolumn packed with Corasil-C_{18} (37 to 50 μm)
[b] Protected by a Guard-Pak CN cartridge

Extraction — I-1. The sample was diluted to 0.25 mℓ with water, spiked with 100 μℓ solution of the internal standard (24 μmol/ℓ in methanol) and extracted with 2 mℓ of chloroform. The organic layer was evaporated under a stream of nitrogen. The residue was dissolved in 25 μℓ chloroform-methanol (9:1) of which 20 μℓ was spotted on a 20 × 20 cm silica gel 60 F_{254} (0.25 mm) thin-layer plate. This plate was developed with ethyl acetate-methanol-ammonia (80:20:1.5) to a height of 10 cm. The caffeine and the internal standard areas were marked under UV and scraped off the plate. The scraping were quantitatively collected and eluted with 40 to 50 μℓ methanol-water (9:1); and 2 to 3 μℓ of this eluate are injected.
I-2. The sample was spiked with 50 μℓ of the internal standard solution (1 μg/mℓ in ethanol), 0.2 mℓ of 6 *M* KOH and 100 mg of sodium sulfate added and the mixture extracted with 4 mℓ of diethyl ether. The ether layer was dried over anhydrous sodium sulfate and evaporated in a stream of nitrogen. The residue was dissolved in 100 μℓ of ethanol and 1 μℓ of this solution was injected.
I-3. Urine (2 mℓ) was made alkaline with 0.2 mℓ of ammonia buffer, pH 9.5, 0.1 mℓ of the internal standard solution (100 μg/mℓ in water) was added, and the mixture was applied to an Extrelut-1 column. After 5 min, the column was eluted with 6 mℓ of dichloromethane-methanol (9:1). The eluate was evaporated under nitrogen at 40°C. The residue was dissolved in 0.2 mℓ of ethyl acetate and 1 μℓ was injected.
I-4. The sample was mixed with 100 μℓ of 0.2 *N* HCl and 75 μℓ of the internal standard solution (5 μg/mℓ in water) and the mixture was extracted with 12 mℓ of chloroform-2-propanol (80:20). The organic phase was evaporated at 50°C under a stream of nitrogen. The residue was reconstituted with 100 μℓ of mobile phase and 25 μℓ were injected.
I-5. The sample was mixed with 1.2 g of ammonium bicarbonate and 100 μℓ of the internal standard solution (20 μg/mℓ in 0.9% sodium chloride solution) and extracted twice with 8 mℓ portions of chloroform-isopropanol (85:15). The combined organic extracts were evaporated under a stream of nitrogen at 50°C. The residue was reconstituted in 500 μℓ of the mobile phase and aliquots of 80 μℓ were injected with an autosampler.
I-6. The sample was treated with an equal volume of 0.15 *M* barium hydroxide. After mixing for 2 min, 1 mℓ of 5% zinc sulfate was added. After mixing and centrifugation aliquots of clear supernatant were injected.
I-7. To 0.5 mℓ of plasma were added 0.2 g of ammonium sulfate, 100 μℓ of an aqueous solution containing 5 nmol of 7-ethyltheophylline, and 5 nmol of 1,3,7-trimethyluric acid and the mixture was extracted with 10 mℓ of chloroform-isopropanol (1:1, v/v). The organic layer was evaporated under a stream of nitrogen at 70°C. The residue was dissolved in 0.5 mℓ of the mobile phase. Aliquots of this solution were injected with an autosampler.
I-8. Extraction is carried out on line with the use of a supercritical fluid extraction technique.

I-9. The sample and 0.1 mℓ of the internal standard solution (10 μg/mℓ in 0.1 *M* phosphate buffer, pH 4) were applied to a prewashed (2 mℓ methanol, 2 mℓ water) 1-mℓ Bond-Elut C_{18} column. After standing for 1 min, the solution was drawn through the column. After 2 min, the column was washed with 2 mℓ of water and eluted with 400 μℓ of acetone. The eluate was evaporated under nitrogen at 55°C, the residue reconstituted in 150 μℓ of mobile phase. Aliquots of 15 μℓ were injected.
I-10. The urine sample (50 μℓ) was treated with 50 μℓ of 0.1 *M* sodium hydroxide. After 10 min at room temp, 50 μℓ of 0.1 *M* HCl and 100 μℓ of the internal standard solution (1 mg/mℓ in water) were added. An aliquot of 20 μℓ of this mixture was injected.
I-11. The urine sample (200 μℓ), 50 μℓ of the internal standard solution (200 mg/ℓ in water), 200 μℓ of 0.1 *M* tetrabutyl ammonium hydrogen sulfate, and 100 μℓ of pH 11 buffer solution (0.1 *M* sodium carbonate-0.1 *M* sodium bicarbonate, 90:10) were vortex mixed before and after the addition of 0.5 g ammonium sulfate. The mixture was extracted by vortex-mixing with 5 mℓ of ethyl acetate-chloroform-isopropanol (45:45:10). The organic layer was evaporated to dryness at 45°C under a stream of air. The residue was dissolved in 500 μℓ of 1% tetrahydrofuran, pH 4.8 and 50 μℓ of this solution was injected.
I-12. The sample was forced through a Sep-Pack C_{18} cartridge which was then washed with 5 mℓ of 0.1 *M* sodium acetate and eluted with 5 mℓ of acetonitrile. The elute was evaporated at 35°C. The residue was dissolved in 200 μℓ of acetonitrile, aliquots of which were spotted on a TLC plate. The plate was developed and visualized under short UV light.

Elution — E-1. Acetonitrile-85% phosphoric acid-water (130:0.5:870)
E-2. Acetonitrile-isopropanol-acetic acid-water (4:4:1:91)
E-3. Methanol-water (30:70)
E-4. Dichloromethane-methanol containing ammonium formate (0.2 g/100 mℓ) and formic acid (15 mℓ/100 mℓ) (975:25)
E-5. Methanol-water (55:45)
E-6. Methanol-1% acetic acid (17:83)
E-7. 0.1% Acetic Acid
E-8. (A) 1% Tetrahydrofuran in 10 m*M* acetate buffer, pH 4.8 (B) 15% Acetonitrile + 1% tetrahydrofuran in 10 m*M* acetate buffer, pH 4.8. Isocratic A for 5 min; stepwise gradient of 5, 10, and 2% increase in B per min; each step being pumped for 5 min.

Solvent — S-1. Dichloromethane-isopropanol-acetic acid (105:15:1)

REFERENCES

1. **Hulshoff, A.,** Determination of caffeine in small plasma samples by gas-liquid chromatography with thin-layer chromatographic sample clean-up, *Anal Lett.*, 12, 1423, 1979.
2. **Hirai, T. and Kondo, H.,** Determination of caffeine in human plasma and urine by gas chromatography using nitrogen phosphorus-detector, *Anal. Sci.*, 1, 191, 1985.
3. **Delbeke, F. T., and Debackere M.,** Simple and rapid method for the determination of caffeine in urine using Extrelut-1 columns, *J. Chromatogr.*, 278, 418, 1983.
4. **Haughey, D. B., Greenberg, R., Schaal, S. F., and Lima, J. J.,** Liquid chromatographic determination of caffeine in biologic fluids, *J. Chromatogr.*, 229, 387, 1982.
5. **Stavric, B., Klassen, R., and Gilbert, S. G.,** Automated high-performance liquid chromatographic assay for monitoring caffeine and its metabolites in biological fluids of monkeys consuming caffeine, *J. Chromatogr.*, 310, 107, 1984.

CAFFEINE (continued)

6. **O'Connell, S. E. and Zurzola, F. J.,** Rapid quantitative liquid chromatographic determination of caffeine levels in plasma after oral dosing, *J. Pharm. Sci.*, 73, 1009, 1984.
7. **Wahllander, A., Renner, E., and Karlaganis, G.,** High-performance liquid chromatographic determination of dimethylxanthine metabolites of caffeine in human plasma, *J. Chromatogr.*, 338, 369, 1985.
8. **Sugiyama, K., Saito, M., Hondo, T., and Senda, M.,** Directly coupled laboratory-scale supercritical fluid extraction-supercritical fluid chromatography, monitored with a multi-wavelength ultraviolet detector, *J. Chromatogr.*, 332, 107, 1985.
9. **Hartley, R., Smith, I. J., and Cookman, J. R.,** Improved high-performance liquid chromatographic method for the simultaneous determination of caffeine and its N-demethylated metabolites in plasma using solid-phase extraction, *J. Chromatogr.*, 342, 105, 1985.
10. **Tang, B. K., Zubovits, T., and Kalow, W.,** Determination of acetylated caffeine metabolites by high-performance exclusion chromatography, *J. Chromatogr.*, 375, 170, 1986.
11. **Scott, N. R., Chakraborty, J., and Marks, V.,** Determination of the urinary metabolites of caffeine and theophylline by high-performance liquid chromatography, *J. Chromatogr.*, 385, 321, 1986.
12. **Scott, N. R., Chakraborty, J., and Marks, V.,** Determination of caffeine, theophylline and theobromine in serum and saliva using high-performance liquid chromatography, *Ann. Clin. Biochem.*, 21, 120, 1984.
13. **Szeto, D. W. and Tee, F. L. S.,** Identification of 1,7-dimethyl-xanthine as a metabolite of caffeine in the dog, *J. High Resol. Chromatogr. Chromatogr. Commun.*, 4, 305, 1981.

CAMAZEPAM

Gas Chromatography

Specimen (mℓ)	Extraction	Column (m × mm)	Packing (mesh)	Oven temp (°C)	Gas mℓ/min	Det.	RT (min)	Internal standard (RT)	Deriv.	Other compounds (RT)	Ref.
Plasma (1)	I-1	1 × 3	3% OV-17 Chromosorb W (80/100)	260	Ar:9 Methane:1 (40)	ECD	10.7	Penfluridol (17)	—	Temazepam (4)	1
Plasma (1)	I-2	1 × 3	3% OV-17 Gas Chrom A (100/120)	300	N_2 (50)	ECD	4	Penfluridol (5)	—	Temazepam (1.8)	2

Extraction — I-1. The sample was treated with the residue after evaporation of 0.4 mℓ of the internal standard solution (1 μg/mℓ in acetone), 2 mℓ of 1 *M* phosphate buffer, pH 7.2 were added and the mixture extracted twice with 8-mℓ portions of diethyl ether. The combined ether extracts were evaporated under a stream of nitrogen at 35°C. The residue was reconstituted in 50 μℓ of acetone and 2 to 3 μℓ of this solution were injected.
I-2. The sample after the addition of 0.5 mℓ of borate buffer, pH 9 was extracted with 6 mℓ of benzene-isopropanol (95:5). An aliquot of 5 mℓ of the organic phase was evaporated at 35°C under vacuum. The residue was dissolved in 100 μℓ of the internal standard solution (250 μg/mℓ in acetone) and 1 to 2 μℓ of this solution was injected.

REFERENCES

1. **Cuisinaud, G., Peillon, E., Ferry, N., Deruaz, D., and Sassard, J.,** Simple and sensitive gas chromatograpic method for the determination of camazepam in human plasma, *J. Chromatogr.*, 178, 314, 1979.
2. **Riva, R., Albani, F., and Baruzzi, A.,** Quantitative determination of camazepam and its metabolite temazepam in man by gas-liquid chromatography with electron-capture detection. *Farmaco Ed. Prat.*, 37, 15, 1982.

CAMPHOR

Gas Chromatography

Specimen (mℓ)	Extraction	Column (m × mm)	Packing (mesh)	Oven Temp (°C)	Gas mℓ/min	Det.	RT (min)	Internal standard (RT)	Deriv.	Other compounds (RT)	Ref.
Dosage	—	1.8 × 3	5% Carbowax-20 *M* Chromosorb W (80/100)	T.P.[a]	N_2 (30)	FID	1.4	Benzyl alcohol (3.6)	—	Menthol (1.8) Methylsalicylate (2.8)	1

[a] Initial temp = 130°C, rate = 6°C/min, final temp = 170°C.

REFERENCES

1. **Tan, H. S. I., Kemper, P. A., and Padron, P. E.,** Gas-liquid chromatographic assay of mixtures of camphor, menthol, and methyl salicylate in ointments, *J. Chromatogr.*, 238, 241, 1982.

CANNABIS[a]

Gas Chromatography

Specimen (mℓ)	Extraction	Column (m × mm)	Packing (mesh)	Oven temp (°C)	Gas (mℓ/min)	Det.	RT[b] (min)	Internal standard (RT)	Deriv.	Other compounds (RT)	Ref.
Urine (10)	I-1	1.8 × 2	3% OV-17 GasChrom Q (100/120)	255	He (30)	FID	7 = 6	Oxyphenbutazone (4.5)	Methyl	—	2
Resin, marihuana (100 mg)	I-2	1.9 × 2	3% OV-1 (NA)	T.P.[c]	He (25)	MS-EI[d]	1 = 4.4 3 = 4.8 4 = 3.2 5 = 5.6 6 = 4.9 7 = 6.1	—	Methyl[e]	—	3
Smoke condensate	I-3	50 × 0.25	OV-101	T.P.[f]	He (NA)	MS-EI	g	—	Trimethylsilyl[h]	—	4
Urine (1)	I-4	25 × 0.3	SE-54	220	He[i]	MS-EI[j]	7 = 2	$[^2H_3]$-Δ^9-tetrahydro-cannabinolic acid	Pentafluoropropyl, pentafluoropropionyl	—	5
Plasma (1)	I-5	2 × 2	1% SE-30 GasChrom Q (100/120)	250	He (30)	MS-EI	1 = 1.2 7 = 2.8 8 = 2.4	$[^2H_2]$cannabinol (1.6)	Trimethylsilyl, methyltrimethylsilyl	—	6, 7
Urine (10)	I-6	1.8 × 3	(NA) OV-1 GasChrom Q (100/120)	T.P.[k]	He (30)	MS-EI	7 = 5.5	—	methyl		8
Urine (5)	I-7	25 × 0.23	SE-54	T.P.[l]	He (NA)	MS-EI	7 = 15.9	—	Methyl	—	9
Plasma (1)	I-8	15 × 0.25	OV-17 (0.15 μm)[m]	T.P.[n]	H_2[o]	ECD	1 = 7.4 7 = 14.4 8 = 9.3	Tetracosanoic acid[p] (10)	Pentafluorobenzyl-trimethylsilyl	—	10

Plasma, urine (1)	I-9	40 × 0.5	3% OV-17 Chromosorb W (G-SCOT)	225	He[q]	MS-EI	1 = 5.1 7 = 5.8 8 = 7.1	[$^2H_3$]-tetrahydro-cannabolic acid	Pentafluoro-propionyl-hexafluoroiso-propyl	8,11-Dehy-dro-xytetra-hydro-cannabinol (9)[r]	11

Liquid Chromatography

Specimen (mℓ)	Extraction	Column (cm × mm)	Packing (μm)	Elution	Flow (mℓ/min)	Det. (nm)	RT (min)	Internal standard (RT)	Other compounds (RT)	Ref.
Urine (10)	I-10	25 × 4.6	Sepralyte-C_8 (5)	E-1	1.6	ABS (214)	7 = 4.2	11-Norcannabinol-9-car-boxylic acid (4.8)	—	12
Urine (2)	I-11	10 × 4.6	Spherisorb-ODS (5)	E-2; grad	4	ABS[s] (215)	1 = 18.4 7 = 14.3 8 = 14	—	—	13
Marihuana-ciga-rette, tar, ash	I-12	25 × 4.6	μ-Bondapak-C_{18} (10)	E-3	1	Electrochem[t]	1 = 14 3 = 12 4 = 9.2 6 = 10.7 7 = 23.2	*n*-Dodecyl-*p*-hydroxy-benzoate (25.3)	Cannabichromene (19.8)	14

Thin-Layer Chromatography

Specimen (mℓ)	Extraction	Plate (Manufacturer)	Layer (mm)	Solvent	Post-separation treatment	Det (nm)	Rf	Internal standard (Rf)	Other compound (Rf)	Ref.
Urine	I-13	20 × 20 cm (An-altech)	Silica gel G (0.25)	S-1	Sp: 0.1% Fast Blue Salt B in 2N NaOH	Visual	7 = 0.1	—	—	15, 16
Urine (20)	I-14	20 × 20 cm (Gelman)	Silica gel glass microfilm (0.25)	S-2	Sp: 0.5% Fast Blue RR in methanol-water (1:1)	Visual	7 = 0.25—0.38	—	—	17

CANNABIS[a] (continued)

Thin-Layer Chromatography

Serum (2)	I-15	20 × 20 (NA)	Silica gel IB2 (NA)	S-3[u]	SP: 4% sodium in methanol	Visual; Fl (430, 500)	1 = 0.68	—	—	18
Urine (10)	I-16	2.5 × 7.5 (Applied Analytical)	Silica gel 60 (NA)	S-4	Sp: 0.5% Fast Blue RR in methanol-water (1:1)	Visual	7 = 0.43—0.5	—	—	19

Note: 1 — Δ^9-Tetrahydrocannabinol; 2 — Δ^8-Tetrahydrocannabinol; 3 — Cannabinol; 4 — Cannabidiol; 5 — Cannabinolic acid; 6 — Cannabidilic acid; 7 — Δ^9-Tetrahydrocannabinolic acid; 8 — 11-Hydroxy-Δ^9-tetrahydrocannabinol.

[a] Determination of cannabinoids has been reviewed.[1]
[b] Different compounds present in cannabis are identified by numbers (see *Note*).
[c] Initial temp= 200°C; initial time = 2 min; rate = 10°C/min; final temp = 260°C; final time = 5 min.
[d] Chemical ionization, with isobutane as the reagent gas was also used.
[e] Retention times of derivatized and underivatized cannabinoid constituents of hashish or marihuana extracts are given.
[f] Initial temp = 50°C; rate = 1°C/min; final temp = 270°C.
[g] Total ion current chromatograms and the molecular weight of different peaks obtained by the analysis of marihuana and tobacco smokes are given.
[h] Nonpolar neutrals are chromatographed without derivatization.
[i] Velocity = 55 cm/sec
[j] Chemical ionization with methane as the reagent gas was also used.
[k] Initial temp = 200°C; rate = 5°C/min; final temp = 270°C.
[l] Initial temp = 75°C; rate = 15°C/min; final temp = 300°C.
[m] Film thickness.
[n] Initial temp = 260°C; rate = 6°C/min; final temp = 300°C; final time = 5 min.
[o] 40 cm/sec.
[p] Pentafluorobenzyl ester of tetracosanoic acid was used as an external standard.
[q] Column head pressure = 16 psig.
[r] Retention time of trimethylsilyl derivative.
[s] Fractions of HPLC eluate are collected every 30 sec which are analyzed by RIA.
[t] Potential = +1.2 V.
[u] Two developments (3 cm, 6 cm) in the same solvent.

Extraction — I-1. The sample was mixed with 8 mℓ of methanol and 2 mℓ of 10 *N* KOH and the mixture was incubated at 50°C for 15 min. After removal from the water bath, 2 mℓ of 1 *M* potassium phosphate buffer, pH 2.5 was added, pH of the mixture adjusted 2 to 2.5 with HCl, and extracted with 20 mℓ of hexane-ethyl acetate (7:1) containing 0.5 μg/mℓ of the internal standard. The organic layer was back extracted into 5 mℓ of 0.5 *N* KOH. The aqueous phase was adjusted to pH 2 to 2.5 with 1 mℓ of 1 *M* potassium phosphate, pH 2.5 buffer and HCl and re-extracted with 15 mℓ of the above extraction solvent without internal standard. The organic layer was evaporated at 50°C under a stream of nitrogen. The residue was dissolved in 70 μℓ of 25% aqueous tetramethyl-ammonium hydroxide-dimethyl sulfoxide (1:20). After 2 min at room temp 5 μℓ of iodomethane was added, mixed, and allowed to stand at room temp for 5 min. The reaction was terminated by the addition of 0.2 mℓ of 0.1 *N* HCl and the mixture extracted with 1 mℓ of iso-octane. The organic layer was evaporated under a stream of nitrogen at 50°C. The residue was dissolved in 20 μℓ of isooctane and 5 μℓ was injected.
I-2. The sample is triturated with 1 mℓ of methanol, placed in ultrasonic bath (20 min), and centrifuged. Aliquots of the supernatant are injected. Alternatively, an aliquot of 25 μℓ of the supernatant is evaporated under a stream of nitrogen, dissolved in 100 μℓ of methyl-8-(dimethylformamide-dimethylacetal, 2 *M* in pyridine). Aliquots of 2 to 4 μℓ of this solution are injected.
I-3. An appropriate amount of smoke is dissolved in dichloromethane which is washed with 0.1 *N* NaOH to separate acids, and then with 0.1 *N* HCl to separate basic fractions. Dichloromethane layer is evaporated. The residue dissolved in cyclohexane and extracted with methanol-water to separate polar neutrals. The cyclohexane layer is washed with nitromethane. Nonpolar constituents remain in cyclohexane and polynuclear aromatic hydrocarbons are extracted by nitromethane.
I-4. The urine sample was hydrolyzed enzymatically, spiked with 40 ng of deutrated analog, and extracted with 4 mℓ of *n*-hexane-diethyl ether (1:1). The organic layer was evaporated under a stream of nitrogen at room temp. The residue was dissolved in 100 μℓ of pentafluoropropionic anhydride-pentafluoropropanol (4:1) and the solution incubated for 15 min at 70°C. After cooling, the excess reagent was evaporated and the residue dissolved in 50 μℓ of cyclohexane containing 60 ng/μℓ Mirex as a GC internal standard. Finally, 1 μℓ was injected.
I-5. The sample was spiked with the internal standard, allowed to equilibrate at room temp for 30 min and extracted with hexane. The organic layer was evaporated and treated with freshly prepared solution of diazomethane in ether. After evaporation of ether, the residue was treated with N,O-*bis*(trimethylsilyl)trifluoroacetamide to prepare the silyl derivative. Alternative methods of extraction have been described for the isolation of metabolites.
I-6. The urine sample was hydrolyzed by refluxing it with 1 mℓ of 10% (W/V) methanolic KOH for 15 min. After cooling the sample was adjusted to pH 3 to 4 with glacial acetic acid and passed through a prewashed (2 mℓ methanol, 5 mℓ water) Sep-Pak-C_{18} cartridge. The cartridge was then washed with 5 mℓ of water, 5 mℓ of acetonitrile-water (4:6) and finally eluted with 2 mℓ of methanol. The eluate was evaporated under nitrogen on a hot water bath and the residue treated with 70 μℓ of aqueous 25% tetramethylammonium hydroxide-dimethylsulfoxide (1:2). After 2 min, 5 μℓ of iodomethane were added and the mixture allowed to stand for 5 min. The mixture was acidified with 0.2 mℓ of 1 *M* acetic acid and extracted twice with 1-mℓ portions of cyclohexane. The combined organic extracts were evaporated, the residue dissolved in 20 μℓ of methanol and a 4- to 5-μℓ aliquot was injected.
I-7. Urine was hydrolyzed with methanolic KOH for 15 min at 50°C. After cooling, 2 mℓ of 0.1 *M* phosphate buffer (pH 7) were added and pH of the urine adjusted to 7 with HCl. A 1- to 5 mℓ aliquot of this hydrolysate was applied to a minicolumn packed with Sepharose 4B coated with antibody for tetrahydrocannabinol. The column was washed with 15 mℓ of water and eluted with 10 mℓ of acetone-water (95:5). The eluate was evaporated with a dry stream of nitrogen. The residue was derivatized as described in I-1 (Reference 2).
I-8. The sample was treated with 200 mg of XAD-2 resin and 170 μℓ of acetonitrile. The solution was adjusted to pH 4 by the addition of 75 μℓ of 1 *M* HCl and the mixture was shaken at room temp for 1 hr at 75 c/min. The supernatant was removed and the resin washed with 4 mℓ of water. The resin was then treated with 4 mℓ of 20% acetonitrile in 0.1 *M* NaOH followed by 100 μℓ of pentafluorobenzylbromide-trichloroethylene (1:9). The reaction mixture was then shaken for 90 min. The resin was filtered, washed with water, and eluted with 10 mℓ of dichloromethane followed with 10 mℓ of diethyl ether. The combined eluates were evaporated at 70°C under a stream of dry nitrogen. The residue was dissolved in 50 μℓ of N,O-*bis*(trimethylsilyl) trifluoroacetamide/trimethylchlorosilane and 200 μℓ of toluene.
I-9. Plasma was incubated overnight with 4000 U of β-glucuronidase, acidified with 1 mℓ KCl/HCl buffer, pH 2.3, prior to extraction twice with 4-mℓ volumes of

CANNABIS[a] (continued)

1.5% isoamyl alcohol in heptane. The combined organic extracts were evaporated under a stream of nitrogen. The residue was treated with hexane (100 μℓ), pentafluoropropionic anhydride (50 μℓ), and hexafluoroisopropanol (50 μℓ). The mixture was heated at 100°C for 10 min. After cooling, the excess reagents and solvent were removed with a stream of nitrogen and the residue dissolved in 20 μℓ of chloroform, aliquots of this solution were injected.

I-10. The urine sample after the addition of the internal standard (500 μℓ of a 2 μg/mℓ solution in ethanol) and 2 mℓ of 2 *N* KOH solution was incubated at 15°C for 15 min. The resulting hydrolyzed urine was then adjusted to pH 5 to 6 with concentrated HCl and passed through a prewashed [methanol (4 mℓ), water (4 mℓ), 1 *N* HCl (1 mℓ)] BondElut THC column. After the sample had passed through, the column was washed with 10 mℓ of 0.1 *N* HCl followed by 25 mℓ of acetonitrile-50 *M* phosphoric acid (1:9) and finally eluted with 1.5 mℓ of acetonitrile. Aliquots of 20 μℓ of this eluate were injected.

I-11. The urine sample is mixed with 2 mℓ of methanol, 2 mℓ of methanol-water, pH 5.5 (1:1). After centrifugation 5 mℓ of the supernatant are injected.

I-12. The samples were extracted with methanol in a Soxhlet extractor for 3 hr. To 1 mℓ of the extract was added 0.1 mℓ of the internal standard solution (130 μg/mℓ in methanol), aliquots of 30 μℓ of this solution were injected.

I-13. A volume of urine was adjusted to pH 4.7 to 6.3 and concentrated by evaporation. The residue was diluted with water to 10 mℓ and extracted twice with 15-mℓ portions of hexane-diethyl ether (65:35). The aqueous phase was processed to isolate the conjugates of THC-11-oic acid. The combined organic extracts were evaporated under a stream of nitrogen at 50°C. The residue was dissolved in 15 mℓ of ether, the ether solution washed with 10 mℓ of 5% sodium bicarbonate, dried over anhydrous sodium sulfate, and evaporated. The residue was dissolved in 30 μℓ of ethanol and applied as a streak on a TLC plate. The same aqueous phase was hydrolyzed enzymatically at 55 to 60°C for 30 min and processed as before.

I-14. The urine is treated with 2 mℓ of 10% methanolic solution of potassium hydroxide at 100°C for 12 min. The pH of the hydrolyzed sample is adjusted to 3 to 4 with 3 mℓ of glacial acetic acid and extracted three times with 15-mℓ portions of cyclohexane-ethyl acetate (96:4). The combined organic extracts are evaporated at 80 to 85°C, the residue dissolved in 40 μℓ of methanol and the entire solution spotted on the plate.

I-15. Methanol (4 mℓ) was added to 2 mℓ of the sample, vortexed, and centrifuged. The clear supernatant was collected and the precipitate rinsed with another 4 mℓ of methanol. The pooled extracts were evaporated to a volume of 3 mℓ in an 80°C water bath with a stream of air. The residual fluid was extracted with 8 mℓ of hexane-isoamyl alcohol (97:3). The hexane layer was evaporated to dryness. The residue was treated with 2 mℓ of Claisen alkali (37 g of KOH dissolved in 20 mℓ water and diluted with 100 mℓ of methanol). The solution was washed with 1 mℓ of hexane. The alkaline solution was concentrated to 0.5 mℓ, diluted with 8 mℓ of water, and extracted with 8 mℓ of hexane. The hexane layer was evaporated. The residue was treated with 10 μℓ of 1 mg/mℓ of 2-*p*-Cl-sulfophenyl-3-phenylindone in acetonitrile. After mixing, 5 μℓ of 0.2 *M* aqueous sodium carbonate were added, and the mixture was incubated for 30 min at 40°C. Acetone (25 μℓ) was added to the reaction mixture, vortexed, centrifuged, and aliquots of the supernatant were spotted on the TLC plate. After chromatography, spots corresponding to tetrahydrocannabinol were scraped and eluted with acetone. The eluates were filtered through membrane filter and the filtrate evaporated with a stream of air in an 80°C water bath. The residue was treated with 1 mℓ of fluorescence reagent (0.2% dibenzo-18 crown-6 in benzene + 1 g of potassium methoxide) and the fluorescence was determined immediately.

I-16. The sample is hydrolyzed by incubating in a boiling water bath for 15 min after the addition of 0.9 mℓ of 10 *M* sodium hydroxide. After cooling, the pH of the solution is adjusted to 1 to 3 by the addition of 0.7 mℓ of concentrated HCl and applied to a prewashed (methanol 6 mℓ, water 6 mℓ) 3-mℓ BondElut-THC column. The column is washed with 0.1 *M* HCl (10 mℓ), 50 m*M* phosphoric acid-acetonitrile (9:1), and eluted with 1 mℓ of acetone. The eluate is extracted with 0.5 mℓ of dichloromethane. The upper layer is discarded. The lower phase is extracted with 0.5 mℓ of hexane. The upper organic layer is collected and evaporated at 60°C with a stream of nitrogen. The residue is dissolved in 10 μℓ of acetone and the entire solution spotted on a TLC plate.

Elution — E-1. Acetonitrile: 50 m*M* phosphoric acid (65:35).
E-2. (A) Acetonitrile: (B) water, pH 3.3 (phosphoric acid). Isocratic 36% (A) from 0 to 10 min; linear gradient from 36 to 70% (A) from 10 to 20 min; isocratic 70% (A) from 20 to 25 min.
E-3. Acetonitrile-methanol-0.02 *N* sulfuric acid (16:7:6).

Solvent — S-1. I development: Acetone-chloroform-triethylamine (80:20:1). II development: Light petroleum-diethyl ether-glacial acetic acid (50:50:15).
S-2. Chloroform-methanol-concentrated ammonium hydroxide (85:15:2).
S-3. Dimethylformamide-water (80:20).
S-4. Ethyl acetate-methanol-water-concentrated ammonium hydroxide (12:5:0.5:1).

REFERENCES

1. **Hawks, R., Ed.,** *Analysis of Cannabinoids,* NIDA Research Monograph, 1982, 42.
2. **Whiting, J. D. and Manders, W. W.,** Confirmation of a tetrahydrocannabinol metabolite in urine by gas chromatography, *J. Anal. Toxicol.*, 6, 49, 1982.
3. **Bjorkman, S.,** Gas chromatography of cannabis constituents, including cannabinoid acids, with on-column alkylation and esteriofication, *J. Chromatogr.*, 237, 389, 1982.
4. **Novotny, M., Merli, F., Wiesler, D., Fencl, M., and Saeed, T.,** Fractionation and capillary gas chromatographic-mass spectrometric characterization of the neurtal components in marijuana and tobacco smoke condensates, *J. Chromatogr.*, 238, 141, 1982.
5. **Karlsson, L., Jonsson, J., Aberg, K., and Roos, C.,** Determination of Δ^9-Tetrahydrocannabinol-11-oic acid in urine as Its pentafluoropropyl-pentafluoropropionyl derivative by GC/MS utilizing netgative ion chemical ionization, *J. Anal. Toxicol.*, 7, 198, 1983.
6. **Harvey, D. J., Leuschner, J. T. A., and Paton, W. D. M.,** Pharmacokinetic studies on delta-1-tetrahydrocannabinol and its major metabolites using metastable ion monitoring, *Spectros. Int. J.*, 3, 195, 1984.
7. **Harvey, D. J., Leuschner, J. T. A. and Paton, W. D. M.,** Measurement of Δ^1-tetrahydrocannabinol in plasma to the low picogram range by GC-MS using metastable ion detection, *J. Chromatogr.*, 202, 83, 1980.
8. **Nakamura, G. R., Stall, W. J., Masters, R. G., and Folen, V. A.,** Analysis of urine for 11-nor-Δ^9-tetrahydrocannabinol-9-carboxylic acid using Sep-PAK cartridges for sample cleanup, *Anal. Chem.*, 57, 1492, 1985.
9. **Lemm, U., Tenczer, J., Baudisch, H., and Krause, W.,** Antibody-mediated extraction of the main tetrahydrocannabinol metabolite, 11-nor-Δ^9-tetrahydrocannabinol-9-carboxylic acid, from human urine and its identification by gas chromatography-mass spectrometry in the sub-nanogram range, *J. Chromatogr.*, 342, 393, 1985.
10. **Rosenfeld, J. M., McLeod, R. A., and Foltz, R. L.,** Solid-supported reagents in the determination of cannabinoids in plasma, *Anal. Chem.*, 58, 716, 1986.
11. **McBurney, L. J., Bobbie, B. A., and Sepp, L. A.,** GC/MS and EMIT analyses for Δ^9-tetrahydrocannabinol metabolites in plasma and urine of human subjects, *J. Anal. Toxicol.*, 10, 56, 1986.
12. **ElSohly, M. A., ElSohly, H. N., Jones, A. B., Dimson, P. A., and Wells, K. E.,** Analysis of the major metabolite of Δ^9-tetrahydrocannabinol in urine II. A HPLC procedure, *J. Anal. Toxicol.*, 7, 262, 1983.
13. **Peat, M. A., Deyman, M. E., and Johnson, J. R.,** High performance liquid chromatography-immunoassay of Δ^9-tetrahydrocannabinol and its metabolites in urine, *J. Forensic Sci.*, 29, 110, 1984.

CANNABIS[a] (continued)

14. **Nakahara, Y. and Sekine, H.,** Studies on confirmation of cannabis use. I. Determination of the cannabinoid contents in marijuana cigarette, tar, and ash using high performance liquid chromatography with electrochemical detection, *J. Anal. Toxicol.*, 9, 121, 1985.
15. **Kanter, S. L., Hollister, L. E., and Zamora, J. U.,** Marijuana metabolites in urine of man XI. Detection of unconjugated and conjugated Δ^9-tetrahydrocannabinol-11-oic acid by thin-layer chromatography, *J. Chromatogr.*, 235, 507, 1982.
16. **Kanter, S. L., Hollister, L. E., and Musumeci, M.,** Marijuana metabolites in urine of man X. Identification of marijuana use by detection of Δ^9-tetrahydrocannabinol-11-oic acid using thin-layer chromatography, *J. Chromatogr.*, 234, 201, 1982.
17. **Kaistha, K. K. and Tadrus, R.,** Semi-quantitative thin-layer mass-screening detection of 11-nor-Δ^9-tetrahydrocannabinol-9-carboxylic acid in human urine, *J. Chromatogr.*, 237, 528, 1982.
18. **Vinson, J. A. and Patel, A. H.,** Detection and quantitation of tetrahydrocannabinol in serum using thin-layer chromatography and fluorometry, *J. Chromatogr.*, 307, 493, 1984.
19. **Kogan, M. J., Newman, E., and Willson, N. J.,** Detection of marijuana metabolite 11-nor-Δ^9-tetrahydrocannabinol-9-carboxylic acid in human urine by bonded-phase adsorption and thin-layer chromatography, *J. Chromatogr.*, 306, 441, 1984.

CANRENONE

Liquid Chromatography

Specimen (mℓ)	Extraction	Column (cm × mm)	Packing (μm)	Elution	Flow (mℓ/min)	Det. (nm)	RT (min)	Internal standard (RT)	Other compounds (RT)	Ref.
Serum, urine (1)	I-1	25 × 2.5	LiChrosorb SI-100 (10)	E-1	2.0	ABS (283)	4.5	—	—	1
Serum (1)	I-2	25 × 4.6	Nucleosil SI-100 (10)	E-2	3.0	ABS (283)	12	Androstadien-17β-ol-3-one (6)	—	2
Plasma, urine (0.2)	I-3	12.5 × 4.6	Spherisorb-ODS-2 (5)[a]	E-3	1.0	ABS (285)	7.8	Spirorenone (5.1)	b	3

[a] Protected by a precolumn.
[b] Spironolactone and metabolites formed after dethioacetylation are detected only at 235 nm.

Extraction — I-1. The sample was vigorously mixed with 200 μℓ of chloroform and centrifuged. Aliquots of 50 to 100 μℓ were injected.
I-2. To the sample are added 100 μℓ of an aqueous solution (1.5 mg/ℓ) of the internal standard and 5 mℓ of carbon tetrachloride. The organic phase is evaporated at 50°C under a gentle stream of nitrogen. The residue is dissolved in 100 μℓ of methanol and 60 μℓ are injected.

I-3. The sample was spiked with 300 ng of the internal standard and extracted twice with 1-mℓ portions of toluene-*n*-hexane (1:1). The combined organic extracts were evaporated under a stream of nitrogen. The residue was dissolved in 250 μℓ of the mobile phase and 200 μℓ were injected.

Elution — E-1. Chloroform-*n*-hexane (50:50).
E-2. Methanol-diisopropyl ether (2.5:97.5).
E-3. Methanol-water (60:40).

REFERENCES

1. **Neurath, G. B. and Ambrosius, D.,** High-performance liquid chromatographic-determination of canrenone, a major metabolite of spironolactone, in body fluids, *J. Chromatogr.*, 163, 230, 1979.
2. **Besenfelder, E. and Endele, R.,** High performance liquid chromatographic method for determination of canrenone in serum, *J. High Resol. Chromatogr. Chromatogr. Commun.*, 4, 419, 1981.
3. **Krause, W., Karras, J. and Jakobs, U.,** Determination of canrenone, the major metabolite of spironolactone, in plasma and urine by high-performance liquid chromatography, *J. Chromatogr.*, 277, 191, 1983.

CAPTOPRIL

Gas Chromatography

Specimen (mℓ)	Extraction	Column (m × mm)	Packing (mesh)	Oven temp (°C)	Gas (mℓ/min)	Det.	RT (min)	Internal standard (RT)	Deriv.	Other compounds (RT)	Ref.
Blood (1)	I-1	1 × 2	2% OV-210 GasChromQ (80/100)	195	He (30)	MS-EI	5.5	Captopril-N-hexyl-maleimide (11)	N-Ethyl-maleimide-hexafluoro-isopropyl	—	1
Plasma, urine (0.1—1)	I-2	2 × 2	3% OV-101 Chromosorb W (100/120)	T.P.[a]	He (30)	MS-EI[b]	9.8	YS-980 (10.5)	N-Ethyl-maleimide-hexafluoro-isopropyl	S-Methyl-captopril (3.4) Captoprildi-sulfide (12.6)	2
Blood, plasma (1)	I-3	1.8 × 2	3% OV-101 Chromosorb W (100/120)	220	Ar: 95-Methane :5 (40)	ECD	3.4	SQ 25, 761 (5.8)	N-Ethyl-maleimide-hexafluoro-isopropyl	—	3

CAPTOPRIL (continued)

Gas Chromatography

Specimen (mℓ)	Extraction	Column (m × mm)	Packing (mesh)	Oven temp (°C)	Gas (mℓ/min)	Det.	RT (min)	Internal standard (RT)	Deriv.	Other compounds (RT)	Ref.
Blood (5)	I-4	15 × 0.22	CP sil 19CB[c] (0.20 μm)[d]	T.P.[a]	He[f]	MSD[g]	5.68, 5.80[b]	4-Fluoro-captopril (5.9, 6.1)[b]	N-Ethyl-maleimide-methyl	S-Benzoyl-captopril (5.2), 4-Fluoro-S-benzoyl-captopril (5.3)[i]	4, 5
Pure compounds	j	10 × 0.22	CP sil 19CB[c] (0.20)[d]	T.P.[k]	He[f]	ECD	5.6, 5.8[b]	—	N-Ethyl-maleimide-methyl	S-Benzoyl-captopril (4.8)	6

Liquid Chromatography

Specimen (mℓ)	Extraction	Column (cm × mm)	Packing (μm)	Elution	Flow (mℓ/min)	Det. (nm)	RT (min)	Internal standard (RT)	Other compounds (RT)	Ref.
Blood[m] (3)	I-5	30 × 3.9	μ-Bondapak-C_{18} (10)	E-1	NA	ABS (254)	7	Thiosalicylic acid (12)	—	7
Plasma (1)	I-6	10 × 5	Hypersil-ODS (5)	E-2	1.0	Electrochem[m]	3	—	—	8
Blood (0.5)	I-7	30 × 3.9	μ-Bondapak-C_{18} (10)	E-3	1.0	Electrochem	10	SA 446 (20)	—	9
Plasma (0.5)	I-8	30 × 3.9	μ-Bondapak-C_{18} (10)	E-4	1.5	Fl (385, 515)	4.5	SQ 25, 233 (7.5)	—	10
Plasma (1)	I-9	30 × 3.9	μ-Bondapak-C_{18} (10)	E-5	1.0	ABS (254)	21	p (32)	—	11

[a] Initial temp = 150°C; initial time = 1 min; rate = 10°C/min; final temp = 290°C.
[b] Chemical ionization was also used with isobutane as the reagent gas.

[c] Cyanopropylphenylmethylpolysiloxane.
[d] Film thickness.
[e] Initial temp = 190°C; initial time = 1 min; rate = 25°C/min; final temp = 285°C; final time = 2.5 min.
[f] Inlet pressure = 7 psi.
[g] Mass selective detector.
[h] Double peaks due to diasteroisomers.
[i] Used as the internal standard for the assay of S-benzoyl captopril.
[j] See I-4 for derivatization.
[k] Initial temp = 180°C; initial time = 1 min; rate = 30°C/min; final temp = 250°C; final time = 5 min.
[l] Inlet pressure = 20 psi.
[m] A separate procedure for the extraction of urine is also described.
[n] Gold-mercury electrode at +0.07 V.
[o] Potential = + 0.9 V.
[p] (4R)-2-(2-Hydroxyphenyl)-3-(3-mercaptopropionyl)-4-thiazolidine-carboxylic acid.

Extraction — I-1. The sample was added to a 0.2% solution of N-ethylmaleimide in 0.1 *M* phosphate buffer and treated twice with 5-mℓ volumes of 10% metaphosphoric acid. After removal of the precipitate by centrifugation at 1600 × *g* for 10 min, the supernatant was extracted three times with 10-mℓ volumes of ethyl acetate. The combined organic layers were evaporated *in vacuo*. The residue was dissolved in 5 mℓ of acetate buffer, pH 6 and washed twice with 5-mℓ volumes of ethyl acetate. The aqueous layer was acidified with 0.1 *N* HCl (5 mℓ) and extracted three times with 5-mℓ volumes of ethyl acetate. The combined organic layers were evaporated *in vacuo*. The residue was spiked with the internal standard, the mixture treated with hexafluoroisopropanol (0.3 mℓ) and trifluoroacetic anhydride (0.05 mℓ) and the resulting solution was removed with a stream of nitrogen, the residue dissolved in 0.2 mℓ of ethyl acetate and a 5-μℓ aliquot of this solution was injected.
I-2. The sample was treated with 1 mℓ of 0.1 *M* phosphate buffer, pH 7.4, 10 μℓ of the internal standard solution (1 mg/mℓ in acetone), and 100 μℓ of N-ethylmaleimide (10 mg/mℓ in water). After 10 min, the mixture was washed with ethyl acetate, acidified with 0.5 mℓ of 2 *M* HCl, saturated with sodium chloride (2 g), and extracted with 10 mℓ of ethyl acetate. The organic layer was evaporated under a stream of nitrogen and the residue treated with 50 μℓ each of hexafluoroisopropanol and perfluorobutyric anhydride. The mixture was incubated at 60°C for 15 min when excess reagents were removed with nitrogen. The residue was reconstituted with 100 μℓ of ethyl acetate, aliquots of this solution were injected.
I-3. Plasma (1 mℓ) was spiked with 800 ng of the internal standard, treated with 0.1 mℓ of a 2% solution of tributylphosphine in methanol, and the mixture incubated at 50°C for 30 min. After cooling, the sample was washed with 10 mℓ of hexane. The aqueous layer was then treated with 0.2 mℓ of 2.5% aqueous solution of N-ethylmaleimide and allowed to stand at room temp for 15 min and the mixture washed with 12 mℓ of benzene. The organic layer was evaporated and treated with hexafluoroisopropanol (100 μℓ) and acetic anhydride (10 μℓ) at 50°C for 1 hr. Excess reagents were removed with a stream of nitrogen, the residue dissolved in 1 mℓ of benzene. Aliquots (5 μℓ) of this solution were injected.
I-4. The sample was treated with 75 mg of N-ethylmaleimide, 100 μℓ acetone solution of the internal standards, 5 mℓ of acetone was added slowly followed by rapid addition of another 10 mℓ of acetone. After centrifugation, acetone solution was evaporated at 65°C under a stream of nitrogen. The residual precipitate was extracted with 10 mℓ of aqueous acetone (15% water in acetone) and the extract combined with the first extract and evaporated. The residual aqueous solution was acidified with 10 mℓ of 0.1 *N* HCl and applied to an activated XAD-2 column. The column was eluted with ethyl acetate which was back extracted into 5% sodium

CAPTOPRIL (continued)

bicarbonate. The aqueous phase was acidified and re-extracted into ethyl acetate which was evaporated. The dry residue was methylated by incubation with 0.1 mℓ of methanolic HCl at 60°C for 10 min. The excess reagent was removed at room temp in a vacuum desiccator. The residue was dissolved in 20 μℓ of acetone and aliquots of 5 μℓ of this solution were injected.

I-5. The sample was treated with 1.5 mℓ of 0.5% solution of *p*-bromophenacylbromide in acetone and immediately vortexed. The mixture was allowed to stand for 5 min, made acidic by addition of 0.3 *N* HCl, and extracted successively with 16 and 8 mℓ of benzene. The combined organic extracts were evaporated *in vacuo*. The residue was dissolved in 4 mℓ of 0.05 *M* phosphate buffer, pH 7 and washed with 6 mℓ of hexane. The aqueous layer was made acidic with 0.1 mℓ of 2 *N* HCl and extracted successively with 6 and 2 mℓ of benzene. The combined extracts were spiked with 0.5 μg of *p*-bromophenacyl adduct of thiosalicylic acid and evaporated, the residue dissolved in 200 μℓ of acetonitrile, and aliquots of 5 to 25 μℓ were injected.

I-6. Plasma proteins are precipitated by the addition of 0.1 mℓ of sulfosalicylic acid (250 mg/mℓ), centrifuged, and aliquots of 10 to 25 μℓ were injected.

I-7. The sample was treated with 20 μℓ of 1% solution of N-(4-dimethylaminophenyl)maleimide in acetone, 0.3 mℓ of 1/30 *M* phosphate buffer (pH 6.8) and 100 ng of the internal standard, vortex mixed, and allowed to stand at 0°C for 30 min. The mixture was then washed twice with 2-mℓ portions of diethyl ether. To the aqueous layer 500 μg of glutathione was added and the solution was kept at 0°C for 20 min. The resulting solution was treated with acetone (3 mℓ) and centrifuged. The precipitate was washed with acetone. The supernatant and washings were combined and concentrated under reduced pressure at room temp to about 1 mℓ. The residual liquid was diluted with 6 mℓ of water and passed through a Sep-Pak C_{18} cartridge. The cartridge was washed with water (2 mℓ) and eluted with acetonitrile (8 mℓ). The eluate was evaporated to dryness *in vacuo* below 40°C. The residue was dissolved in methanol (200 μℓ) and an aliquot of this solution was injected.

I-8. The sample was treated with an equal volume of 10% trichloroacetic acid containing 5 μg of the internal standard. The mixture was vortexed and centrifuged at 0°C. To 50 μℓ of the supernatant was added a mixture 100 μℓ of borate buffer (2.5 *M*, pH 9.5) containing 4 m*M* Na_2EDTA, 50 μℓ of a 0.4 mg/mℓ solution of ammonium 7-fluorobenzo-2-oxa-1,3-diazole-4-sulfonate in the borate buffer, and 10 μℓ of a solution of tri-*n*-butylphosphine in dimethylacetamide. The solution was vigorously mixed, allowed to stand at 60°C for 1 hr, and an aliquot of 30 μℓ of the cooled reaction mixture was injected.

I-9. To the sample were added 2 mℓ of 0.1 *M* phosphate buffer, pH 6 and 0.5 mℓ of 0.5% N-(4-benzoylphenyl)maleimide in acetone solution. The tube was vortex mixed for 15 sec and allowed to stand at room temp for 10 min. To the mixture 2 mℓ of 0.5 *M* phosphate buffer (pH 7) and 0.1 mℓ of the N-(4-benzoylphenyl)maleimide adduct of the internal standard in 0.1 mℓ of acetone were added. The mixture was washed twice with 4 mℓ of diethyl ether, acidified with 0.5 mℓ of 6 *M* HCl, and extracted with 7 mℓ of chloroform. The organic layer was evaporated at 40°C under a stream of air. The residue was dissolved in 2 mℓ of 0.5 *M* phophate buffer and washed twice with 4 mℓ of diethyl ether. The aqueous layer was acidified with 6 *M* HCl and re-extracted with 7 mℓ of chloroform. The organic layer was evaporated to dryness. The residue was dissolved in 100 μℓ of methanol and a 20-μℓ aliquot of this solution was injected.

Elution — E-1. Acetonitrile-water-acetic acid (48:51.5:0.5).
E-2. Methanol-0.1 *M* KH_2PO_4, pH 2 (35:65).
E-3. Acetonitrile-0.8% $NH_4H_2PO_4$, pH 3 (1:2).
E-4. Methanol-1% phosphoric acid (35:65).
E-5. Acetonitrile-methanol-1% acetic acid (45:11:75).

REFERENCES

1. **Matsuki, Y., Fukuhara, K., Ito, T., Ono, H., Ohara, N., Yui, T., and Nambara, T.,** Determination of captopril in biological fluids by gas-liquid chromatography, *J. Chromatogr.*, 188, 177, 1980.
2. **Drummer, O. H., Jarrott, B., and Louis, W. J.,** Combined gas chromatographic-mass spectrometric procedure for the measurement of captopril and sulfur-conjugated metabolites of captopril in plasma and urine, *J. Chromatogr.*, 305, 83, 1984.
3. **Bathala, M. S., Weinstein, S. H., Meeker, F. S., Jr., Singhvi, S. M., and Migdalof, B. H.,** Quantitative determination of captopril in blood and captopril and its disulfide metabolites in plasma by gas chromatography, *J. Pharm. Sci.*, 73, 340, 1984.
4. **Jemal, M., Ivashkiv, E., and Cohen, A. I.,** Simultaneous determination of captopril and S-benzoyl captopril in human blood by capillary gas chromatography-mass selective detection, *Biomed. Mass Spectrom.*, 12, 664, 1985.
5. **Cohen, A. I., Ivashkiv, E., McCormick, T., and McKinstry, D. N.,** Identification and determination of the S-methyl metabolite of captopril in human plasma by selected-ion monitoring gas chromatography-mass spectrometry, *J. Pharm. Sci.*, 73, 1493, 1984.
6. **Jemal, M. and Cohen, A. I.,** N-ethylmaleimide as a new electrophoric derivatizing reagent for the gas chromatography of thiols, *Anal. Chem.*, 57, 2407, 1985.
7. **Kawahara, Y., Hisaoka, M., Yamazaki, Y., Inage, A., and Morioka, T.,** Determination of captopril in blood and urine by high-performance liquid chromatography, *Chem. Pharm. Bull.*, 29, 150, 1981.
8. **Perrett, D. and Drury, P. L.,** The determination of captopril in physiological fluids using high performance liquid chromatography with electrochemical detection, *J. Liq. Chromatogr.*, 5, 97, 1982.
9. **Shimada, K., Tanaka, M., Nambara, T., Imai, Y., Abe, K., and Yoshinaga, K.,** Determination of captopril in human blood by high-performance liquid chromatography with electrochemical detection, *J. Chromatogr.*, 227, 445, 1982.
10. **Toyo'Oka, T., Imai, K., and Kawahara, Y.,** Determination of total captopril in dog plasma by HPLC after prelabelling with ammonium 7-fluorobenzo-2-oxa-1,3-diazole-4-sulphonate (SBD-F), *J. Pharm. Biomed. Anal.*, 2, 473, 1984.
11. **Hayashi, K., Miyamoto, M., and Sekine, Y.,** Determination of captopril and its mixed disulphides in plasma and urine by high-performance liquid chromatography, *J. Chromatogr.*, 338, 161, 1985.

CARAMIPHEN

Gas Chromatography

Specimen (mℓ)	Extraction	Column (m × mm)	Packing (mesh)	Oven temp (°C)	Gas mℓ/min	Det.	RT (min)	Internal standard (T)	Deriv.	Other compounds (RT)	Ref.
Blood (5)	I-1	1.8 × 4	3% OV-17 Chromosorb W (80/100)	270	He (60)	NPD	1.7	Diphenyl-pyraline (2.7)	—	—	1

CARAMIPHEN (continued)

Extraction — I-1. The sample was spiked with a 0.5-mℓ aliquot of an aqueous solution of the internal standard (0.5 μg/mℓ), pH adjusted to 9 with 5% sodium hydroxide, and extracted twice with 1 mℓ aliquots of hexane. The combined organic extracts were evaporated under a nitrogen stream. The residue was dissolved in 25 μℓ of methanol and aliquots of 2 μℓ of this solution were injected.

REFERENCE

1. **Levandoski, P. and Flanagan, T.,** Use of nitrogen-specific detector for GLC determination of caramiphen in whole blood, *J. Pharm. Sci.*, 69, 1353, 1980.

CARBAMAZEPINE

Gas Chromatography

Specimen (mℓ)	Extraction	Column (m × mm)	Packing (mesh)	Oven temp (°C)	Gas (mℓ/min)	Det.	RT (min)	Internal standard (RT)	Deriv.	Other compounds (RT)	Ref.
Plasma (1)	I-1	1.8 × 4	3% OV-17 Varaport (80/100)	270	N_2 (70)	FID	5	Cyheptamide (4)	—	CBZ-E[a] (6.5)	1
Serum (1)	I-2	1 × 2	1% Cyclohexane dimethanol succinate Diatomite C.Q. (100/120)	250	He (45)	FID	5.5	Dehydroepiandrosterone (6.8)	—	—	2

Liquid Chromatography

Specimen (mℓ)	Extraction	Column (cm × mm)	Packing (μm)	Elution	Flow (mℓ/min)	Det. (nm)	RT (min)	Internal standard (RT)	Other compounds (RT)	Ref.
Plasma (0.5)	I-3	15 × 4.6	Supelco LC-18 (5)	E-1	2.5	ABS (212)	4.7	Cyheptamide (7.6)	CBZ-E (2.1)	3
Plasma (0.254)	I-4	25 × 4.6	LiChrosob RP-8 (10)	E-2	1.8	ABS (215)	5.9	10-Methoxycarbamazepine (7.4)	CBZ-E (3.9)	4

Plasma (0.1)	I-5	30 × 2.9	μ-Bondapak phenyl (10)[b]	E-3	NA	ABS (254)	5.7	Lorazepam (6.7) N-Demethyldiazepam (8.9)	CBZ-E (4.1)5
Plasma (0.5)	I-6	25 × 4.6	LiChroCart RP-8 (5)[c]	E-4	1.2	ABS (215)	6.6	Clonazepam (8.8)	CBZ-E (4.8)6
Plasma (1)	I-7	30 × 3.9	μ-Bondapak-C_{18} (10)	E-5	1.0	ABS (215)	8.2	10-Methoxycarbamaze-pine (10.3)	CBZ-E (5.8)7

[a] Carbamazepine epoxide.
[b] Protected by a 70 × 6 mm precolumn packed with Corasil C_{18} (37 to 50 μm).
[c] Protected by a 75 × 4.6 mm precolumn packed with C_{18}-silica (30 μm).

Extraction — I-1. The sample was spiked with the residue after evaporation of 40 μℓ of the internal standard solution (0.2 mg/mℓ in acetone). The sample was alkalinized with 0.2 mℓ of 1.5 *N* NaOH, dichloromethane (1 mℓ) was added and after shaking, 2 mℓ of petroleum ether was added. The sample was swirled without actually mixing the three layers. Petroleum ether was discarded, 6 mℓ of dichloromethane added and the tubes shaken on a horizontally rotating device. After centrifugation, the dichloromethane layer was evaporated at 35°C by a nitrogen stream. The residue was dissolved with 20 μℓ of acetone and aliquots of 1 to 1.5 μℓ of this solution were injected.
I-2. The sample is spiked with the residue after evaporation of 1 mℓ of the internal standard solution (0.7 mg/100 mℓ in chloroform), 1 mℓ of 5 *M* sodium hydroxide added, and extracted with 5 mℓ of chloroform. The chloroform layer is evaporated at 60°C under a stream of nitrogen. The residue is dissolved in 20 μℓ of chloroform and a 3-μℓ aliquot is injected.
I-3. The sample is spiked with the residue after evaporation of 15 μℓ of the internal standard solution (0.25 mg/mℓ in methanol), 1 mℓ of phosphate buffer (0.1 *M*, pH 7.4) is added, and the mixture extracted with 10 mℓ of chloroform. The chloroform layer is evaporated, the residue dissolved in 200 μℓ of mobile phase, and aliquots of 10 μℓ of this solution are injected.
I-4. The sample after the addition of the internal standard (4 μg) was made alkaline with 250 μℓ of 4 *N* sodium hydroxide and extracted with 2 mℓ of dichloromethane. The organic layer was evaporated at 40°C under a stream of nitrogen. The residue was dissolved in 100 μℓ of acetonitrile and aliquots of 10 μℓ were injected.
I-5. The sample was spiked with 50 μℓ of an aqueous solution of the 2 internal standards (lorazepam 10 μg/mℓ, desmethyldiazepam 15 μg/mℓ), 100 μℓ of an aqueous saturated solution of tribasic sodium phosphate added, and the mixture extracted with 2 mℓ of chloroform. The organic phase was evaporated at 42°C under reduced pressure and the residue dissolved in 100 μℓ of the mobile phase. Aliquots of 50 μℓ of this solution were injected with an autosampler.

CARBAMAZEPINE (continued)

Elution — E-1. Methanol-water (45:55).
E-2. Acetonitrile-water (35:65).
E-3. Water-acetonitrile-methanol (62:35:3).
E-4. Acetonitrile-water (40:60).
E-5. Methanol-water (55:45).

REFERENCES

1. **Ranise, A., Benassi, E., and Besio, G.,** Rapid gas chromatographic method for the determination of carbamazepine and unrearranged carbamazepine-10,11, epoxide in human plasma, *J. Chromatogr.*, 222, 120, 1981.
2. **Cocks, D. A., Dyer, T. F., and Edgar, K.,** Simple and rapid gas-liquid chromatographic method for estimating carbamazeine in serum, *J. Chromatogr.*, 222, 496, 1981.
3. **Sawchuk, R. J. and Cartier, L. L.,** Simultaneous liquid-chromatographic determination of carbamazepine and its epoxide metabolite in plasma, *Clin. Chem.*, 28, 2127, 1982.
4. **Elyas, A. A., Ratnaraj, N., Goldberg, V. D., and Lascelles, P. T.,** Routine monitoring of carbamazepine and carbamazepine-10,11-epoxide in plasma by high-performance liquid chromatography using 10-methoxycarbamazepine as internal standard, *J. Chromatogr.*, 231, 93, 1982.
5. **Garsela, D. M. and Rocci, M. L., Jr.,** Liquid chromatographic microassay for carbamazepine and its 10,11-epoxide in plasma, *J. Pharm. Sci.*, 73, 1874, 1984.
6. **Chan, K.,** Simultaneous determination of carbamazepine and its epoxide metabolite in plasma and urine by high-performance liquid chromatography, *J. Chromatogr.*, 342, 341, 1985.
7. **Hooper, W. D., King, A. R., Patterson, M., Dickinson, R. G., and Eadie, M. J.,** Simultaneous plasma carbamazepine and carbamazepine-epoxide concentrations in pharmacokinetic and bioavailability studies, *Ther. Drug Monit.*, 7, 36, 1085.

CARBENICILLIN

Liquid Chromatography

Specimen (mℓ)	Extraction	Column (cm × mm)	Packing (μm)	Elution	Flow (mℓ/min)	Det. (nm)	RT (nm)	Internal standard (RT)	Other Compounds (RT)	Ref.
Dosage	—	15 × 4.6	Spherisorb-ODS (5)	E-1	1.3	ABS (254)	12	—	Penicillin G (27)	1

Serum (0.2)	I-1	15 × 4.6	Develosil-ODS (5)[a]	E-2	1.0	ABS (328)	4	—	Ticarcillin (4) Sulbenicillin (6)	2

[a] Column temp = 40°C.

Extraction — I-1. The sample after the addition of 200 μℓ of 10 *M* urea solution was ultrafiltered using an Amicon MPS-1 micropartition system with YMT membranes. To a 200-μℓ aliquot of the ultrafiltrate, 200 μℓ of 2 *M* 1,2,4-triazole reagent (pH 9 and containing $HgCl_2$) were added. The mixture was incubated at 60°C for 10 min. A 30- to 90-μℓ aliquot of the cooled reaction mixture was injected.

Elution — E-1. Methanol-0.005 *M* KH_2PO_4, pH 3.35 (35:65) containing 0.1% tetramethylammonium bromide.
E-2. Acetonitrile-0.0001 *M* phosphate buffer containing 0.005M tetrabutylammonium bromide and sodium thiosulfate (1:1:8).

REFERENCES

1. **Twomey, P. A.,** High-performance liquid chromatographic analysis of carbenicillin and its degradation products, *J. Pharm. Sci.*, 70, 824, 1981.
2. **Haginaka, J. and Wakai, J.,** High-performance liquid chromatographic assay of carbenicillin, ticarcillin and sulbenicillin in serum and urine using pre-column reaction with 1,2,4-triazole and mercury(II) chloride, *Analyst*, 110, 1185, 1985.

CARBIDOPA

Liquid Chromatography

Specimen (mℓ)	Extraction	Column (cm × mm)	Packing (μm)	Elution	Flow (mℓ/min)	Det. (nm)	RT (min)	Internal standard (RT)	Other compounds (RT)	Ref.
Plasma (1)	I-1	25 × 3.2	Spherisorb-ODS (5)[a]	E-1	1.0	Electrochem[b]	9	Dihydroxybenzyl-amine (4)	Levodopa (2.5) 3,4-Dihydroxy-phenyl-acetic acid (6)	1

[a] Protected by a 30 × 3.2 mm precolumn.
[b] Working electrode potential = +0.70 V.

Extraction — I-1. The sample was spiked with 50 ng of the internal standard and treated with 100 mg of acid washed alumina and 1 mℓ of 0.5 *M* Tris-EDTA, pH 8.6. After mixing for 5 min, the supernatants were discarded and the alumina washed twice with 10 mℓ of water and eluted with 0.5 mℓ of 0.2 *M* perchloric acid and 20 μℓ of the eluate were injected.

CARBIDOPA (continued)

Elution — E-1. Methanol-0.1 *M* NaH_2PO_4, 20 m*M* citric acid, 1.25 m*M* sodium octaine sulfonic acid, 0.15 m*M* sodium EDTA (8:92), pH 3.2.

REFERENCE

1. **Nissinen, E. and Taskinen, J.,** Simultaneous determination of carbidopa, levodopa and 3,4-dihydroxyphenyl-acetic acid using high-performance liquid chromatography with electrochemical detection, *J. Chromatogr.*, 231, 459, 1982.

CARBINOXAMINE

Gas Chromatography

Specimen (mℓ)	Extraction	Column (m × mm)	Packing (mesh)	Oven (temp °C)	Gas (mℓ/min)	Det.	RT (min)	Internal standard (RT)	Deriv.	Other compounds (RT)	Ref.
Serum (2)	I-1	10 × 0.25	SE-30 (0.25 μm)[a]	T.P.[b]	He[c]	NPD	3.9	N-Ethylhydrocodone (6.7) Brompheniramine (4.1)	—	Hydrocodone (6.3)	1

[a] Film thickness.
[b] Initial temp= 185°C; initial time = 1.1 min; rate = 25°C/min for 0.9 min; 10°C/min; final temp = 250°C; final time = 1 min.
[c] Linear flow rate = 30 to 40 cm/sec.

Extraction — I-1. The sample was treated with 1 mℓ of the solution of internal standards (40 ng/mℓ of each) in 0.01 *N* HCl, 1 mℓ of 2 *N* KOH and the mixture was extracted with 6 mℓ of dichloromethane 2-propanol (9:1). The organic phase was back extracted into 2 mℓ of 0.1 *N* sulfuric acid. The aqueous phase was made alkaline with 0.3 mℓ of 2 *N* KOH and extracted with 2 mℓ of benzene. The organic phase was evaporated at 35 to 40°C with a stream of air. The residue was constituted with 40 μℓ of 5% methanol in *n*-nonyl alcohol. An aliquot of 2 to 3 μℓ of this solution was injected.

REFERENCE

1. **Hoffman, D. J., Leveque, M. J., and Thomson, T.,** Capillary GLC assay for carbinoxamine and hydrocodone in human serum using nitrogen-sensitive detection, *J. Pharm. Sci.*, 72, 1342, 1983.

CARBOPROST

Liquid Chromatography

Specimen (mℓ)	Extraction	Column (cm × mm)	Packing (μm)	Elution	Flow (mℓ/min)	Det. (nm)	RT (min)	Internal standard (RT)	Other compounds (RT)	Ref.
Dosage	I-1	30 × 3.9	μ-Porasil (10)	E-1	1.8	ABS (254)	13	Gauifenesin (8)	5-*Trans*-isomer of carboprost (18)	1[a]

[a] An altenative procedure of chromatographic determination of carboprost without derivatization with refractive index detection is described.

Extraction — I-1. The sample was treated with 2 mℓ of 0.5 *M* citrate buffer, pH 4 and extracted with 20 mℓ of dichloromethane. A 4-mℓ aliquot of the organic layer was evaporated. The residue was treated with 80 μℓ of a 20 mg/mℓ of α-bromo-2-acetonaphthone in acetonitrile and 60 μℓ of a 10 μℓ/mℓ of diisopropylethylamine in acetonitrile solutions. The mixture was incubated at 30 to 35°C for 15 min. The solution was evaporated to dryness and the residue constituted with 2 mℓ of a 7 mg/mℓ of the internal standard in the mobile phase. Aliquots of this solution were injected.

Elution — E-1. Dichloromethane-1,3-butanediol-water (496:3.5:0.25).

REFERENCE

1. **Brown, L. W. and Carpenter, B. E.,** Comparison of two high-pressure liquid chromatographic assays for carboprost, a synthetic prostaglandin, *J. Pharm. Sci.*, 69, 1396, 1980.

CARBOQUONE

Liquid Chromatography

Specimen (mℓ)	Extraction	Column (cm × mm)	Packing (μm)	Elution	Flow (mℓ/min)	Det. (nm)	RT (min)	Internal standard (RT)	Other compounds (RT)	Ref.
Plasma, ascites (2)	I-1	30 × 3.9	μ-Bondapak-C_{18} (10)	E-1	1.5	ABS (340)	3	2,5-Diethyleneimino-3,6-dimethylbenzoquinone (4.5)	—	1

Extraction — I-1. The sample was mixed with 3 mℓ of an aqueous solution of the internal standard and extracted with 10 mℓ of chloroform. The organic layer was evaporated in a rotary evaporator after the addition of 2 drops of ethylene glycol. The residue was dissolved with 200 μℓ of methyl alcohol and an aliquot of 10 to 20 μℓ of this solution was injected.

Elution — E-1. Acetonitrile-water (36:64)

REFERENCE

1. **Hisaoka, M., Morioka, T., and Yagita, A.,** Determination of carboquone in plasma and ascites by high performance liquid chromatography after intravenous and intraperitoneal administration in man, *Gann*, 73, 161, 1982.

CARISOPRODOL

Gas Chromatography

Specimen (mℓ)	Extraction	Column (m × mm)	Packing (mesh)	Oven temp (°C)	Gas mℓ/min	Det.	RT (min)	Internal standard (RT)	Deriv.	Other compounds (RT)	Ref.
Plasma (1)	I-1	2 × 2	GP 3% SP2100 DB Supelcoport (100/120)	180	He (60)	NPD	4.5	Tybamate (9.5)	—	—	1[a]

[a] The procedure was developed in two different laboratories with some differences in the extraction procedure and chromatographic parameters.

Extraction — I-1. Plasma was mixed with the residue obtained after the evaporation of 100 μℓ of the internal standard solution (100 μg/mℓ in methanol) and extracted with 5 mℓ of chloroform. The organic phase was evaporated under nitrogen, the residue dissolved in 0.5 mℓ of methanol, and aliquots of this solution were injected.

REFERENCE

1. **Kucharczyk, N., Segelman, F. H., Kelton, E., Summers, J., and Sofia, R. D.**, Gas chromatographic determination of carisoprodol in human plasma, *J. Chromatogr.*, 377, 384, 1986.

CARMINOMYCIN

Liquid Chromatography

Specimen (mℓ)	Extraction	Column (cm × mm)	Packing (mg)	Elution	Flow (mℓ/min)	Det. (nm)	RT (min)	Internal standard (RT)	Other compounds (RT)	Ref.
Serum (2)	I-1	30 × 3.9	μ-Bondapak-C_{18} (10)	E-1	1.9	Fl (470[a], 560)	5.8	Adriamycin (2.6)	Carminomycinol (4)	1
Plasma (1)	I-2	25 × 4.6	μ-Bondapak-C_{18} (10)	E-2	2	Fl (490, 550)	10.9[b]	4′-Epiadriamycin (5.8)[b]	Carminomycinol (8.5)[b]	2

[a] An interference filter (No. 5-60) at 380 to 480 nm was used. The excitation maximum is at 470 nm.
[b] Capacity factors.

Extraction — I-1. The sample was spiked with 100 μℓ of a methanolic solution of the internal standard (1 μg/mℓ) and washed with 5 mℓ of hexane. The aqueous layer was then extracted with 5 mℓ of chloroform. The organic layer was washed with 1 mℓ of 0.1 *M* phosphate buffer, pH 7.5, and then evaporated under a stream of nitrogen at 37°C. The residue was dissolved in 100 μℓ of methanol, 80 μℓ of which were injected.
I-2. Plasma was spiked with the internal standard (final concentration = $4.3 \times 10^{-7}M$) and extracted with 5 mℓ of chloroform-isopropanol (4:1). The organic layer was evaporated with air at 35°C. The residue was dissolved in 150 μℓ of methanol, of this solution, 100 μℓ was injected.

Elution — E-1. Acetonitrile-0.1 *M* ammonium acetate buffer, pH 4 (40:60).
E-2. Acetonitrile-0.01 *M* phosphate buffer, pH 4 (2:3).

REFERENCES

1. **Fandrich, S. E. and Pittman, K. A.**, Analysis of carminomycin in human serum by fluorometric high-performance liquid chromatography, *J. Chromatogr.*, 223, 155, 1981.

CARMINOMYCIN (continued)

2. **Lankelma, J., Penders, P. G. M., McVie, J. G., Leyva, A., Ten Bokkel-Huinink, W. W., De Planque, M. M., and Pinedo, H. M.,** Plasma concentrations of carminomycin and carminomycinol in man, measured by high pressure liquid chromatography, *Eur. J. Cancer Clin. Oncol.*, 18, 363, 1982.

CARPIPRAMINE

Liquid Chromatography

Specimen (mℓ)	Extraction	Column (cm × mm)	Packing (μm)	Elution	Flow (mℓ/min)	Det. (nm)	RT (min)	Internal standard (RT)	Other compounds (RT)	Ref.
Plasma (0.3—3)	I-1	15 × 3.9	Zorbax-SIL (5)	E-1	0.4	ABS (250)	12.6	Chlorpromazine (9.7)	—	1

Extraction — E-1. The sample after the addition of 1 mℓ of 10% NaOH and water (2 mℓ) was extracted with 25 mℓ of *n*-heptane containing 1.5% isoamyl alcohol. The organic layer was back extracted with 5 mℓ of 10 M HCl. To 4.5 mℓ of the aqueous layer, 1 mℓ of an aqueous solution containing the internal standard (200 ng/mℓ) was added. The solution was made alkaline with 0.5 mℓ of 40% sodium hydroxide and extracted with 100 μℓ of chloroform. A 30-μℓ volume of the chloroform extract was injected.

Elution — E-1. Dichloromethane-methanol-0.2% aqueous ammonia (89.8:10:0.2).

REFERENCE

1. **Sadanaga, T., Hikida, K., Tameto, K., and Nakanishi, M.,** Determination of carpipramine in plasma by high-performance liquid chromatography, *J. Chromatogr.*, 183, 246, 1980.

CARPROFEN

Gas Chromatography

Specimen (mℓ)	Extraction	Column (m × mm)	Packing (mesh)	Oven temp (°C)	Gas mℓ/min	Det.	RT (min)	Internal standard (RT)	Deriv.	Other compounds (RT)	Ref.
Blood (1)	I-1	1.5 × 2	3% OV-17 Gas Chrom Q (100/120)	280	Iso-butane (NA)	MS-CI	2.3	[$^2H_3$] Carprofen	Methyl	—	1

Liquid Chromatography

Specimen (mℓ)	Extraction	Column (cm × mm)	Packing (μm)	Elution	Flow (mℓ/min)	Det. (nm)	RT (min)	Internal standard (RT)	Other compounds (RT)	Ref.
Blood, urine (5)	I-2	25 × 4.6	Partisil (10)[a]	E-1	2.0	ABS (254)	R,S[b] = 6.4 S,S = 7.6	—	—	2
Dosage	—	NA	Chromegabond-C_{18} (10)	E-2	NA	ABS (254)	32	Benzophenone (15.5)	c	3
Urine (10)	I-3	25 × 4	Hibar RP-18 (5)[d,e]	E-3; grad	0.8	ABS (254)	30.5	N-phenylanthranilic acid (27.2)	Diflunisal (14.9) Tolmetin (14.9) Tiaprofenic acid (14.9) Sulindae (18.8) Indoprofen (19) Zomepirac (20) Ketoprofen (21.8) Nifluminic acid (22.6) Naproxen (23.7) Fenbufen (26.4) Pirprofen (28) Flubiprofen (29.8)	4

CARPROFEN (continued)

Liquid Chromatography

Specimen (mℓ)	Extraction	Column (cm × mm)	Packing (μm)	Elution	Flow (mℓ/min)	Det. (nm)	RT (min)	Internal standard (RT)	Other compounds (RT)	Ref.
									Fenoprofen (29.8) Diclofenac (29.8) Diclofenac (29.8) Indomethacin (30.8) Lonazolac (31.4) Flupnamic acid (32.5) Ibuprofen (34.4)	

Thin-Layer Chromatography

Specimen (mℓ)	Extraction	Plate (Manufacturer)	Layer (mm)	Solvent	Post-separation treatment	Det. (nm)	Rf	Internal standard (Rf)	Other compounds (Rf)	Ref.
Urine (5)	I-4	20 × 20 cm (Merck)	Silica gel 60 (0.25)	S-1	SP: 10% Ferric chloride-1% potassium ferricyanide (1:2)	Visual	0.49	—	Zomepirae (0.32) Diclofenac (0.55) Mefanamie acid (0.7) Ketoprofen (0.1)[f]	5

[a] The column was protected by a guard column packed with Pellosil (38 μm) silica gel.
[b] Diastereomers after derivatization with S-(-)-α-methylbenzylamine.
[c] Separation of known impurities and possible degradation products of carprofen is shown.
[d] Protected by a 30 × 4 mm guard column packed with Pherisorb RP-8 (30 to 40 μm).
[e] Column temp = 35°C.
[f] This drug does not produce blue color with spray reagent.

Extraction — I-1. The sample was diluted with 1 mℓ of water, 1 nmol of an aqueous solution of the internal standard, and 0.3 mℓ of 5 *M* acetate buffer, pH 4.5 were added. The solution was then extracted with 10 mℓ of benzene. An aliquot of 8 mℓ of the organic layer was evaporated, the residue was dissolved in 100 μℓ of methanol, treated with 1 mℓ of ethereal diazomethane, and the solution was allowed to stand at room temp for 1 hr. This solution was evaporated and the residue was reconstituted in 50 μℓ of chloroform, aliquots of which were injected.
I-2. The sample was adjusted to pH 5 with 2 mℓ of 1 *M* acetate buffer and extracted twice with 12-mℓ portions of ether. The combined ether extracts were evaporated with nitrogen at 30 to 40°C. The residue was dissolved in 100 μℓ of ethanol and spotted on a silica gel 60 F_{254} TLC plate as a 3-cm wide streak. The plate was developed in chloroform-acetic acid (90:10). The band corresponding to carprofen (Rf = 0.54) was scraped off the plate and, after the addition of 3 mℓ of 0.2 *M* acetate buffer (pH 5), was extracted twice with 5-mℓ volumes of ether. The combined ether extracts were evaporated. The residue was treated with 0.1 mℓ of the solution of 1,1-carbonyldiimidazole (32.5 mg/mℓ in chloroform). The mixture was allowed to stand at room temp for 5 min, then 10 μℓ of acetic acid and 25 μℓ of (5)-(-)-α-methylbenzylamine were added stepwise, and the solution well mixed with a vortex mixer and allowed to react at room temp for 20 min. Then 3 mℓ of 0.2 *N* ammonium hydroxide was added and the mixture extracted with 5 mℓ of hexane. A 4.5-mℓ aliquot of hexane layer was evaporated under nitrogen at 30 to 40°C. After drying thoroughly in a vacuum desiccator for 5 min, the residue was dissolved in 0.5 to 2 mℓ of the HPLC mobile phase and a 20-μℓ aliquot was injected.
I-3. To a 10-mℓ aliquot of the urine sample, 1 mℓ of a solution of 100 μg/mℓ of the internal standard in 0.05 *M* boric acid buffer was added. The mixture was then acidified with 1 mℓ of 1 *M* HCl and extracted three times with 20-mℓ portions of diethyl ether. The organic layers were combined and evaporated. The residue was reconstituted in 5 mℓ of a mixture of acetonitrile-water (1:1). Of these solutions 10 to 50 μℓ were injected.
I-4. A 5-mℓ volume of urine was hydrolyzed enzymatically by incubation at 40 to 45°C overnight at pH 5.5. The hydrolyzed urine after adjusting the pH to 4 was extracted twice with 10-mℓ portions of diethyl ether. The combined ether extract was evaporated at room temperature with a stream of nitrogen. The residue was dissolved in 50 μℓ of isopropanol and 5 μℓ of the resulting solution was spotted.

Elution — E-1. Dichloromethane containing 0.75% methanol.
E-2. Methanol-1% acetic acid (65:35).
E-3. (A) Acetonitrile; (B) 0.05 *M* acetate buffer (pH 4.5). Gradient from 25% (A) to 55% (A) in 30 min.

Solvent — S-1. Chloroform-methanol-water (70:30:2).

REFERENCES

1. **Hodshon, B. J., Garland, W. A., Perry, C. W., and Bader, G. J.,** Determination of carprofen in blood by gas chromatography chemical ionization mass spectrometry, *Biomed. Mass Spectrom.*, 6, 325, 1979.
2. **Stoltenborg, J., Puglisi, C. V., Rubio, F., and Vane, F. M.,** High-performance liquid chromatographic determination of stereoselective disposition of carprofen in humans, *J. Pharm. Sci.*, 70, 1207, 1981.
3. **Ross, A. J., Del Mauro, M. D., Sokoloff, H. D., and Casey, D. L.,** Analysis of carprofen dosage forms and drug substance by high-performance liquid chromatography, *J. Pharm. Sci.*, 73, 1211, 1984.
4. **Battista, H. J., Wehinger, G., and Henn, R.,** Separation and identification of non-steroidal antirheumatic drugs containing a free carboxyl function using high-performance liquid chromatography, *J. Chromatogr.*, 345, 77, 1985.
5. **Dettwiler, M., Rippstein, S., and Jeger, A.,** A rapid sensitive determination of carprofen and zomepirac using thin-layer chromatography and gas chromatography-mass spectrometry, *J. Chromatogr.*, 244, 153, 1982.

CARPRONIUM CHLORIDE

Gas Chromatography

Specimen (mℓ)	Extraction	Column (m × mm)	Packing (mesh)	Oven temp (°C)	Gas (mℓ/min)	Det.	RT (min)	Internal standard (RT)	Deriv.	Other compounds (RT)	Ref.
Plasma (2)	I-1	1 × 4	15% FFAP Anakrom SD (90/100)	140	He (50)	MS-EI	8.5[a]	[$^2H_9$]Carpronium chloride	Isopropyl	—	1

[a] Retention time of the major pyrolytic product, N,N-dimethyl-β-aminobutyrate.

Extraction — I-1. The sample was spiked with an aqueous solution of the internal standard and 80 μℓ of 60% perchloric acid were added. After centrifugation the supernatant was evaporated to dryness *in vacuo*. The residue was dissolved in 2 mℓ of 0.1 *M* acetate buffer, pH 5. Iodine reagent (0.2 mℓ) was added to the solution and the iodine complex was extracted with 4 mℓ of 1,2-dichloroethane. The extract was evaporated *in vacuo*, the residue dissolved in 2 mℓ of methanol. Anion exchange resin (Dowex 1 × 10, Cl,10 mg) was added to the solution and mixed for a few minutes. The methanolic supernatant was evaporated. The residue was dissolved in 2 mℓ of isopropanol-HCl reagent and the mixture incubated at 70°C for 1 hr. The residue was dissolved in 50 μℓ of acetonitrile. Aliquots of 2 to 5 μℓ of this solution were applied on the flash heater filament of the pyrolyzer.

REFERENCE

1. **Sano, M., Ohya, K., and Shintani, S.,** Analysis of drugs by pyrolysis, II. An improved method for the determination of carpronium chloride in plasma by selected ion monitoring, *Biomed. Mass Spectrom.*, 7, 1, 1980.

CATECHOLAMINES[a,b]

Liquid Chromatography

Specimen (mℓ)	Extraction	Column (cm × mm)	Packing (μm)	Elution	Flow (mℓ/min)	Det. (nm)	RT (min)	Internal standard (RT)	Other compounds (RT)	Ref.
Plasma, urine (2, 0.5)	I-1	15 × 4.6	Technicon LC-8 (5)	E-1	NA	Electrochem	NE[c] = 4.5 E[c] = 5.5 DA[c] = 12	DHBA[c] (7.5)	—	7

Plasma (1)	I-2	25 × 4	Yanaco ODS-T (10)	E-2	1.0	Electro-chem	NE = 6 E = 7.5 DA = 13.5	DHBA (9.2)	—	8
Urine (1)	I-3	10 × 4.6	Analytichem-SCX (5)[d]	E-3	1.0	Electro-chem	NE = 6 E = 10 DA = 10.8	DHBA (7.5)	—	9
Urine (4.5)	I-4	30 × 3.9	μ-Bondapak-C_{18} (10)	E-4	2.0	Electro-chem	NE = 4 E = 65 DA = 12	DHBA (8.5)	—	10
Plasma (0.5)	I-5	15 × 4.6	TSK ODS-120T (5)	E-5	1.0	Fl (345,485)	NE = 3.2 E = 5 DA = 6.6	Isoproterenol (8.4)	—	11, 12

[a] A large number of procedures for the assay of catecholamines have been published. See References 1 to 5 for recent reviews of analytical procedures for catecholamines.

[b] In the published procedures, catecholamines have been determined with electrochemical detection in the oxidation mode. In a recent report catecholamines have been determined in the combined oxidation/reduction mode.[6]

[c] DA = dopamine, DHBA = dihydroxybenzylamine, E = epinephrine, NE = norepinephrine.

[d] Protected by a Brownlee LiChrosorb RP-18 Precolumn.

Extraction — I-1. The plasma sample was spiked with 100 μℓ of a solution of the internal standard (7 ng/mℓ); 1 mℓ of 2 *M* NH_4OH-NH_4Cl buffer, pH 8.5 containing 0.2% (w/v) diphenylborate-ethanolamine and 0.5% (w/v) of EDTA. The mixture was extracted with 5 mℓ of *n*-heptane-*n*-octanol (19:1) containing 0.25% (w/v) tetraoctylammonium bromide. A 4-mℓ aliquot of the organic layer was mixed with 2 mℓ of *n*-octanol and 250 μℓ of 0.08 *M* acetic acid mixed for 2 min, centrifuged, and 200 μℓ of the aqueous phase were injected.

I-2. Boric acid gel (5 mg; Affi-Gel 601) was activated by allowing it to swell in 1 mℓ of water overnight, then washing it successively with 1 mℓ of 1 *M* HCl, 1 mℓ of water, 1 mℓ of 0.1 *M* NaOH, and finally with two 1-mℓ portions of water. To the activated gel 10 μℓ of 25 m*M* $Na_2S_2O_5$ in water, 5 μℓ of 100 ng/mℓ of the internal standard in 10 m*M* HCl and 1 mℓ of the plasma sample were added. The mixture was shaken for 10 min, supernatant was discarded, gel was washed with 1 mℓ of water, and eluted with 100 μℓ of 0.75 *M* acetic acid. Aliquots of 50 μℓ of the eluate were injected.

I-3. A phenylboronic acid column was conditioned by washing with 1 mℓ of methanol and 1 mℓ of 0.1 *M* HCl. A primary-secondary amine column was attached above the phenylboronic acid column and was conditioned in succession with 2 mℓ of methanol, 4 mℓ of NH_4OH (3 mℓ/ℓ), and finally 4 mℓ of 5 m*M* phosphate buffer, pH 8.5. The sample was adjusted to pH 5 ± 0.5 with 0.3% NH_4OH, 75 μℓ of 1 mg/ℓ of the internal standard added, and the mixture applied to the upper column. The sample was allowed to pass through and the columns were rinsed with 4- and 2-mℓ portions of phosphate buffer, pH 8.5. The amine column was removed and the phenylboronic column was washed with 1 mℓ of methanol, 1 mℓ of acetonitrile-pH 8.5 phosphate buffer (1:1) and the column was eluted with 1 mℓ of 0.1 *M* HCl. Aliquots of 20 μℓ of this eluate were injected.

I-4. The sample was treated with 0.2 mℓ of reducing agent (1.9 g of sodium metabisulphite + 1.76 g ascorbic acid/100 mℓ of water), 1 mℓ of the internal standard solution (600 μg/ℓ in 0.1*M* HCl), and 4.5 mℓ of Tris buffer (pH 8.7). Alumina (200 mg) was added and mixed for 20 min. After centrifugation the supernatant was discarded and alumina was washed with 10 mℓ of barbitone buffer, pH 8.5 and eluted with 0.5 mℓ of 0.5 *M* HCl. The clear supernatant was collected and washed with 8 mℓ of ethyl acetate. Aliquots of 50 μℓ of the aqueous phase were injected.

CATECHOLAMINES[a,b] (continued)

I-5. The sample was spiked with 25 μℓ of 10 pmol/mℓ of the internal standard solution and 0.5 mℓ of 0.2 *M* lithium phosphate buffer (pH 5.8) and passed through a prewashed (10 mℓ of water) Toyopak SCX cartridge. The cartridge was then successively washed twice with 5 mℓ of water and once with 50% acetonitrile. The cartridge was then eluted with 300 μℓ of 0.6 *M* KCl-acetonitrile (1:1) containing 0.6 m*M* potassium ferricyanide. The eluate, 50 μℓ of 1,2-diphenylethylenediamine (21 mg/mℓ in ethanol) was added and the mixture incubated at 37°C for 40 min. The reaction mixture was cooled in ice water and a 100 μℓ aliquot of the mixture was injected.

Elution — E-1. Methanol-water (25:75) containing 0.05 *M* sodium acetate, 0.15 *M* acetic acid, 0.01% (w/v) sodium dodecylsulfate, 0.01% (w/v) sodium chloride, and 0.01% (w/v) EDTA.
E-2. Acetonitrile-0.1 *M* citrate buffer, pH 5 containing 6 m*M* sodium 1-octanesulfonate, and 2 m*M* Na_2 EDTA (60:940).
E-3. 0.1 *M* phosphate buffer, pH 3.5 containing 100 mg/ℓ Na_2 EDTA.
E-4. Sodium dihydrogen phosphate (15.6 g), EDTA (40 mg), PIC B8 (1 vial), and 70 mℓ methanol diluted to 1ℓ with water.
E-5. Acetonitrile-methanol-50 m*M* Tris buffer, pH 7 (5:1:4).

1. **Allenmark, S.,** Analysis of catecholamines by HPLC, *J. Liq. Chromatogr.*, 5 (Suppl. 1), 1, 1982.
2. **Tjaden, U. R., de Jong, J., and van Valkenburg, C. F. M.,** Gradient elution of biogenic amines and derivatives in reversed phase ion-pair partition chromatography with electrochemical and fluorometric detection, *J. Liq. Chromatogr.*, 6, 2255, 1983.
3. **Raum, W. J.,** Methods of plasma catecholamine measurement including radioimmunoassay, *Am. J. Physiol.*, 246, E4—E12, 1984.
4. **Hjemdahl, P.,** Catecholamine measurements by high-performance liquid chromatography, *Am. J. Physiol.*, 247, E14—E20, 1984.
5. **Bouloux, P., Perrett, D. and Besser, G. M.,** Methodological considerations in the determination of plasma catecholamines by high-performance liquid chromatography with electrochemical detection, *Ann. Clin. Biochem.*, 22, 194, 1985.
6. **Goldstein, D. S., Stull, R., Zimlichman, R., Levinson, P. D., Smith, H., and Keiser, H. R.,** Simultaneous measurement of DOPA, DOPAC, and catecholamines in plasma by liquid chromatography with electrochemical detection, *ESA Rev.*, 2, 1, 1986.
7. **Smedes, F., Kraak, J. C., and Poppe, H.,** Simple and fast solvent extraction system for selective and quantitative isolation of adrenaline, noradrenaline and dopamine from plasma and urine, *J. Chromatogr.*, 231, 25, 1982.
8. **Maruta, K., Fujita, K., and Ito, S.,** Liquid chromatography of plasma catecholamines, with electrochemical detection, after treatment with boric acid gel, *Clin. Chem.*, 30, 1271, 1984.
9. **Wu, A. H. B. and Gornet, T. G.,** Preparation of urine samples for liquid-chromatographic determination of catecholamines: bonded-phase phenylboronic acid, cation-exchange resin, and alumina adsorbents compared, *Clin. Chem.*, 31, 298, 1985.
10. **Davidson, D. F. and Fitzpatrick, J.,** A simple, optimised and rapid assay for urinary free catecholamines by HPLC with electrochemical detection, *Ann. Clin. Biochem.*, 22, 297, 1985.
11. **Mitsui, A., Nohta, H., and Ohkura, Y.,** High-performance liquid chromatography of plasma catecholamines using 1,2-diphenylethylenediamine as precolumn fluorescence derivatization reagent, *J. Chromatogr.*, 344, 61, 1985.

12. **Nohta, H., Mitsui, A., and Ohkura, Y.,** Spectrofluorimetric determination of catecholamines with 1,2-diphenylethyl-enediamine, *Anal. Chim. Acta,* 165, 171, 1984.

CEFACLOR

Liquid Chromatography

Specimen (mℓ)	Extraction	Column (cm × mm)	Packing (μm)	Elution	Flow (mℓ/min)	Det. (nm)	RT (min)	Internal standard (RT)	Other compounds (RT)	Ref.
Serum (1)	I-1	25 × 4.6	LiChrosorb RP-18 (10)	E-1	2.0	ABS (254)	2.9	—	—	1
Plasma, urine (0.1)	I-2	30 × 3.9	μ-Bondapak-C_{18} (10)	E-2	2.0	ABS (254)	3.8	Cephaloglycin (6)	—	2

Extraction — I-1. The sample was treated with an equal volume of 1 *M* perchloric acid—4.5% dioxane in acetonitrile (85:15) with constant mixing. After centrifugation, 300 μℓ of the clear supernatant were injected.
I-2. The sample was treated with 2.5 μg of an aqueous solution of the internal standard (1 mg/mℓ) and 200 μℓ of methanol. The mixture was vortexed and centrifuged. The supernatant was evaporated under a gentle stream of nitrogen. The residue was dissolved in 100 μℓ of mobile phase and 50 μℓ of the solution were injected.

Elution — E-1. Methanol-dioxane-0.05 *M* citrate buffer, pH 4 (13:7:80).
E-2. Methanol-0.5% acetic acid (20:80).

REFERENCES

1. **Ullmann, U. and Diekmann, H. W.,** Dunnschicht- und Hochdruck-Flussigkeitschromatographie mit Cefaclor in Urin und Serum, *Infection,* 6, 5554, 1979.
2. **Nahata, M. C.,** Determination of cefaclor by high-performance liquid chromatography, *J. Chromatogr.,* 228, 429, 1982.

CEFAMANDOLE

Liquid Chromatography

Specimen (mℓ)	Extraction	Column (cm × mm)	Packing (μm)	Elution	Flow (mℓ/min)	Det. (nm)	RT (min)	Internal standard (RT)	Other compounds (RT)	Ref.
Serum (0.5)	I-1	30 × 3.9	μ-Bondapak-C_{18} (10)[a]	E-1	3.0	ABS (254)	5	—	—	1

[a] A radial pak (10 × 10.8 cm) packed with 10 μm-C_{18} silica with a modified mobile phase was also used (Retention time of cefamandole = 3 min).

Extraction — I-1. The sample was extracted with dichloromethane-acetonitrile mixture. The organic layer was evaporated. The residue was dissolved in the mobile phase and aliquots of 50 μℓ of this solution were injected.

Elution — E-1. Acetonitrile-0.1 *M* sodium phosphate buffer, pH 6 (15:85).

REFERENCE

1. **Bawdon, R. E., Leveno, K. J., Quirk, J. G., Cunningham, F. G., and Guss, S. P.,** High pressure liquid chromatographic assay of cefamandole in serum following intravenous and intraperitoneal administration, *J. Liq. Chromatogr.*, 6, 2747, 1983.

CEFAZOLIN

Liquid Chromatography

Specimen (mℓ)	Extraction	Column (cm × mm)	Packing (μm)	Elution	Flow (mℓ/min)	Det. (nm)	RT (min)	Internal standard (RT)	Other Compounds (RT)	Ref.
Serum (0.5)	I-1	30 × 3.9	μ-Bondapak-C_{18} (10)	E-1	1.0	ABS (272)	6.5	8-Chlorotheophylline (11)	—	1
Serum (0.5)	I-2	30 × 3.9	μ-Bondapak-phenyl (10)	E-2	2.0	ABS (270)	6	Moxalactam (9)	—	2

Extraction — I-1. The sample was spiked with 15 μg of the internal standard and treated with 0.2 mℓ of 10% sodium tungstate, 0.3 mℓ of 5 *N* sulfuric acid, and 0.3 mℓ of water. After thorough mixing and centrifugation 30 μℓ of the clear supernatant was injected.
I-2. The sample was spiked with 300 μℓ of 0.1 *M* ammonium acetate buffer containing the internal standard and treated with 3 mℓ of 2 propanol. The supernatant was washed with 3 mℓ of 2 propanol. The supernatant was washed with 3 mℓ of chloroform-isoamyl alcohol (96:4). Finally, 100 μℓ of the upper aqueous layer was injected.

Elution — E-1. Methanol-0.05 *M* acetic acid (pH 3) (35:65).
E-2. Acetonitrile-water-0.1 *M* phosphate buffer, pH 7.6 containing 0.5 *M* tetrabutylammonium phosphate (22:77:1).

REFERENCES

1. **Miller, K. W., McCoy, H. G., Chan, K. K. H., Fischer, R. P., Lindsay, W. G., Seifert, R. D., and Zaske, D. E.,** Effect of cardiopulmonary bypass on cefazolin disposition, *Clin. Pharmacol. Ther.*, 27, 550, 1980.
2. **Polk, R. E., Kline, B. J., and Markowitz, S. M.,** Cefazolin and moxalactam pharmacokinetics after simultaneous intravenous infusion, *Antimicrob. Agents Chemother.*, 20, 576, 1981.

CEFMENOXIME

Liquid Chromatography

Specimen (mℓ)	Extraction	Column (cm × mm)	Packing (μm)	Elution	Flow (mℓ/min)	Det. (nm)	RT (min)	Internal standard (RT)	Other compounds (RT)	Ref.
Plasma (1)	I-1	30 × 4	μ-Bondapak-C_{18} (10)	E-1	2.0	ABS (254)	6	*p*-Nitro-acid (3.2) *p*-Anisic acid (7)	—	1
Serum (0.5)	I-2	30 × 3.9	μ-Bondapak-CN (10)	E-2	2.5	ABS (254)	3.4	*p*-Anisic acid (5.6)	—	2
Plasma (0.2)	I-3	30 × 3.9	μ-Bondapak phenyl (10)	E-3	2.0	ABS (254)	9.2	*p*-Anisic acid (13)	—	3
Serum (0.5)	I-4	30 × 3.9	μ-Bondapak-C_{18} (10)	E-4	2.0	ABS (254)	5.6	Cefoxitin (7.3)	—	4

Extraction — I-1. The sample is treated with one fifth the volume of the internal standard containing 4% sodium dodecyl sulfate and the mixture is filtered through Amicon ultrafiltration system with CF25 or CF50A membrane cones. Aliquots of 90 μℓ the ultrafiltrate are injected.
I-2. The sample was spiked with 100 μℓ of solution of the internal standard (24 mg/ℓ in water) and the mixture treated with 100 μℓ of perchloric acid. After mixing and centrifugation, 25 μℓ of the clear supernatant were injected.

CEFMENOXIME (continued)

I-3. The sample was treated with 0.2 mℓ of acetonitrile containing the internal standard (0.8 μg/mℓ). The mixture was vortexed, centrifuged, and the supernatant was evaporated under nitrogen to 0.1 mℓ. Aliquots of 25 to 50 μℓ of the solution were injected.
I-4. The sample was treated with 0.1 mℓ of the internal standard solution (1 mg/mℓ in 0.05 *M* phosphate buffer, pH 6.1), 0.1 mℓ of phosphate buffer and 2.5 mℓ of acetonitrile. After centrifugation, the supernatant was washed with 4 mℓ of dichloromethane. The upper aqueous phase was collected and 5 to 60 μℓ of this were injected.

Elution — E-1. Acetonitrile-0.2 *M* acetate buffer, pH 5.3 (13:87).
E-2. 0.1 *M* acetate buffer, pH 3.8.
E-3. Acetonitrile-0.2% phosphoric acid (14:86).
E-4. Acetonitrile-0.05 *M* ammonium acetate buffer containing 0.005 *M* tetrabutylammonium hydrogen sulfate (20:80).

REFERENCES

1. **Granneman, G. R. and Sennello, L. T.,** A very precise high-performance liquid chromatographic procedure for the determination of cefmenoxime, a new cephalosporin antibiotic, in plasma, *J. Chromatogr.*, 229, 149, 1982.
2. **Reitberg, D. P. and Schentag, J. J.,** Liquid-chromatographic assay of cefmenoxime in serum and urine, *Clin. Chem.*, 29, 1415, 1983.
3. **Noonan, I. A., Gambertoglio, J. G., Barriere, S. L., Conte, J. E., Jr., and Lin, E. T.,** High-performance liquid chromatographic determination of cefmenoxime (AB-50912) in human plasma and urine, *J. Chromatogr.*, 273, 458, 1983.
4. **Smith, I. L., Swanson, D. J., Welage, L. S., DeAngelis, C., Boudinot, S. A., and Schentag, J. J.,** Determination of cefmenoxime in human serum by ion-pair reverse-phase high performance liquid chromatography, *Anal. Lett.*, 18, 1077, 1985.

CEFMETAZOLE

Liquid Chromatography

Specimen (mℓ)	Extraction	Column (cm × mm)	Packing (μm)	Elution	Flow (mℓ/min)	Det. (nm)	RT (min)	Internal standard (RT)	Other compounds (RT)	Ref.
Serum (0.05)	I-1	30 × 3.9	μ-Bondapak-C_{18} (10)	E-1	1.0	ABS (254)	7.2	Barbital (12)	—	1

Extraction — I-1. The sample was mixed with an equal volume of the internal standard solution (2.7 m *M* in methanol containing 5% trichloroacetic acid). The mixture was kept in an ice bath for 10 min and then centrifuged. Aliquots of 10 to 20 μℓ of the resulting supernatant were injected without delay.

Elution — E-1. Acetonitrile-0.005 *M* citrate buffer (pH 5.4).

REFERENCE

1. **Sekine, M., Sasahara, K., Kojima, T., and Morioka, T.,** High-performance liquid chromatographic method for determination of cefmetazole in human serum, *Antimicrob. Agents Chemother.*, 21, 740, 1982.

CEFONICID

Liquid Chromatography

Specimen (mℓ)	Extraction	Column (cm × mm)	Packing (μm)	Elution	Flow (mℓ/min)	Det. (nm)	RT (min)	Internal standard (RT)	Other compounds (RT)	Ref.
Plasma (0.5)	I-1	25 × 4.6	Ultrasphere-ODS (5)	E-1[a]	NA	ABS (265)	6	Cefazolin (9.2)	—	1
Plasma (0.2)	I-2	30 × 3.9	μ-Bondapak-C_{18} (10)	E-2	2	ABS (254)	13.5	Cephalothin (11.5)	—	2

[a] A different mobile phase is used for higher sensitivity.

Extraction — I-1. The sample was mixed with 50 μℓ of a solution of the internal standard (55 mg/mℓ in 0.05 *M* phosphate buffer, pH 6.5) and applied to a prewashed (methanol 5 mℓ, water 5 mℓ) BondElut 1-mℓ C_{18} column. After the sample had passed through the column was washed with 500 μℓ of 0.05 *M* phosphate buffer, pH 6.5, and then eluted with 300 μℓ of methanol. Aliquots of 15 μℓ of the eluate were injected.
I-2. The sample was deproteinized with 0.4 mℓ of the internal standard solution (12.5 μg/mℓ in acetonitrile-water 75:25). After mixing and centrifugation 5 to 20 μℓ of the clear supernatant were injected.

Elution — E-1. Acetonitrile-methanol-0.05 *M* phosphate buffer containing 0.01 *M* triethylamine, pH 7.2 (9:6:85).
E-2. Acetonitrile-0.1% phosphoric acid containing 0.3% tetrabutylammonium hydrogen sulfate (25:75).

CEFONICID (continued)

REFERENCES

1. **Brendel, E., Zschunke, M., and Meineke, I.,** High-performance liquid chromatographic determination of cefonicid in human plasma and urine, *J. Chromatogr.*, 339, 359, 1985.
2. **Phelps, R., Zurlinden, E., Conte, J. E., Jr., and Lin, E.,** High-performance liquid chromatographic determination of cefonicid in human plasma, serum and urine, *J. Chromatogr.*, 375, 111, 1986.

CEFOPERAZONE

Liquid Chromatography

Specimen (mℓ)	Extraction	Column (cm × mm)	Packing (μm)	Elution	Flow (mℓ/min)	Det. (nm)	RT (min)	Internal standard (RT)	Other compounds (RT)	Ref.
Serum (2)	I-1	25 × 4.6	μ-Bondapak-C_{18} (10)	E-1	1.5	ABS (228)	2.3	—	—	1
Serum, tissue (0.1)	I-2	30 × 3.9	μ-Bondapak-phenyl	E-2	2.7	ABS (254)	10.5	Cefoxitin (5.8)	—	2
Serum, urine (1)	I-3	25 × 4.6	μ-Bondapak-C_{18} (10)	E-3; grad	2.0	ABS (254)	23	—	a	3
Plasma, urine, CSF (0.1)	I-4	30 × 3.9	μ-Bondapak-C_{18} (10)	E-4	2.0	ABS (254)	10	Hydrochlorthiazide (6)	—	4
Serum, urine (0)	I-5	30 × 3.9	μ-Bondapak-C_{18} (10)	E-5	2.5	ABS (228)	8	Cephalothin (11.5)	—	5

[a] Separation of different degradation products of cefoperazone is shown.

Extraction — I-1. The sample was applied to a prewashed (methanol, water) Sep-Pak C_{18} cartridge. The cartridge was washed with water and then eluted with 2 × 1 mℓ of methanol-water (1:1). Aliquots of 25 μℓ of the combined eluate were injected.

I-2. The sample was treated with an equal volume of the internal standard solution (75 μg/mℓ in methanol). After mixing and centrifugation, 20 μℓ of the clear supernatant were injected.

I-3. The sample is treated with an equal volume of methanol, vortexed, and centrifuged. Aliquots of 20 μℓ of the clear supernatant are injected.

I-4. The plasma sample was treated with an equal volume of a methanolic solution of the internal standard (100 μg/mℓ). After vortexing and centrifugation, aliquots of 10 to 15 μℓ were injected. Urine and CSF samples were treated in a similar manner but the concentration of the internal standard solution was different.
I-5. The sample was treated with 2 volumes of methanol containing the internal standard (15 μg/mℓ). After vortexing and centrifugation, 20 μℓ of the clear supernatant were injected.

Elution — E-1. Methanol-water (1:1).
E-2. Acetonitrile-water (20:80) containing 0.005 *M* tetrabutyammonium phosphate.
E-3. (A) 0.0012 *M* Triethylamine + 0.042 *M* acetic acid; (B) Acetonitrile + water (24:76) containing 0.0012 *M* triethylamine + 0.042 *M* acetic acid. Gradient from 75% (A) to 60% (A) during 15 min.
E-4. Acetonitrile-1 *M* triethylamine in acetonitrile-1 *M* acetic acid-water (120:1.2:2.8:876).
E-5. Acetonitrile-10 m*M* phosphate buffer, pH 4.5 (15:85).

REFERENCES

1. **Dupont, D. G. and DeJager, B. L.,** High-performance liquid chromatographic determination of cefoperazone in serum, *J. Liq. Chromatogr.*, 4, 123, 1981.
2. **Muder, R. R., Diven, W. F., Yu, V. L., and Johnson, J.,** Determination of cefoperazone concentration in serum and muscle tissue with a versatile high-pressure liquid chromatographic method, *Antimicrob. Agents Chemother.*, 22, 1076, 1982.
3. **Dokladalova, J., Quercia, G. T., and Stankewich, J. P.,** High-performance liquid chromatographic determination of cefoperazone in human serum and urine, *J. Chromatogr.*, 276, 129, 1983.
4. **Hwang, P. T. R. and Meyer, M. C.,** High-performance liquid chromatographic micro-assay for cefoperazone In human plasm, urine and CSF, *J. Liq. Chromatogr.*, 6, 743, 1983.
5. **Schachter, S. H., Spino, M., Tesoro, A., and MacLeod, S. M.,** High performance liquid chromatographic analysis of cefoperazone in serum and urine, *J. Liq. Chromatogr.*, 7, 2421, 1984.

CEFORANIDE

Liquid Chromatography

Specimen (mℓ)	Extraction	Column (cm × mm)	Packing (μm)	Elution	Flow (mℓ/min)	Det. (nm)	RT (min)	Internal standard (RT)	Other compounds (RT)	Ref.
Plasma (0.5)	I-1	30 × 3.9	μ-Bondapak-C_{18} (10)	E-1	2.6	ABS (254)	5.4	Cephalexin (9)	—	1

CEFORANIDE (continued)

Extraction — I-1. The sample was spiked with an aqueous solution of the internal standard solution (60 μg/mℓ) and the mixture treated with 1 mℓ of acetonitrile and 0.1 mℓ of 6% aqueous trichloroacetic acid. After mixing and centrifugation, the aqueous phase was washed with 5 mℓ of dichloromethane. Aliquots of the aqueous phase were then injected.

Elution — E-1. Methanol-0.05 *M* ammonium acetate buffer, pH 4 (10:90).

REFERENCE

1. **Dajani, A. S., Thirumoorthi, M. C., Bawdon, R. E., Buckle, J. A., Pfeffer, M., Van Harken, D. R., and Smyth, R. D.,** Pharmacokinetics of intramuscular ceforanide in infants, children, and adolescents, *Antimicrob. Agents Chemother.*, 21, 282, 1982.

CEFOTAXIME

Liquid Chromatography

Specimen (mℓ)	Extraction	Column (cm × mm)	Packing (μm)	Elution	Flow (mℓ/min)	Det. (nm)	RT (min)	Internal standard (RT)	Other compounds (RT)	Ref.
Serum (1)	I-1	25 × 4.6	LiChrosorb RP-8 (5)	E-1	1.0	ABS (310)	9.5	—	Desacetyl-cefotaxime (7)	1
Serum (0.1—0.3)	I-2	25 × 4	LiChrosorb RP-18 (7)[a]	E-2	1.5	ABS (254)	4.8	—	Desacetyl-cefotaxime (7.5)	2
Plasma, urine (1)	I-3	10 × 3	Spherisorb-ODS (5)	E-3	1.1	ABS (262)	13.5	—	Desacetyl-cefotaxime (3.9)[b]	3
Plasma (0.2)	I-4	25 × 4	Partisil PSX ODS (10)	E-4	2.0	ABS (254)	7	—	Desacetyl-cefotaxime (3)	4
Plasma, urine (1)	I-5	30 × 4.5	μ-Bondapak-C_{18} (10)[a]	E-5	1.3	ABS (254)	8.7	—	Desacetyl-cefotaxime (3.5)	5

[a] Protected by a 22 × 4 mm guard column packed with Corasil C_{18} 37 to 50 μm.
[b] Two minor metabolites elute at 6.2 and 6.9 min.

Extraction — I-1. The sample was mixed with 0.1 mℓ of 70% trichloroacetic acid. After centrifugation an aliquot of 100 μℓ of the supernatant was injected.
I-2. The sample was mixed with an equal volume of 0.4 *M* perchloric acid. The mixture was kept at 0°C for 15 min and centrifuged for 2 min. The clear supernatant was removed and incubated for 10 min at 25°C. The solution was buffered with 4 *M* sodium acetate to a final pH 4.5. Aliquots of 20 to 25 μℓ were injected.
I-3. The sample was mixed with 8 mℓ of chloroform-acetone (1:3). After centrifugation a measured volume of the upper aqueous layer was freeze dried. The residue was reconstituted in 100 μℓ of the mobile phase. After centrifugation, 20 μℓ of the supernatant were injected.
I-4. The sample was treated with an equal volume of trichloroacetic acid (60 g/ℓ). After mixing and centrifugation 20 μℓ of the supernatant were injected.
I-5. The sample was mixed with 3 mℓ of acetonitrile. After centrifugation 1 mℓ of the supernatant was washed with 1 mℓ of chloroform-butanol-1 (3:1). After centrifugation a 20-μℓ aliquot of the upper aqueous layer was injected.

Elution — E-1. Methanol-2 m*M* H_3PO_4 (72:28).
E-2. Acetonitrile-methanol-water (10:7:83) containing 20 m*M* sodium dihydrogenphosphate.
E-3. Methanol-water-acetic acid (12:87:1).
E-4. Methanol-0.05 *M* heptanesulfonic acid, pH 3.5 (25:75).
E-5. Acetonitrile-0.007 *M* phosphoric acid (15:85).

REFERENCES

1. **Bergan, T. and Solberg, R.,** Assay of cefotaxime by high-pressure liquid chromatography, *Chemotherapy*, 27, 155, 1981.
2. **Kees, F., Strehl, E., Seeger, K., Seidel, G., Dominiak, P., and Grobecker, H.,** Comparative determination of cefotaxime and desacetyl cefotaxmine in serum and bile by biosassay and high-performance liquid chromatography, *Arzneim. Forsch.*, 31, 362, 1981.
3. **Dell, D., Chamberlain, J., and Coppin, F.,** Determination of cefotaxime and desacetylcefotaxime in plasma and urine by high-performance liquid chromatography, *J. Chromatogr.*, 226, 431, 1981.
4. **Hammond, E. M., Legge, M., and MacLean, A. B.,** Determination of cefotaxime in plasma by high pressure liquid chromatogaphy, *Med. Lab. Sci.*, 41, 299, 1984.
5. **Yost, R. L. and Derendorf, H.,** Rapid chromatographic determination of cefotaxime and its metabolite in biological fluids, *J. Chromatogr.*, 341, 131, 1985.

CEFOTETAN

Liquid Chromatography

Specimen (mℓ)	Extraction	Column (cm × mm)	Packing (μm)	Elution	Flow (mℓ/min)	Det. (nm)	RT min	Internal standard (RT)	Other compounds (RT)	Ref.
Plasma, urine (0.2)	I-1	12.5 × 4	LiChrosorb RP-18 (5)	E-1	1.0	ABS (280)[a]	A = 13.4[b] B = 14.8	—	Iothalamic acid (2.7)	1

CEFOTETAN (continued)

[a] Absorbance was also monitored at 254 nm.
[b] Epimers R and S.

Extraction — I-1. A sample of plasma was mixed with one volume of 0.1 *M* NaH_2PO_4 and 2 volumes of acetonitrile. The mixture was allowed to stand at 4°C for 15 min and centrifuged. The clear supernatant was washed with 2 mℓ of dichloromethane. After centrifugation, an aliquot of 5 to 50 μℓ of the upper aqueous layer was injected.

Elution — E-1. Acetonitrile-water (75:925) containing 5.5 g of $NaH_2PO_4.H_2O$, 1.8 g $Na_2HPO_4.2H_2O$ and 20 mg of tetrabutylammonium bromide.

REFERENCE

1. **Kees, F., Grobecker, H., and Naber, K. G.,** High-performance liquid chromatographic analysis of cefotetan epimers in human plasma and urine, *J. Chromatogr.*, 305, 363, 1984.

CEFOXITIN

Liquid Chromatography

Specimen (mℓ)	Extraction	Column (cm × mm)	Packing (μm)	Elution	Flow (mℓ/min)	Det. (nm)	RT (min)	Internal standard (RT)	Other compounds (RT)	Ref.
Serum, CSF	I-1	30 × 3.9	μ-Bondapak-C_{18} (10)[a]	E-1	2.0	ABS (238)	4.8	—	—	1
Serum (0.5)	I-2	30 × 3.9	μ-Bondapak-C_{18} (10)	E-2	2.0	ABS (235)	3.2	—	Cephalothin (5.8)	2
Serum (1)	I-3	30 × 3.9	μ-Bondapak-C_{18} (10)[b]	E-3	1.0	ABS (254)	5.5	Cephalothin (9)	—	3

[a] Protected by a guard column packed with Corasil-C_{18}.
[b] Protected by a Whatman guard column packed with Co:Pell ODS.

Extraction — I-1. The sample was treated with 3 volumes of anhydrous ethanol. After mixing and centrifugation, 20 μℓ of the supernatant were injected.

I-2. The serum sample was treated with an equal volume of freshly prepared 5% trichloroacetic acid in methanol. The sample was mixed and allowed to stand on ice for 30 min; and then centrifuged. Aliquots of 25 μℓ of the clear supernatant were injected.
I-3. The sample was spiked with 0.2 mℓ of the internal standard solution (500 μg/mℓ in 1% phosphate buffer, pH 6) and the pH of the mixture adjusted to 3 by the addition of 0.5 mℓ of 0.4 *M* HCl. The mixture was then applied to a prewashed (2 mℓ methanol, 2 mℓ water) 1-mℓ Baker-C_{18} disposable column. After the sample had passed through, the column was washed with 1 mℓ of 0.1 *M* HCl and then eluted with 0.5 mℓ of methyl alcohol. A 20-μℓ aliquot of this eluate was injected.

Elution — E-1. Methanol: 0.03% ammonium carbonate (15:85).
E-2. Acetonitrile-acetic acid-0.005 *M* potassium dihydrogen phosphate (25:0.5:74.5).
E-3. Acetonitrile-acetic acid-water (29:70:1).

REFERENCES

1. **Torchia, M. G. and Danzinger, R. G.,** Reversed-phase high-performance liquid chromatographic assay for cefoxitin in proteinaceous biological samples, *J. Chromatogr.*, 181, 120, 1980.
2. **Wheeler, L. A., De Meo, M., Kirby, B., Jerauld, R. S., and Finegold, S. M.,** High-performance liquid chromatographic assay for measurement of cefoxitin in serum, *J. Chromatogr.*, 183, 357, 1980.
3. **Purser, C., Baltar, A., Ho, I. K., and Hume, A. S.,** New rapid method of analysis of cefoxitin in serum and bone, by high-performance liquid chromatography, *J. Chromatogr.*, 311, 135, 1984.

CEFPIMIZOLE

Liquid Chromatography

Specimen (mℓ)	Extraction	Column (cm × mm)	Packing (μm)	Elution	Flow (mℓ/min)	Det. (nm)	RT (min)	Internal standard (RT)	Other compounds (RT)	Ref.
Plasma, urine (1)	I-1	25 × 4.6	Supelcosil LC-18 (5)[a]	E-1	NA	ABS (254)	12	Acetophenone (19)	—	1

[a] Protected by a 50 × 2.1 mm guard column packed with Co:Pell-ODS, (35 μm).

Extraction — I-1. The sample was diluted with 1 mℓ of an aqueous solution containing 0.01 *M* EDTA and 0.05 *M* tetrabutylammonium hydroxide, pH 5, and treated with 4 mℓ of acetonitrile. After mixing, the mixture was allowed to stand for 2 hr at 4°C. The supernatant was collected and the precipitate washed with 2 mℓ of acetonitrile-above mentioned diluent (75:25). The washings and the original supernatant were combined and washed with 0.2 mℓ of dichloromethane. The aqueous phase was collected and mixed with 50 μℓ of a 200 μg/mℓ solution of acetophenone in methanol and aliquots of this solution were injected.

CEFPIMIZOLE (continued)

Elution — E-1. Methanol-water (350:650) containing 1 m*M* of EDTA, 5 m*M* of tetrabutylammonium hydroxide, adjusted to pH 6 with acetic acid.

REFERENCE

1. **Lakings, D. B. and Wozniak, J. M.,** High-performance liquid chromatographic methods for the determination of cefpimizole in plasma and urine, *J. Chromatogr.*, 308, 261, 1984.

CEFRADINE

Liquid Chromatography

Specimen (mℓ)	Extraction	Column (cm × mm)	Packing (μm)	Elution	Flow (mℓ/min)	Det. (nm)	RT (min)	Internal standard (RT)	Other compounds (RT)	Ref.
Plasma (0.5)	I-1	30 × 4	μ-Bondapak-C_{18} (10)	E-1	2.0	ABS (254)	5.6	—	Cephalexin (4.2)	1
Serum (1)	I-2	20 × 4.6	Hypersil-ODS (5)	E-2	1.0	ABS (264)	7.5	—	Cephalexin (6)	2

Extraction — I-1. The sample (0.5 mℓ of plasma) was mixed with 1.5 mℓ of methanol. After centrifugation, aliquots of 20 μℓ of the supernatant were injected. I-2. The sample was treated with an equal volume of 10% trichloroacetic acid. After mixing and centrifugation, aliquots of 50 μℓ of the supernatant were injected.

Elution — E-1. Methanol-0.01 *M* phosphate buffer, pH 6.8 (20:80).
E-2. Methanol-water-3.86% sodium acetate-4% acetic acid (250:732:15:3).

REFERENCES

1. **Hayashi, Y.,** High-performance liquid chromatographic micro-assay for Cefradine in biological fluids, *Jpn. J. Antibiotics*, 34, 440, 1981.
2. **Clarke, G. S. and Robinson, M. L.,** HPLC analysis of Cepharadine in human serum and urine, *J. Clin. Hosp. Pharm.*, 8, 373, 1983.

CEFROXADINE

Liquid Chromatography

Specimen (mℓ)	Extraction	Column (cm × mm)	Packing (μm)	Elution	Flow (mℓ/min)	Det. (nm)	RT (min)	Internal standard (RT)	Other compounds (RT)	Ref.
Serum (1)	I-1	15 × 4.6	LiChrosorb RP-8 (5)	E-1	1.0	ABS (280)	6.5	—	—	1

Extraction — I-1. The sample (1 mℓ of serum) was treated with 0.1 mℓ of freshly prepared 70% (w/v) trichloroacetic acid. After centrifugation, aliquots of the clear supernatant were injected to fill a 100-μℓ loop.

Elution — E-1. Methanol-2 m*M* H_3PO_4 (28:72).

REFERENCE

1. **Bergan, T. and Solberg, R.,** Comparison of microbiological and high pressure liquid chromatographic assays of the new cephalosporin cefroxadine, *Methods Find. Exp. Clin. Pharmacol.*, 3, 179, 1981.

CEFSULODIN

Liquid Chromatography

Specimen (mℓ)	Extraction	Column (cm × mm)	Packing (μm)	Elution	Flow (mℓ/min)	Det. (nm)	RT (min)	Internal standard (RT)	Other compounds (RT)	Ref.
Dosage	—	25 × 4.6	Zorbax-C_8 (10)[a]	E-1	1.5	ABS (254)	7	Phenoxyacetic acid (18.5)	b	1
Serum, urine (0.5)	I-1	15 × 4	Nucleosil-C_{18} (5)[c]	E-2	0.8	ABS (254)	7.5	Cefacetrile (11.5)	Cefotiam (8)[d] Cefmenoxime (8)[d]	2
Plasma (1)	I-2	30 × 3.9	μ-Bondapak-C_{18} (10)	E-3	2.0	ABS (254)	3.5	*p*-Fluoromethylbenzyl-amine (11)	—	3
Plasma, urine (0.2)	I-3	15 × 4.7	LiChrosorb RP-8 (5)[e]	E-4	1.0	ABS (254)	8	Cephalexin (16)	Cefotiam (26)[f]	4

CEFSULODIN (continued)

Liquid Chromatography

Specimen (mℓ)	Extraction	Column (cm × mm)	Packing (μm)	Elution	Flow (mℓ/min)	Det. (nm)	RT (min)	Internal standard (RT)	Other compounds (RT)	Ref.
Serum (0.2)	I-4	30 × 3.9	μ-Bondapak-C_{18} (10)	E-5	1.5	ABS (280)	3.8	Cefazolin (6.3)	Cefotiam (8.3)	5
Plasma, CSF (0.05)	I-5	30 × 3.9	μ-Bondapak-C_{18} (10)	E-6	1.5	ABS (280)	4.8	—	—	6

[a] Protected by a 30 × 4.6 mm Brownlee RP-8 precolumn.
[b] Separation of possible degradation products and manufacturing impurities is shown.
[c] Protected by a 10 × 4 mm precolumn packed with Nucleosil C_{18} (5 μm).
[d] Different chromatographic conditions for the determination of these drugs are described.
[e] Protected by a 10 × 0.47 cm precolumn packed with Co:Pell-ODS.
[f] Conditions for the determination of cefuroxime, cefotoxime, cefroxadin, and cephalexin are described.

Extraction — I-1. The sample (0.5 mℓ of serum) was spiked with the internal standard and mixed with 1 mℓ of methanol. After centrifugation, the sample was filtered through a 0.5-μm Millipore filter. Aliquots of the filtrate were injected.
I-2. The sample was stabilized prior to storage by dilution with an equal volume of 1 *M* phosphate buffer, pH 6. The diluted sample was thoroughly mixed with 50 μℓ of an aqueous solution (8 mg/mℓ) of the internal standard and filtered through Amicon ultrafiltration system. Aliquots of 70 μℓ of the ultrafiltrate were injected.
I-3. The sample was treated with 20 μℓ of 0.45 *N* H_3PO_4 and 100 μℓ of methanol. After mixing and centrifugation aliquots of the clear supernatant were injected.
I-4. The serum sample was treated with an equal volume of acetonitrile containing 50 μg/mℓ of the internal standard. After mixing and centrifugation, the supernatant was filtered through a 0.45-μm membrane filter. A 10-μℓ aliquot of the filtrate was injected.
I-5. Plasma (50 μℓ) from the experimental rat was diluted with 50 μℓ of pooled human plasma and treated with 150 μℓ of methanol. The samples were mixed, centrifuged, and aliquots of the supernatant were injected. Samples of CSF were collected directly in the mini vial of the autoinjector and aliquots of 7 μℓ injected.

Elution — E-1. Acetonitrile-methanol-water (15:35:950) containing 0.02 *M* ammonium acetate, pH 4.1.
E-2. Acetonitrile-0.1 *M* acetate buffer, pH 4.4 (8:92).
E-3. Acetonitrile-0.02 *M* ammonium acetate, pH 4.2 (4.5:95.5).
E-4. Methanol-0.1 *M* tetrabutylammonium hydrogen sulfate in 0.09 *M* K_3PO_4 (15:85).
E-5. Methanol-water (35:65) containing 0.005 *M* tetrabutylammonium phosphate.
E-6. Acetonitrile-methanol, 0.02 *M* ammonium acetate-acetic acid, pH 4.1 (15:35:46:4).

REFERENCES

1. **Elrod, L., Jr., White, L. B., Wimer, D. C., and Cox, R. D.,** Determination of cefsulodin sodium [D(-)-SCE-129] by high-performance liquid chromatography, *J. Chromatogr.*, 237, 515, 1982.
2. **Itakura, K., Mitani, M., Aoki, I., and Usui, Y.,** High performance liquid chromatographic assay of cefsulodin, cefotiam and cefmenoxime in serum and urine, *Chem. Pharm. Bull.*, 30, 622, 1982.
3. **Granneman, G. R. and Sennello, L. T.,** Precise high-performance liquid chromatographic procedure for the determination of cefsulodin, a new antipseudomonal cephalosporin antibiotic, in plasma, *J. Pharm. Sci.*, 71, 1112—1115, 1982.
4. **Lecaillon, J. B., Rouan, M. C., Souppart, C., Febvre, N., and Juge, F.,** Determination of cefsulodin, cefotiam, cephalexin, ceflotaxime, desacetyl-cefotaxime, cefuroxime and cefroxadin in plasma and urine by high-performance liquid chromatography, *J. Chromatogr.*, 228, 257, 1982.
5. **Yamamura, K., Nakao, M., Yamada, J. I., and Yotsuyanagi, T.,** Simultaneous determinations of cefsulodin and cefotiam in serum and bone marrow blood by high-performance liquid chromatography, *J. Pharm. Sci.*, 72, 958, 1983.
6. **Mohler, J., Meulemans, A., Vicart, P., and Vulpillat, M.,** The "on line" determination of cefsulodin by high performance liquid chromatography in CSF and plasma of rats, *J. Liq. Chromatogr.*, 9, 189, 1986.

CEFTAZIDIME

Liquid Chromatography

Specimen (mℓ)	Extraction	Column (cm × mm)	Packing (μm)	Elution	Flow (mℓ/min)	Det. (nm)	RT (min)	Internal standard (RT)	Other compounds (RT)	Ref.
Serum, urine (0.15)	I-1	30 × 4	MicroPak C_{18} (10)[a,b]	E-1[c]	1.0	ABS (257)	9	—	—	1
Serum (0.1)	I-2	30 × 3.9	μ-Bondapak-C_{18} (10)[d]	E-2[e]	1.2	ABS (255)	5	8-Chlorotheophylline (10.7)	—	2
Serum, urine, CSF (0.1)	I-3	30 × 3.9	μ-Bondapak-C_{18} (10)[d]	E-3	2.0	ABS (254)	11	Hydrochlorthiazide (6)	—	3

[a] Protected by a guard column (40 × 4 mm) packed with Vydac RP-18 (40 μm) packing.
[b] Column temp = 250°C.
[c] An alternative gradient system is described to get sharper peaks.
[d] Protected by a guard column packed with Corasil C_{18} (37 to 50 μm).
[e] A different mobile phase and flow rate is described for the analysis of urine.

CEFTAZIDIME (continued)

Extraction — I-1. The serum sample was treated with an equal volume of cold methanol. After vortexing, the mixture was allowed to stand on ice for 5 min and then centrifuged. Aliquots of the clear supernatant (20 μℓ) were injected.
I-2. The serum sample was spiked with an equal volume of an aqueous solution (30 μg/mℓ) of the internal standard and the proteins were precipitated by vortexing with one volume of methanol. After centrifugation an aliquot of 20 μℓ of the clear supernatant was injected.
I-3. The serum sample was spiked with an equal volume of a methanolic solution of the internal standard (100 μg/mℓ) and treated with 100 μℓ of methanol. After mixing and centrifugation aliquots of 10 to 15 μℓ of the clear supernatant were injected.

Elution — E-1. Methanol-50 m*M* ammonium dihydrogen phosphate containing 117 μ*M* perchloric acid (20:80).
E-2. Methanol-150 m*M* KH_2PO_4, pH 6.5 (18:82).
E-3. Acetonitrile-water-acetic acid (120:200:20), pH 4.

REFERENCES

1. **Myers, C. M. and Blumer, J. L.,** Determination of ceftazidime in biological fluids by using high-pressure liquid chromatography, *Antimicrob. Agents Chemother.*, 24, 343, 1983.
2. **Leeder, J. S., Spino, M., Tesoro, A. M., and MacLeod, S. M.,** High-pressure liquid chromatographic analysis of ceftazidime in serum and urine, *Antimicrob. Agents Chemother.*, 24, 720, 1983.
3. **Hwang, P. T. R., Drexler, P. G., and Meyer, M. C.,** High-performance liquid chromatographic determination of ceftazidime in serum, urine, CSF and peritoneal dialysis fluid, *J. Liq. Chromatogr.*, 7, 979, 1984.

CEFTIZOXIME

Liquid Chromatography

Specimen (mℓ)	Extraction	Column (cm × mm)	Packing (μm)	Elution	Flow (mℓ/min)	Det. (nm)	RT (min)	Internal standard (RT)	Other compounds (RT)	Ref.
Serum, bile, urine (0.5)	I-1	30 × 4	μ-Bondapak-phenyl (10)[a]	E-1[b]	2.0	ABS (254)	5	—	—	1
Serum (0.5)	I-2	30 × 3.9	μ-Bondapak-C_{18} (10)	E-2	2.0	ABS (270)	4.2	—	—	2
Serum (0.2)	I-3	30 × 3.9	μ-Bondapak-C_{18} (10)	E-3	1.5	ABS (310)	6	Cefotaxime (9)	—	3

Serum (0.5)	I-4	30 × 3.9	μ-Bondapak-phenyl (10)	E-4	1.5	ABS (254)	6	Cefotaxime (8.5)	—	4

[a] Protected by a 50 × 2 mm guard column packed with Corasil-phenyl (37 to 50 μm).

[b] Different mobile phases are used for the analysis of bile and urine samples. For bile and urine absorbance is monitored at 280 nm.

Extraction — I-1. The sample (0.5 mℓ of serum) was treated with 0.1 mℓ of acetonitrile and 0.1 mℓ of 0.2 *M* phosphate buffer, pH 2.6. After 10 min, the mixture was filtered through a 0.5-μm membrane filter and 10 μℓ of the filtrate were injected.
I-2. The sample was applied to a DEAE-Sephadex A-25 column (3 mℓ bed volume) conditioned with pH 7.2 phosphate buffer/saline. The column was washed with 4 mℓ of buffered saline and eluted with 5 mℓ of 1 *M* sodium chloride. An aliquot of 100 μℓ of this eluate was injected.
I-3. The sample was spiked with 100 μℓ of an aqueous solution of the working internal standard solution and the proteins precipitated with 1 mℓ of acetonitrile. After centrifugation, the supernatant was washed with 1.5 mℓ of dichloromethane and 10- to 20-μℓ aliquots of the upper aqueous layer were injected.
I-4. The sample (0.3 mℓ) was spiked with 75 μℓ of an aqueous solution of the internal standard (1 mg/mℓ) and was treated with 0.5 mℓ of acetonitrile. After mixing and centrifugation aliquots of 10 to 20 μℓ of the clear supernatant were injected.

Elution — E-1. Acetonitrile-0.02 *M* phosphate buffer, pH 2.6 (13:87).
E-2. Acetonitrile-acetic acid-water (13:1.5:85.5), pH 2.8.
E-3. Acetonitrile-2.8% acetic acid (13:87).
E-4. Acetonitrile-0.02 *M* phosphate buffer, pH 2.6 (13:87).

REFERENCES

1. **Suzuki, A., Noda, K., and Noguchi, H.,** High-performance liquid chromatographic determination of ceftizoxime, a new cephalosporin antibiotic, in rat serum, bile and urine, *J. Chromatogr.*, 182, 448, 1980.
2. **Fasching, C. E., Peterson, L. R., Bettin, K. M., and Gerding, D. N.,** High-pressure liquid chromatographic assay of ceftizoxime with an anion-exchange extraction technique, *Antimicrob. Agents Chemother.*, 22, 336, 1982.
3. **McCormick, E. M., Echols, R. M., and Rosano, T. G.,** Liquid chromatographic assay of ceftizoxime in sera of normal and uremic patients, *Antimicrob. Agents Chemother.*, 25, 336, 1984.
4. **LeBel, M., Ericson, J. F., and Pitkin, D. H.,** Improved high-performance liquid chromatographic (HPLC) assay method for ceftizoxime, *J. Liq. Chromatogr.*, 7, 961, 1984.

CEFTRIAXONE

Liquid Chromatography

Specimen (mℓ)	Extraction	Column (cm × mm)	Packing (μm)	Elution	Flow (mℓ/min)	Det. (nm)	RT (min)	Internal standard (RT)	Other compounds (RT)	Ref.
Plasma, saliva, urine (0.25)	I-1	25 × 4	LiChrosorb-NH_2 (5)	E-1	1.5	ABS (274)	9	—	—	1
Serum, urine, CSF (0.2)	I-2	25 × 4.5	LiChrosorb RP-8 (5)	E-2	1.6	ABS (280)	7	—	—	2
Serum, tissue	I-3	30 × 3.9	μ-Bondapak-C_{18} (10)	E-3	2.3	ABS (254)	7.5	—	Cefazolin (5)[a]	3

[a] The mobile phase and the flow rate are different for this drug

Extraction — I-1. The sample (0.25 mℓ) was diluted with 0.75 mℓ of water, treated with 2 mℓ of acetonitrile, mixed, and centrifuged. Aliquots of 20 to 50 μℓ of the supernatant were injected.
I-2. A sample of 200 μℓ of serum or of 1:10 diluted urine was mixed with 200 μℓ of acetonitrile. After mixing and centrifugation, the supernatant was washed with 500 μℓ of dichloromethane. A 10-μℓ aliquot of the upper aqueous phase was injected. Samples of CSF were injected directly after centrifugation.
I-3. Protein precipitation. Details are described in earlier papers (*Antimicrob. Agents Chemother.*, 22, 999, 1982; *Am. J. Obstet. Gynecol.*, 144, 546, 1982).

Elution — E-1. Acetonitrile-water-10% ammonium carbonate (70:26:4).
E-2. Acetonitrile-water-1 *M* phosphate buffer, pH 7.5 (400:592.5:7.5) containing 2.73 g of hexadeyl trimethyl ammonium bromide.
E-3. Methanol-0.1 *M* sodium phosphate buffer, pH 8 (10:90).

REFERENCES

1. **Ascalone, V. and Dal Bo, L.,** Determination of ceftriaxone, a novel cephalosporin, in plasma, urine and saliva by high-performance liquid chromatography on an NH_2 bonded-phase column, *J. Chromatogr.*, 273, 357, 1983.
2. **Bowman, D., Aravind, M. K., Miceli, J. N., and Kauffman, R. E.,** Reversed-phase high-performance liquid chromatographic method to determine ceftriaxone in biological fluids, *J. Chromatogr.*, 309, 209, 1984.
3. **Bawdon, R. E., Hemsell, D. L., and Hemsell, P. G.,** Serum and pelvic tissue concentrations of ceftriaxone and cefazolin at hysterectomy, *J. Liq. Chromatogr.*, 7, 2011, 1984.

CEFUROXIME

Liquid Chromatography

Specimen (mℓ)	Extraction	Column (cm × mm)	Packing (μm)	Elution	Flow (mℓ/min)	Det. (nm)	RT (min)	Internal standard (RT)	Other compounds (RT)	Ref.
Plasma (0.1)	I-1	15 × 4.6	LiChrosorb RP-8 (5)	E-1	1.2	ABS (278)	5	—	—	1
Plasma, urine	I-2	30 × 3.9	μ-Bondapak-C_{18} (10)	E-2	1.5	ABS (270)	NA	—	—	2
Serum (1)	I-3	30 × 3.9	μ-Bondapak-C_{18} (10)[a]	E-3	1.5	ABS (254)	5.4	—	—	3

[a] Column temp = 40°C.

Extraction — I-1. The sample (0.1 mℓ of plasma) was mixed with 0.4 mℓ of 0.33 *N* perchloric acid. After mixing and centrifugation, 100 μℓ of the clear supernatant were injected.
I-2. The sample was mixed and incubated with an equal volume of 60% methanol-0.2 *M* sodium acetate for 2 min at 60°C. After centrifugation, 25 μℓ of the supernatant were injected.
I-3. The serum sample was treated with an equal volume of 6% trichloroacetic acid. After centrifugation, the pH of the supernatant was adjusted to 6 with 6% sodium bicarbonate. Aliquots of this solution were injected.

Elution — E-1. Methanol-0.067 *M* KH_2PO_4 (75:425).
E-2. Methanol-0.01 *M* sodium acetate (20:80).
E-3. Acetonitrile-0.01 *M* acetate buffer, pH 4.8 (11:89).

REFERENCES

1. **Hekster, Y. A., Baars, A. M., Vree, T. B., van Klingeren, B., and Rutgers, A.,** Comparison of high performance liquid chromatography and microbiological assayn in the determination of plasma cefuroxime concentrations in rabbits, *J. Antimicrob. Chemother.*, 6, 65, 1980.
2. **Bundtzen, R. W., Toothaker, R. D., Nielson, O. S., Madsen, P. O., Welling, P. G., and Craig, W. A.,** Pharmacokinetics of cefuroxime in normal and impaired renal function: comparison of high-pressure liquid chromatography and microbiological assays, *Antimicrob. Agents Chemother.*, 19, 443, 1981.
3. **Brisson, A. M. and Fourtillan, I.,** Pharmacocinetique du cefuroxime chez l'homme, *Therapie*, 36, 143, 1981.

CELIPROLOL

Liquid Chromatography

Specimen (mℓ)	Extraction	Column (cm × mm)	Packing (μm)	Elution	Flow (mℓ/min)	Det. (nm)	RT (min)	Internal standard (RT)	Other compounds (RT)	Ref.
Blood, plasma urine (1)	I-1	25 × 4.6	Spherisorb-ODS (5)	E-1	1.0	ABS (237)	12.1	Acebutolol (9)	Metabolite[a] (7.8)	1
Plasma, urine (1)	I-2	30 × 4	Micro-Pac MCH-C_{18} (10)[b]	E-2	NA	Fl (265, 418)	4.78	Acebutolol (4) ST 1412 (6.8)	—	2

[a] Unidentified metabolite.

[b] Protected by a (30 × 4.6 mm) precolumn packed with spheri-RP-18 (5 μm).

Extraction — I-1. The plasma was mixed with 1 mℓ of 0.01 *M* phosphate buffer (pH 6) containing 1 μg of the internal standard and 200 μℓ of 2 *M* sodium hydroxide. The mixture was extracted with 10 mℓ of ethyl acetate. The organic layer was back extracted into 150 μℓ of 0.01 *M* sulfuric acid. Aliquots of the aqueous phase were injected with an autosampler. Blood samples were pretreated with 2 mℓ of acetonitrile. The supernatant was evaporated and the residue subjected to above extraction procedure. Urines were extracted with diethyl ether rather than ethyl acetate.

I-2. The sample was mixed with 50 μℓ of an aqueous solution of the internal standard, 1 mℓ of water and 0.2 mℓ of 4 *N* sodium hydroxide. The mixture was extracted with 4 mℓ of ethyl acetate. A 3-mℓ aliquot of the organic layer was back extracted into 100 μℓ of 0.01 *N* sulfuric acid. Aliquots of 50 μℓ of the aqueous phase were injected.

Elution — E-1. Acetonitrile-0.1 *M* phosphate buffer, pH 4-water (44:6:39).

E-2. Acetonitrile-0.1 *M* phosphate buffer, pH 3.3-water (55:6:39).

REFERENCES

1. **Buskin, J. N., Upton, R. A., Sorgel, F., Williams, R. L., Lang, E., and Benet, L. Z.,** Specific and sensitive assay of celiprolol in blood, plasma and urine using high-performance liquid chromatography, *J. Chromatogr.*, 230, 454, 1982.
2. **Hippmann, V. D. and Takacs, F.,** Eine quantitative Methode zur Bestimmung von Celiprolol im biologischen Material mit Hilfe der Hochleistungsflussig-chromatographie unter Verwendung eines Fluoreszenz-Detektors, *Arzneim. Forsch.*, 33, 8, 1983.

CEPHALEXIN

Liquid Chromatography

Specimen (mℓ)	Extraction	Column (cm × mm)	Packing (μm)	Elution	Flow (mℓ/min)	Det. (nm)	RT (min)	Internal standard (RT)	Other compounds (RT)	Ref.
Dosage	I-1	25 × 4.6	Amino Sil-x-1 (13)[a]	E-1	0.75	ABS (425)	11.6	*o*-Nitrophenol (5.5)	Lysine (3.9)	1
Plasma, saliva, urine (0.1)	I-2	30 × 3.9	μ-Bondapak-C_{18} (10)	E-2	2.0	ABS (254)	3	Cephaloglycin (4)	—	2
Serum (0.02)	I-3	10 × 8	Radial-Pak C_{18} (10)	E-3	2.0	ABS (254)	7.5	—	—	3
Plasma (0.5)	I-4	25 × 4	Nucleosil-C_{18} (5)[b]	E-4	NA	Fl (355, 435)	7	Methylanthranilate (10)	c	4
Plasma (1)	I-5	30 × 4	Altech-C_{18} (10)[d]	E-5	2	ABS (254)	6.6	μ-Hydroxyethyltheophylline (15.6)	e	5

[a] Column temp = 50°C.
[b] Column temp = 55°C.
[c] Procedures for the determination of ampicillin, amoxicillin, and cephradine are described.
[d] Protected by a guard column packed with 30 to 38 μm C_{18} silica.
[e] Procedures for the determination of cefadroxil in plasma and urine are described.

Extraction — I-1. A solution of lysine salt of cephalexin was adjusted to pH 10 and treated with an aqueous solution (0.4%) of 2,4,6-trinitrobenzenesulfonic acid. The mixture was kept in the dark for 1 hr at room temperature. The pH of the mixture was adjusted to 4.8 with acetate buffer, 0.1% aqueous solution of the internal standard was added and an aliquot of 1 μℓ of this solution was injected.
I-2. The sample was treated with 200 μℓ of methanol containing 2.5 μg/mℓ or 25 μg/mℓ of the internal standard. After vortexing and centrifugation, the supernatant was evaporated at 40°C under a stream of nitrogen. The residue was dissolved in 75 μℓ of the mobile phase and an aliquot of 50 μℓ was injected.
I-3. A 20-μℓ aliquot of serum was treated with 100 μℓ of methanol. After mixing and centrifugation, a 90-μℓ aliquot of the supernatant was mixed in the injection syringe with an equal volume of 0.01 *M* sodium 2-pentanesulfonate (pH 2.5 with acetic acid). The whole mixture was then injected.
I-4. The sample (0.5 mℓ of plasma) was diluted to 4 mℓ with water and treated with 3 mℓ of 10% trichloroacetic acid. Three milliliters of the supernatant was treated with 2 mℓ of 0.1 *M* disodium hydrogen citrate solution. One milliliter of 0.5% (w/v) H_2O_2 prepared in 0.1 *M* disodium hydrogen citrate was added and the mixture heated in a boiling water bath for 70 min. After cooling, the reaction mixture was extracted with 7 mℓ of acetone-chloroform (2:3). A 5-mℓ aliquot of the organic extract was evaporated. The residue was dissolved in 100 μℓ of methanol containing the internal standard, an aliquot of 20 μℓ of this solution was injected.
I-5. The sample is treated with 0.1 mℓ of an aqueous solution of the internal standard and 1 mℓ of acetonitrile. After mixing and centrifugation a 20-μℓ aliquot of the supernatant was injected.

CEPHALEXIN (continued)

Elution — E-1. Methanol-water (5:40) containing 1% citric acid.
E-2. Methanol-water (20:80) containing 0.5% acetic acid.
E-3. Acetonitrile-water (30:70) containing 0.005 *M* sodium 2-propanesulfonate, pH 3.
E-4. Methanol-water (3:2).
E-5. Methanol-0.1 *M* monobasic sodium phosphate, pH 4.5 (19:81).

REFERENCES

1. **Fabregas, J. L. and Beneyto, J. E.,** Simultaneous determination of cephalexin and lysine in their salt using high-performance liquid chromatography of derivatives, *J. Pharm. Sci.*, 69, 1378, 1980.
2. **Nahata, M. C.,** High-performance liquid chromatographic determination of cephalexin in human plasma, urine and saliva, *J. Chromatogr.*, 225, 532, 1981.
3. **Tsutsumi, K., Kubo, H., and Kinoshita, T.,** Determination of serum cephalexin by high performance liquid chromatography, *Anal. Lett.*, 14, 1735, 1981.
4. **Miyazaki, K., Ohtani, K., Sunada, K., and Arita, T.,** Determination of ampicillin, amoxicillin, cephalexin, and cephradine in plasma by high-performance liquid chromatography using fluorometric detection, *J. Chromatogr.*, 276, 478, 1983.
5. **Welling, P. G., Selen, A., Pearson, J. G., Kwok, F., Rogge, M. C., Ifan, A., Marrero, D., Craig, W. A., and Johnson, C. A.,** A pharmacokinetic comparison of cephalexin and cefadroxil using HPLC assay procedures, *Biopharm. Drug Dispos.*, 6, 147, 1985.

CEPHALOGLYCIN

Liquid Chromatography

Specimen (mℓ)	Extraction	Column (cm × mm)	Packing (μm)	Elution	Flow (mℓ/min)	Det. (nm)	RT (min)	Internal standard (RT)	Other compounds (RT)	Ref.
Urine	I-1	25 × 4.6	LiChrosorb RP-18 (10)[a]	E-1	1.4	ABS (254)	21	—	Deacetylcephaloglycin (8) Deacetylcephaloglycin lactone (38)	1

[a] Protected by a 20 × 4.6 mm precolumn packed with LiChrosorb RP-2.

Extraction — I-1. A 5- to 20-μℓ aliquot of the filtered urine was injected directly.

Elution — E-1. Water-methanol (4:1) containing 4 m*M* sodium *n*-heptylsulfonate and 10 m*M* $NH_4H_2PO_4$.

REFERENCE

1. **Haginaka, J., Nakagawa, T., and Uno, T.,** Chromatographic analysis and pharmacokinetic investigation of cephaloglycin and its metabolites in man, *J. Antibiotics*, 33, 236, 1980.

CEPHALORIDINE

Liquid Chromatography

Specimen (mℓ)	Extraction	Column (cm × mm)	Packing (μm)	Elution	Flow (mℓ/min)	Det. (nm)	RT (min)	Internal standard (RT)	Other compounds (RT)	Ref.
Urine (0.05)	I-1	15 × 3.9	Nova-PakC_{18} (5)	E-1	0.8	ABS (254)	9.2	—	—	1

Extraction — I-1. A 50-μℓ aliquot of the centrifuged urine sample was treated with 50 μℓ of 2 *N* trichloroacetic acid and 350 μℓ of water. The mixture was mixed, allowed to stand in ice for 15 min, and centrifuged. The supernatant (200 μℓ) was treated with 100 mg of Dowex-1-chloride and mixed for 1 min and filtered through a Millipore (0.45 μm) filter. An aliquot of the filtrate was injected.

Elution — E-1. Acetonitrile-0.1 *M* phosphate buffer, pH 7.5 (10:90).

REFERENCE

1. **Ishihara, A., Sudo, J., and Tanabe, T.,** Separation of cephaloridine from the rat urine contaminated by feces and diet pellets in high-performance liquid chromatography, *J. Toxicol. Sci.*, 10, 1, 1985.

CEPHALOSPORINS

Liquid Chromatography

Specimen (mℓ)	Extraction	Column (cm × mm)	Packing (μm)	Elution	Flow (m:li/min)	Det. (nm)	RT[a] (min)	Internal standard (RT)	Other compounds (RT)	Ref.
Plasma (1)	I-1	30 × 3.9	μ-Bondapak-C_{18} (10)	E-1	1.5	ABS (1 = 270) (2 = 275) (4 = 240) (5 = 234) (6 = 245) (7 = 254) (8 = 240)	1 = 6.2 2 = 4.2 4 = 8.4 5 = 3.4 6 = 3.3 7 = 3.1 8 = 240	—	Deacetylcephalothin (2.7)	1
Serum	I-2	10 × 8	Radial-Pak-C_{18} (10)	E-2	3.0	ABS (254)	1 = 8.2 2 = 3.6 6 = 4.2 9 = 6.7	p-Nitroacetanilide (12.4)	Chloramphenical (9.5)	2
Plasma (0.15)	I-3	15 × 4.7	LiChrosorb RP-18 (5)[b]	E-3	1.2	ABS (254)	c	—	—	3
Plasma (0.5)	I-4	15 × 4.6	Ultrasphere-ODS (5)[d]	E-4	2.0	ABS (254)	c	Cephapirin	—	4
Plasma (0.3)	I-5	30 × 4	μ-Bondapak-C_{18} (10)[e]	E-5	2.5	ABS (254)	c	8-Chloro-theophylline[f]	—	5

[a] Numbers refer to different cephalosporins — See *Note*.
[b] Protected by a 10 × 0.47 cm precolumn packed with Co:Pell ODS, 30 to 38 μm.
[c] Unrelated retention times, as a different mobile phase is used for a particular drug.
[d] Column temp = 45°C.
[e] Protected by a 40 × 4 mm precolumn packed with 10 μm C_{18}-silica.
[f] In some cases an alternative antibiotic was used as the internal standard.

Note: 1—Cefamandole, 2—Cefazolin, 3—Cefonicid, 4—Cefoperazone, 5—Cefotaxime, 6—Cefoxitin, 7—Cefuroxime, 8—Cephalothin, 9—Cephapirin.

Extraction — I-1. The plasma sample (1 mℓ) was mixed with 0.5 mℓ of 0.4 *M* HCl. The mixture was extracted with 7 mℓ of chloroform-1-pentanol (3:1). The organic layer was back extracted into 350 μℓ of phosphate buffer, pH 7 and aliquots of 10 to 50 μℓ of the upper aqueous phase were injected.

I-2. The sample was mixed with an equal volume of the internal standard solution (25 mg/100 mℓ in acetonitrile). The mixture was vortexed, allowed to stand at room temp, again mixed, and centrifuged. Aliquots of 25 μℓ of the clear supernatant were injected.
I-3. A 150 μℓ of plasma sample was treated with 50 μℓ of 10% aqueous (w/v) solution of trichloroacetic acid. After mixing and centrifugation a 20- to 40-μℓ aliquot of the clear supernate was injected.
I-4. The sample (0.5 mℓ of plasma) was treated with 1 mℓ of acetonitrile containing the internal standard. After vortexing and centrifugation, aliquots of 10 μℓ of the supernatant were injected.
I-5. The sample (0.3 mℓ of plasma) was combined with an equal volume of 70% methanol containing the internal standard-30% 0.1 *M* sodium acetate, pH 5.2. The mixture was vortexed and incubated at −20°C for 10 min. After centrifugation, an aliquot of 10 μℓ of the supernatant was injected.

Elution — E-1. Methanol-0.01 *M* acetate buffer, pH 4.8 (15:85).
E-2. Methanol-0.75% acetic acid (30:70), pH 5.5 with triethylamine.
E-3. Methanol-water-1.8 *M* H_2SO_4 (x:y:0.2%). x = 15 to 40% depending upon drug.
E-4. Acetonitrile-0.01 *M* sodium dihydrogen phosphate (x:100 − x) x = 6 to 11% depending upon drug.
E-5. Acetonitrile-methanol-0.01 *M* sodium acetate (x:y:100 − x + y) (x:y = 96:4), x + y = 9 to 20%.

REFERENCES

1. **Brisson, A. M. and Fourtillan, J. B.,** Determination of cephalosporins in biological material by reversed-phase liquid column chromatography, *J. Chromatogr.*, 223, 393, 1981.
2. **Danzer, L. A.,** Liquid-chromatographic determination of cephalosporins and chloramphenical in serum, *Clin. Chem.*, 29, 956, 1983.
3. **Rouan, M. C., Abadie, F., Leclerc, A., and Juge, F.,** Systemic approach to the determination of cephalosporins in biological fluids by reversed-phase liquid chromatography, *J. Chromatogr.*, 275, 133, 1983.
4. **Nygard, G. and Khalil, S. K. W.,** An isocratic HPLC method for the determination of cephalosporins in plasma, *J. Liq. Chromatogr.*, 7, 1461, 1984.
5. **Signs, S. A., File, T. M., and Tan, J. S.,** High-pressure liquid chromatographic method for analysis of cephalosporins, *Antimicrob. Agents Chemother.*, 26, 652, 1984.

CEPHAMYCIN C

Thin-Layer Chromatography

Specimen (mℓ)	Extraction	Plate (Manufacturer)	Layer (mm)	Solvent	Post-separation treatment	Det. (nm)	Rf	Internal standard (Rf)	Other compounds (Rf)	Ref.
Fermentation broth	I-1	20 × 20 cm (Merck)	Silica gel F254 (0.25)	S-1	—	Reflectance (273)	NA	—	—	1

Extraction — I-1. A 10-μℓ volume of glacial acetic acid was applied onto each origin. Aliquots of broth samples taken at various stages of the fermentation were centrifuged and 5-μℓ volumes of the supernatant were applied to acetic acid treated origins.

CEPHAMYCIN C (continued)

Solvent — S-1. Ethanol-acetic acid-ammonium hydroxide (6:3:1).

REFERENCE

1. **Treiber, L. R.,** Quantitative analysis of cephamycin C in fermentation broths by means of thin-layer spectrodensitometry, *J. Chromatogr.*, 213, 129, 1981.

CEPHAPIRIN

Liquid Chromatography

Specimen (mℓ)	Extraction	Column (cm × mm)	Packing (μm)	Elution	Flow (mℓ/min)	Det. (nm)	RT (min)	Internal standard (RT)	Other compounds (RT)	Ref.
Serum (0.5)	I-1	30 × 4	μ-Bondapak-C_{18} (10)	E-1	1.5	ABS (270)	3.7	—	Desacetyl-cephapirin (1.8) Cefoxitin (7.7) Cefotaxime (7.8) Desacetylcefotax-ime (3.3)	1

Extraction — I-1. The sample (0.5 mℓ of serum) was applied to a Sephadex DEAE-A-25 resin column (bed volume of 3 mℓ). The column had been washed prior to sample application with phosphate buffered saline (pH 7.2). After the sample had passed through, the column was washed with 4.5 mℓ of above mentioned buffered saline. The column was eluted with 5 mℓ of 1 *M* sodium chloride.

Elution — E-1. Acetonitrile-0.15% acetic acid, pH 2.8 (13:87).

REFERENCE

1. **Fasching, C. E. and Peterson, L. R.,** Anion-exchange extraction of cephapirin, cefotaxime, and cefoxitin from serum for liquid chromatography, *Antimicrob. Agents Chemother.*, 21, 628, 1982.

CETIEDIL

Gas Chromatography

Specimen (mℓ)	Extraction	Column (m × mm)	Packing (mesh)	Oven temp (°C)	Gas (mℓ/min)	Det.	RT (min)	Internal standard (RT)	Deriv.	Other compounds (RT)	Ref.
Plasma (2)	I-1	1.5 × 2	2% OV-101 Chromosorb W (100/120)	252	He (20)	NPD	1.7	Papaverine (3.3)	—	—	1
Plasma (1)	I-2	25 × 3.2	CP Sil 8CB (0.12 μm)[a]	T.P.[b]	He (3)	NPD	5.8	Bepridil (6.2)	—	—	2

[a] Film thickness.
[b] Initial temp = 185°C; rate = 10°C/min; final temp = 265°C.

Extraction — I-1. The sample was treated with 4 mℓ of borate buffer (pH 7) and the mixture extracted with 20 mℓ of *n*-heptane-2-propanol (95:5) containing 5 ng/mℓ of the internal standard. The organic layer was back extracted into 3 mℓ of 1 *M* HCl. The acidic extract was washed with 5 mℓ of *n*-heptane, made basic with 0.5 mℓ of 10 *M* KOH and extracted with 4 mℓ of dichloromethane. The organic layer was evaporated, the residue dissolved in 10 μℓ of 2-propanol and an aliquot was injected.

I-2. The sample was treated with 2 mℓ of methanol containing 10 ng/mℓ of the internal standard. After mixing and centrifugation the supernatant was applied to a prewashed (3 mℓ of 0.04 M ammonium acetate in methanol and 6 mℓ of methanol) 500 mg BondElut-C_{18} column. The column was then eluted three times with 0.5-mℓ aliquots of 0.02 M ammonium acetate in methanol. To the eluate 100 μℓ of 0.58% ammonium hydroxide was added and extracted with 6 mℓ of hexane. The organic layer was evaporated under a gentle stream of nitrogen. The residue was reconstituted with 50 μℓ of toluene-methanol (9:1) solution and aliquots of 3 μℓ were injected with an autoinjector.

REFERENCES

1. **Henderson, J. D., Mankad, V. N., Glenn, T. M., and Cho, Y. W.,** Gas chromatographic analysis of cetiedil, a candidate antisickling agent, in human plasma with nitrogen-sensitive detection, *J. Pharm. Sci.*, 73, 1748, 1984.
2. **Holland, M. L. and Ng, K. T.,** Automated capillary gas chromatographic assay using nitrogen-phosphorus detection for the determination of cetiedil in plasma, *J. Chromatogr.*, 345, 178, 1985.

CGP 6258

Gas Chromatography

Specimen (mℓ)	Extraction	Column (m × mm)	Packing (mesh)	Oven temp (°C)	Gas (mℓ/min)	Det.	RT (min)	Internal standard (RT)	Deriv.	Other compounds (RT)	Ref.
Plasma (1)	I-1	1.5 × 4	3% OV-225 Supelcoport (80/100)	220	N_2 (40)	ECD	6	CGP 7726 (8.5)	Pentafluoropropyl; heptafluoro-butyric	—	1

Extraction — I-1. The sample (1 mℓ of plasma), 0.5 mℓ of the internal standard solution (0.675 nmol in 0.01*M* NaOH), 1 mℓ of 1 *M* HCl, and 5 mℓ of toluene were mixed and centrifuged. The organic phase was evaporated to dryness under a stream of nitrogen. The residue was treated with 0.1 mℓ of heptafluorobutyric anhydride and 0.5 mℓ of pentafluoropropanol. The mixture was incubated at 70°C for 1 hr. The excess reagents were removed under a stream of nitrogen at 40°C. The residue was dissolved in 1 mℓ of *n*-heptane and aliquots of 3 to 5 μℓ were injected.

REFERENCE

1. **Degen, P. H. and Schneider, W.,** Simultaneous derivatisation of carboxyl and hydroxyl groups of a new antiphlogistic drug for its determination by electron-capture gas chromatography, *J. Chromatogr.*, 277, 361, 1983.

CGS 13945

Liquid Chromatography

Specimen (mℓ)	Extraction	Column (cm × mm)	Packing (μm)	Elution	Flow (mℓ/min)	Det. (nm)	RT (min)	Internal standard (RT)	Other compounds (RT)	Ref.
Plasma (1)	I-1	15 × 4.6	Ultrasphere-C_8 (5)	E-1	2	ABS (254)	(11)	CGS 13748 (7)	CGS 13934 (4)	1

Extraction — I-1. The sample was spiked with 2 μg of the internal standard (solvent removed), pH adjusted to 3 with dilute phosphoric acid and extracted with 4 mℓ of dichloromethane. The organic layer was evaporated under reduced pressure at 37°C. The residue was reconstituted in 500 μℓ of the mobile phase and 50 to 100 μℓ of this solution were injected on the column.

Elution — E-1. Acetonitrile-tetrahydrofuran-aqueous phosphoric acid, pH 2 (15:15:70).

REFERENCE

1. **Rakhit, A. and Tipnis, V.,** Liquid-chromatographic determination of an angiotensin convering enzyme inhibitor, CGS 14945, and its active metabolite (CGS 13934) in plasma, *Clin. Chem.*, 30, 1237, 1984.

CGS 5391B

Liquid Chromatography

Specimen (mℓ)	Extraction	Column (cm × mm)	Packing (μm)	Elution	Flow (mℓ/min)	Det. (nm)	RT (min)	Internal standard (RT)	Other compounds (RT)	Ref.
Plasma (0.5)	I-1	25 × 4.6	Zorbax C_8 (10)	E-1	2.0	ABS (254)	NA[a]	CGS 5089A[a]	—	1

[a] The internal standard elutes prior to the drug. Total run time = 7 min.

Extraction — I-1. The sample was treated with 1 mℓ of 0.1 *M* citrate-phosphate buffer, pH 5 containing 10 μg/mℓ of the internal standard. The mixture was extracted with 10 mℓ of chloroform. The organic extract was evaporated at 45 to 55°C with a stream of nitrogen. The residue was dissolved in 2 mℓ of acetonitrile by vortexing for 5 sec and allowing the mixture to stand at room temperature prior to chromatography. Extracts of rat plasma were chromatographed immediately.

Elution — E-1. Acetonitrile-0.05 *M* phosphate buffer, pH 2.9 (35:65).

REFERENCE

1. **Thompson, T. A., Borman, C. H., Vermeulen, J. D., and Rosen, R.,** Assay of the anti-inflammatory compound CGS 5391B in blood plasma by automated HPLC, *J. Liq. Chromatogr.*, 4, 2015, 1981.

CHLORAMBUCIL

Liquid Chromatography

Specimen (mℓ)	Extraction	Column (cm × mm)	Packing (μm)	Elution	Flow (mℓ/min)	Det. (nm)	RT (min)	Internal standard (RT)	Other compounds (RT)	Ref.
Plasma	I-1	25 × 4.6	Partisil PXS-ODS (10)[a]	E-1	1.5	ABS (254)	14.5	—	Phenylacetic mustard (7) Monohydroxy-chlorambucil (6)	1
Plasma (0.1)	I-2	10 × 8	Radial-Pak-C_{18} (10)	E-2	1.0	ABS (263)	8	—	—	2
Plasma (3)	I-3	25 × 4.6	Spherisorb-ODS (5)[b]	E-3	1.3	ABS (260)	4	—	—	3

[a] Protected by a 50 × 3.9 mm guard column packed with Co:Pell-ODS (30 to 32 μm).
[b] Protected by a 30 × 4.6 mm guard column packed with Spherisorb-ODS (5 μm).

Extraction — I-1. Aliquots of plasma are injected directly.
I-2. The sample was treated with 4 volumes of acetonitrile, vortexed, and centrifuged. The supernatant was rapidly frozen in a solution of dry ice-acetone bath and again centrifuged. Aliquots (200 μℓ) of the upper clear supernatant was collected and a volume of 75 μℓ was injected immediately.
I-3. Plasma sample (3 mℓ) was treated with 132 μℓ of cold concentrated perchloric acid (2°C). The mixture was vortexed and centrifuged at −6°C. The supernatant was passed through a C_{18} Sep-Pak cartridge. The cartridge was washed with 10 mℓ of 15% methanol in water and eluted with 2 mℓ of methanol. The eluate was stored at −20°C prior to chromatography.

Elution — E-1. Methanol-0.02 *M* KH_2PO_4 (50:50).
E-2. Acetonitrile-0.2% acetic acid (65:35).
E-3. Methanol-water (80:20).

REFERENCES

1. **Zakaria, M. and Brown, P. R.,** Rapid assay for plasma chlorambucil and phenyl acetic mustard using reversed-phase liquid chromatography, *J. Chromatogr.*, 230, 381, 1982.
2. **Ahmed, A. E., Koenig, M., and Farrish, H. H., Jr.,** Studies on the quantitation of chlorambucil in plasma by reversed-phase high-performance liquid chromatography, *J. Chromatogr.*, 223, 392, 1982.
3. **Adair, C. G., Burns, D. T., Crockard, A. D., and Harriott, M.,** Determination of chlorambucil in plasma using reversed-phase high-performance liquid chromatography, *J. Chromatogr.*, 342, 447, 1985.

CHLORAMPHENICOL[a]

Gas Chromatography

Specimen (mℓ)	Extraction	Column (m × mm)	Packing (mesh)	Oven temp (°C)	Gas (mℓ/min)	Det.	RT (min)	Internal standard (RT)	Deriv.	Other compounds (RT)	Ref.
Tissue (5 g)	I-1	1.8 × 2	3% OV-1 Gas Chrom A (100/120)	220	N_2 (30)	ECD	6	Thiamphenicol (12)[b]	Trimethylsilyl	—	2

Liquid Chromatography

Specimen (mℓ)	Extraction	Column (cm × mm)	Packing (μm)	Elution	Flow (mℓ/min)	Det. (nm)	RT (min)	Internal standard (RT)	Other compounds (RT)	Ref.
Plasma (0.25)	I-2	25 × 4	LiChrosorb RP-18 (10)[c]	E-1	1—25	ABS (275)	9	Benzocaine (15)	CH-1-MS[d] (4.2)	3
Plasma (0.05)	I-3	25 × 4	μ-Bondapak-C_{18} (10)	E-2	2.5	ABS (254)	8.8	N- (β-Hydroxy-*p*-Nitrophenethyl) acetamide (5.2)	—	4
Plasma (0.4)	I-4	25 × 4	LiChrosorb-Cg (10)	E-3	2.5	ABS (280)	6	Ethylparaben (15)	CH-3-MS (9)	5
Serum, CSF, Urine (0.05)	I-5	N/A	ODS-HC-SI2-X-(NA)[f]	E-4	1.5	ABS (272)	3.52	5-Ethyl-*p*-tolyl-barbituric acid (5)	CH-1-MS (2.2) CH-3-MS (2.5)	6, 7[g]
Dosage	—	25 × 3.2	Partisil-ODS (10)	E-5	2.0	ABS (254)	17	—	2-Amino- (*p*-nitrophenyl)-1,2-propandiol	9
Serum (0.05)	I-6	30 × 3.9	μ-Bondapak-C_{18} (10)	E-6	1.5	ABS (254)	3—9	N-Acetylchlor-amphenicol (1.9)	CH-3-MS (5.8)	10
Plasma (0.05)	I-7	30 × 4	μ-Bondapak-C_{18} (10)	E-7	1.5	ABS (278)	15.5	Thiamphenicol (5)	CH-1-MS (8) CH-3-MS (12)	11
Plasma (0.03)	I-8	30 × 4	μ-Bondapak-C_{18} (10)[h]	E-8	1.5	ABS (278)	8.5	5-Ethyl-5-*p*-tolylbarbituric acid (14.5)	CH-1-MS (12.5) CH-3-MS (16.5)	12

CHLORAMPHENICOL[a] (continued)

Plasma (0—1)	I-9	25 × 4	LiChrosorb RP-18 (5)[i]	E-9	1.0	ABS (210)	7.9	Phenaceten (6.8)	Phenobarbital (5.8) Carbamazepine (8.6) Phenytoin (12.9)	13
Serum (1)	I-10	30 × 4	Micropak C-18 (10)	E-10	2.0	ABS (278)	5	2,4-Dinitroacetanilide (10)	—	14
Serum (0.2)	I-11	10 × 8	Radial Pak-C_{18} (5)	NA	2.0	ABS (278)	4.3	Mephensin (6.6)	—	15

Thin-Layer Chromatography

Specimen (mℓ)	Extraction	Plate (Manufacturer)	Layer (mm)	Solvent	Post-separation treatment	Det. (nm)	Rf	Internal standard (Rf)	Other compounds (Rf)	Ref.
Dosage	—	20 × 20 cm (Merck)	Silica gel G F_{254} (0.25)	S-1	—	Reflectance (250)	0.9	—	1-(4′-Nitrophenyl)-2-aminopropane-1,3-diol (0.25)	16

[a] Chromatographic methods for the determination of chloramphenicol in food and tissues have been recently reviewed.[1]
[b] The retention times of the drug and the internal standard increased with the run.
[c] Column temperature = 50°C.
[d] CH-1-MS = Chloramphenicol-1-monosuccinate; CH-3-MS = Chloramphenicol-3-monosuccinate (major component).
[e] A Perkin-Elmer Column.
[f] Column temperature = 50°C.
[g] This procedure is subject of interference by dobutamine.[8]
[h] Column temperature = 37°C.
[i] Column temperature = 45°C.

Extraction — I-1. The sample was homogenized with 2 mℓ of the internal standard solution (50 μg/mℓ) in methanol. The homogenate was extracted twice with 10-mℓ portions of ethyl acetate. The combined extracts were evaporated under a stream of nitrogen at 60°C. The residue was dissolved in 0.2 mℓ of methanol and 2.8 mℓ of 1 *N* HCl. The solution was then washed three times with 1.5-mℓ portions of light petroleum. The aqueous phase was then applied to a Sep-Pak cartridge prewashed with 2 mℓ of methanol, 5 mℓ of water. After the sample had passed through, the cartridge was eluted with two 3-mℓ volumes of methanol-1 *N* HCl (40:60). The combined eluate was evaporated under a stream of dry nitrogen at 60°C. The residual aqueous phase was extracted twice with 2-mℓ volumes of ethyl acetate. The pooled extracts were evaporated under a stream of nitrogen. The residue was then incubated with 400 μℓ of Trisil at 35°C for 30 min. The excess reagent was evaporated under a stream of nitrogen, the residue dissolved in 1-mℓ of benzene and aliquots of this solution were injected.
I-2. The plasma sample (stored at pH 4) was spiked with 25 μℓ of the solution of the internal standard (60 μg/mℓ in methanol) and the mixture treated with 50 μℓ of a 30% solution of trichloroacetic acid. After vortexing and centrifugation, a portion of the supernate (90 μℓ) was injected.
I-3. The sample was treated with an equal volume (50 μℓ) of acetonitrile containing 5 μg of the internal standard. The mixture was vortex mixed, centrifuged and aliquots of 20 to 30 μℓ were injected.
I-4. The sample was spiked with 50 μℓ of a solution of the internal standard (20 mg%), 1 mℓ of citrate-phosphate buffer, pH 3 was added, and the mixture extracted with 5 mℓ of anhydrous diethyl ether. The organic layer was evaporated at room temperature under a gentle stream of air. The residue was reconstituted with 100 μℓ of the mobile phase and 20 μℓ aliquots were injected.
I-5. The sample was diluted 1:10 with water, 100 μℓ of 1 *N* sodium acetate added to the diluted sample and extracted with 1 mℓ of ethyl acetate containing the internal standard (20 mg/ℓ). The organic phase was evaporated at 40°C under nitrogen. The residues were sequentially reconstituted with 50 μℓ of methanol and 10 μℓ were injected.
I-6. The sample (50 μℓ of serum) was treated with 100 μℓ of acetonitrile containing 10 μg of the internal standard. After vortexing and centrifugation, 50 μℓ of the supernate was injected.
I-7. The sample was mixed with 100 μℓ of 1 *M* sodium acetate buffer, pH 4.6 and 50 μℓ of an aqueous solution of the internal standard (1 mg/mℓ). The mixture was extracted with 1 mℓ of ethyl acetate. The organic layer was evaporated at room temperature with a stream of air. The residue was reconstituted with 100 μℓ of the mobile phase and portions of this solution were injected.
I-8. The sample was extracted with 100 μℓ of ethyl acetate containing (60 mg/ℓ) of the internal standard. An aliquot of 400 μℓ of the organic phase was evaporated at room temperature under nitrogen. The residue was dissolved in 100 μℓ of the mobile phase and 80 μℓ were injected.
I-9. The sample was mixed with 200 μℓ of the internal standard solution (6 μg/mℓ in acetonitrile) and 500 μℓ of 0.1 *M* phosphate buffer (pH 6.5). The supernate was extracted with 3 mℓ of diethyl ether. The upper organic phase was evaporated, the residue dissolved in 100 μℓ of mobile phase, and 10 μℓ of the solution was injected.
I-10. The sample was mixed with an equal volume of the internal standard solution (100 μg/mℓ in acetonitrile). After 4 sec of vortex mixing, 0.4 g of salt mix (8 g NaCl + 2 g Na_3PO_4) was added. After vortex mixing for 1 min, a 10-μℓ aliquot of the upper layer was injected.
I-11. A 200-μℓ aliquot of the sample was mixed with 1 mℓ of the internal standard solution (30 mg/ℓ in methanol). After mixing and centrifugation, a 25 μℓ of the clear supernate was injected.

Elution — E-1. Acetonitrile-0.05 *M* sodium acetate, pH 5.7 (22:78).
E-2. Methanol-water (20:80), containing 0.01 *M* monobasic-ammonium phosphate adjusted to pH 3 with HCl.
E-3. Methanol-0.05% aqueous phosphoric acid (4:6).
E-4. Acetonitrile-0.1 *N* sodium acetate (15:85), pH 6.4.
E-5. 50 m*M* potassium dihydrogen phosphate, pH 4.5.

CHLORAMPHENICOL[a] (continued)

E-6. Methanol-water-acetic acid (37:62:1).
E-7. Acetonitrile-0.05 *M* sodium acetate buffer, pH 5.3 (20:80).
E-8. Acetonitrile-10 μmol phosphate buffer, pH 2.7 (25:75).
E-9. Tetrahydrofuran-methanol-water (65:33.5:60).
E-10. Acetonitrile-methanol-0.1 *M* phosphate buffer, pH 3.25 (10:22.5:67.5).

Solvent — S-1. Ethyl acetate-formic acid-water (10:2:8), upper layer.

REFERENCES

1. **Allen, E.,** Review of chromatographic methods for chloramphenicol residues in milk, eggs, and tissues from food-producing animals, *J. Assoc. Off. Anal. Chem.*, 68, 990, 1985.
2. **Nelson, J. R., Copeland, K. F. T., Forster, R. J., Campbell, D. J., and Black, W. D.,** Sensitive gas-liquid chromatographic method for chloramphenicol in animal tissues using electron-capture detection, *J. Chromatogr.*, 276, 438, 1983.
3. **Burke, J. T., Wargin, W. A., and Blum, M. R.,** High-pressure liquid chromatographic assay for chloramphenicol, chloramphenicol-3-monosuccinate, and chloramphenicol-1-monosuccinate, *J. Pharm. Sci.*, 69, 909, 1980.
4. **Gal, J., Marcell, P. D., and Tarascio, C. M.,** High-performance liquid chromatographic micro-assay for chloramphenicol in human blood plasma and cerebrospinal fluid, *J. Chromatogr.*, 181, 123, 1980.
5. **Oseekey, K. B., Rowse, K. L., and Kostenbauder, H. B.,** High-performance liquid chromatographic determination of chloramphenicol and its monosuccinate ester in plasma, *J. Chromatogr.*, 182, 459, 1980.
6. **Aravind, M. K., Miceli, J. N., Kauffman, R. E., Strebel, L. E., and Done, A. K.,** Simultaneous measurement of chloramphenicol and chloramphenicol succinate by high-performance liquid chromatography, *J. Chromatogr.*, 221, 176, 1980.
7. **Aravind, M. K., Miceli, J. N., Done, A. K., and Kauffman, R. E.,** Determination of chloramphenicol-glucuronide in urine by high-performance liquid chromatography, *J. Chromatogr.*, 232, 461, 1982.
8. **Powel, M. B., Robinson, C. A., and Furner, R. L.,** Interference with high performance liquid chromatographic chloramphenicol assay in a patient receiving dobutamine, *Ther. Drug Monit.*, 7, 121, 1985.
9. **Lee, M. G., Dawes, S., and Mannion, P.,** A rapid and specific method for measuring chloramphenicol degradation in aqueous solution following autoclaving, *J. Chromatogr. Sci.*, 19, 96, 1981.
10. **Nahata, M. C. and Powell, D. A.,** Simultaneous determination of chloramphenicol and its succinate ester by high-performance liquid chromatography, *J. Chromatogr.*, 223, 247, 1981.
11. **Velagapudi, R., Smith, R. V., Ludden, T. M., and Sagraves, R.,** Simultaneous determination of chloramphenicol and chloramphenicol succinate in plasma using high-performance liquid chromatography, *J. Chromatogr.*, 228, 423, 1982.
12. **Soldin, S. J., Golas, C., Rajchgot, P., Prober, C. G., and MacLeod, S. M.,** The high performance liquid chromatographic measurement of chloramphenicol and its succinate esters in serum, *Clin. Biochem.*, 16, 171, 1983.

13. **Kushida, K., Chiba, K., and Ishizaki, T.,** Simultaneous liquid chromatographic determination of chloramphenicol and antiepileptic drugs (phenobarbital, phenytoin, carbamazepine, and primidone) in plasma, *Ther. Drug Monit.*, 5, 127, 1983.
14. **Ryan, F. J., Austin, M. A., and Mathies, J. C.,** Simple and precise method for liquid chromatographic determination of chloramphenicol in serum using a phase separation extraction, *Ther. Drug Monit.*, 6, 465, 1984.
15. **Davidson, D. F. and Fitzpatrick, J.,** Rapid column-chromatographic analysis of chloramphenicol in serum, *Clin. Chem.*, 32, 701, 1986.
16. **Pietta, P.,** Simplified thin-layer chromatographic method for the simultaneous determination of chloramphenicol and 1-(4′-nitrophenyl)-2-aminopropane-1,3-diol, *J. Chromatogr.*, 177, 177, 1979.

CHLORDIAZEPOXIDE

Liquid Chromatography

Specimen (mℓ)	Extraction	Column (cm × mm)	Packing (μm)	Elution	Flow (mℓ/min)	Det. (nm)	RT (min)	Internal standard (RT)	Other compounds (RT)	Ref.
Plasma (1—2)	I-1	25 × 4	LiChrosorb RP-18 (10)	E-1	2.0	ABS (260)	7.8	Nitrazepam (6.3)	Demoxepam (4) Desmethylchlordiazepoxide (5)	1
Brain tissue	I-2	30 × 4	μ-Bondapack-C_{18} (10)	E-2	2.0	ABS (254)	4.5	Diazepam (5.2)	Desmethylchlordiazepoxide (3.5)	2
Plasma (0.2)	I-3	10 × 4	LiChrosorb RP-18 (10)	E-3	2.0	ABS (240)	2.8	Diazepam (5.8)	Demoxepam (1.5) Desmethylchlordiazepoxide (2) Oxazepam (2.2)	3
Plasma (1)	I-4	30 × 3.9	μ-Bondapack-C_{18} (10)	E-4	2.0	ABS (254)	11	Chlordesmethyldiazepam (14)	Demoxepam (6) Desmethylchlordiazepoxide (8)	4

Extraction — I-1. The sample was adjusted to pH 9 with 0.1 *N* sodium hydroxide; diluted with 5 mℓ of water and extracted twice with 7-mℓ portions of diethyl ether. The combined ether extract was evaporated in a stream of nitrogen. The residue was dissolved in 100 μℓ of the mobile phase containing 10 μg of the internal standard. The solution was washed with 100 μℓ of *n*-hexane. Aliquots of 20 μℓ of the lower aqueous phase were injected.
I-2. The brain homogenate was spiked with 10 μℓ of 0.1 mg/mℓ of diazepam in methanol, made alkaline with 5 mℓ of 0.01 *N* NaOH, and centrifuged. The supernate was extracted with 10 mℓ of heptane containing 1.5% isoamyl alcohol. The organic layer was evaporated under a stream of nitrogen. The residue was dissolved in 50 μℓ of methanol and an aliquot of 10 μℓ was injected.

CHLORDIAZEPOXIDE (continued)

I-3. The sample was spiked with 10 μℓ of an aqueous solution of the internal standard (3 μg/mℓ) and extracted with 1 mℓ of ether. The ether layer was evaporated with a stream of air, the residue dissolved in 0.2 mℓ of the mobile phase, and 0.1 mℓ was injected.
I-4. A methanolic solution of the internal standard was evaporated to get a 10 μg amount. This was treated with the sample and the mixture was extracted with 3 to 5 mℓ of benzene containing 1.5% isoamyl alcohol. The organic layer was evaporated at 40°C under reduced pressure. The residue was dissolved in 100 μℓ of methanol of which 25 to 30 μℓ was injected.

Elution — E-1. Acetonitrile-0.1% ammonium carbonate (31:69).
E-2. Methanol-water (96:34).
E-3. Acetonitrile-methanol-0.01 *M* sodium acetate (20:20:60).
E-4. Acetonitrile-methanol-water-1 *M* sodium acetate (225:225:550:1).

REFERENCES

1. **Ascalone, V.,** Determination of chlordiazepoxide and its metabolites in human plasma by reverse-phase high-performance liquid chromatography, *J. Chromatogr.*, 181, 141, 1980.
2. **Greizerstein, H. B. and McLaughlin, I. G.,** The high-pressure liquid chromatographic determination of chlordiazepoxide and its N-demethyl metabolite in mouse brain, *J. Liq. Chromatogr.*, 3, 1023, 1980.
3. **Vree, T. B., Baars, A. M., Hekster, Y. A., and Van Der Kleijn, E.,** Simultaneous determination of chlordiazepoxide and its metabolites in human plasma and urine by means of reversed-phase high-performance liquid chromatography, *J. Chromatogr.*, 224, 519, 1981.
4. **Divoll, M., Greenblatt, D. J., and Shader, R. I.,** Liquid chromatographic determination of chlordiazepoxide and metabolites in plasma, *Pharmacology*, 24, 261, 1982.

CHLORHEXIDINE

Liquid Chromatography

Specimen (mℓ)	Extraction	Column (cm × mm)	Packing (μm)	Elution	Flow (mℓ/min)	Det. (nm)	RT (min)	Internal standard (RT)	Other compounds (RT)	Ref.
Urine (10)	I-1	30 × 4	μ-Bondapac-C_{18} (10)	E-1	1.5	ABS (260)	9	3-Bromobenzophen-one (10)	—	1, 2
Contact lens washing	I-2	24 × 4.6	Regis RP-6 (5)	E-2	1.0	ABS (220)	9	—	*p*-Chloroaniline (4)	3

Extraction — I-1. The sample was passed through a prewashed (2 mℓ methanol, 2 mℓ water) Sep-Pak C_{18} cartridge. After the sample had passed through, the cartridge was washed with 2 mℓ of water which was then eluted with 2 mℓ of methanol. The chromatographic standard was added to the eluate to give a final concentration of 10 μg/mℓ. Aliquots of this solution were injected.
I-2. Lens washings were injected directly.

Elution — E-1. Methanol-acetate buffer, pH 5 (60:40) containing 100 mg/ℓ pentadecafluorooctanoic acid.
E-2. Acetonitrile-0.5% phosphoric acid + 0.22% sodium 1-heptane-sulfonate (60:40).

REFERENCES

1. **Gaffney, M. H., Cooke, M., and Simpson, R.,** Improved method for the determination of chlorhexidine in urine, *J. Chromatogr.*, 306, 303, 1984.
2. **Huston, C. E., Wainwright, P., Cooke, M., and Simpson, R.,** High-performance liquid chromatographic method for the determination of chlorhexidine, *J. Chromatogr.*, 237, 457, 1982.
3. **Stevens, L. E., Durrwachter, J. R., and Helton, D. O.,** Analysis of chlorhexidine sorption in soft contact lenses by catalytic oxidation of [^{14}C] chlorhexidine and by liquid chromatography, *J. Pharm. Sci.*, 75, 83, 1986.

CHLORMETHIAZOLE

Gas Chromatography

Specimen (mℓ)	Extraction	Column (m × mm)	Packing (mesh)	Oven temp (°C)	Gas (mℓ/min)	Det.	RT (min)	Internal standard (RT)	Deriv.	Other compounds (RT)	Ref.
Plasma (0.05—0.5)	I-1	1.5 × 3	5% OV-7 Gas Chrom Q (100/120)	145	N_2 (30)	NPD	6	Bromethiazole (9)	—	5-Acetyl-4-methylthiazide 5-(1-Hydroxyethyl)-4-methylthiazole (6)	1
Serum (1)	I-2	1.5 × 2	1% SP1000 Supelcoport (100/120)	95	He (22)	NPD	2	Bromethiazole (3.6)	—	—	2
Plasma, serum (0.2—1)	I-3	1 × 3	8% OV-17 Gas Chrom Q (80/100)	170	N_2 (30)	NPD	3	Quinaldine (4)	—	5-Acetyl-4-methylthiazole (2.2) 5-(1-Hydroxyethyl)-4-methylthiazole (2.5)	3

CHLORMETHIAZOLE (continued)

Liquid Chromatography

Specimen (mℓ)	Extraction	Column (cm × mm)	Packing (μm)	Elution	Flow (mℓ/min)	Det. (nm)	RT (min)	Internal standard (RT)	Other compounds (RT)	Ref.
Plasma (0.4)	I-4	25 × 4.6	Ultrasphere-ODS (5)	E-1	2.0	ABS (254)	4.2	2-Amino-4-methylthiazole (1.9)	—	4
Plasma (0.5)	I-5	30 × 3.9	μ-Bondapak-C_{18} (10)[a]	E-2	1.8	ABS (254)	9	Carbamazepine (11.5)	—	5

[a] Protected by a guard column packed with Corasil-C_{18}.

Extraction — I-1. The sample was mixed with 25 μℓ of an aqueous solution of the internal standard (0.1 mg/mℓ), 200 μℓ of water, 100 μℓ of phosphate buffer (pH 7.0, 0.1 *M*), and the mixture was extracted with 5 mℓ of diethyl ether. The organic layer was back extracted into 500 μℓ of 1 *M* HCl. The aqueous phase was made alkaline with 100 μℓ of 10 *M* NaOH and re-extracted with 1 mℓ of ether. After centrifugation, 5 to 10 μℓ of the ethereal extract were injected.
I-2. To the sample were added 1 mℓ of an aqueous solution of the internal standard (10 mg/ℓ), 0.5 mℓ of 10% sodium carbonate solution. The mixture was extracted twice with 5-mℓ portions of diethyl ether. The combined extracts were evaporated with a stream of nitrogen to about 20 μℓ of which 1 μℓ was injected.
I-3. To the sample were added, 1 mℓ of an aqueous solution of the internal standard (1 μg/mℓ), 0.5 mℓ of 5 *M* sodium hydroxide and the mixture was extracted with 5 mℓ of diethyl ether. The organic layer was back extracted into 2 mℓ of 1 *M* HCl. The aqueous layer was concentrated to about 100 μℓ and a 5-μℓ aliquot of this was injected.
I-4. The sample after the addition of 1 mℓ of an aqueous solution of the internal standard was deproteinized with 25 μℓ of phosphotungstic acid reagent for 90 min at room temp and centrifuged. Aliquots of 20 μℓ of the clear supernate were injected.
I-5. The sample was diluted with an equal volume of methanol containing the internal standard (10 μg/mℓ). A 25-μℓ aliquot of the supernate was injected.

Elution — E-1. Acetonitrile-0.25 *M* KH_2PO_4, pH 4.6 (45:55).
E-2. Methanol-water (45:55).

REFERENCES

1. **Tsuei, S. E., Thomas, J., and Nation, R. L.,** Simultaneous quantitation of chlormethiazole and two of its metabolites in blood and plasma by gas-liquid chromatography, *J. Chromatogr.*, 182, 55, 1980.
2. **Heipertz, R. and Reimer, Ch.,** A rapid gas-chromatographic method for the quantitative determination of clomethiazole in human serum, *Clin. Chim. Acta*, 110, 131, 1981.

3. **Witts, D. J., Arnold, K., and Exton-Smith, A. N.,** The plasma levels of chlormethiazole and two of its metabolites in elderly subjects after single and multiple dosing, *J. Pharm. Biomed. Anal.*, 1, 311, 1983.
4. **Kim, C. and Khanna, J. M.,** Determination of chlormethiazole in blood by high performance liquid chromatography, *J. Liq. Chromatogr.*, 6, 907, 1983.
5. **Hartley, R., Becker, M., and Leach, S. F.,** Determination of chlormethiazole in plasma by high-performance liquid chromatography, *J. Chromatogr.*, 276, 471, 1983.

CHLORMEZANONE

Gas Chromatography

Specimen (mℓ)	Extraction	Column (m × mm)	Packing (mesh)	Oven temp (°C)	Gas mℓ/min	Det.	RT (min)	Internal standard (RT)	Deriv.	Other compounds (RT)	Ref.
Plasma (0.1—1)	I-1	1 × 3	5% EGS Gas Chrom Q (80/100)	100	N_2 (40)	ECD	4.2	*p*-Bromobenzal-dehyde (8.1)	Hydrolysis	—	1

Extraction — I-1. The sample after the addition of 1 mℓ of acetate buffer (pH 4.5) was extracted with 6 mℓ of toluene. A 5-mℓ volume of the toluene extract was evaporated to dryness. The residue was dissolved in 1 mℓ of 0.1 *N* NaOH and extracted with 2 mℓ of hexane containing (0.5 μg/mℓ) of the internal standard. The hexane layer was collected after freezing the aqueous layer in dry ice-acetone bath. Isoamyl alcohol (1 mℓ) was added to hexane extract and the mixture concentrated to about 1 mℓ. Aliquots of 1 to 2 μℓ were injected.

REFERENCE

1. **Ohya, K., Shintani, S., Suzuki, W., and Sano, M.,** Sensitive and selective methods for the determination of chlormezanone in plasma by electron-capture gas chromatography, *J. Chromatogr.*, 221, 67, 1980.

10-CHLORO-5-(2-DIMETHYLAMINO ETHYL)-7H-INDOLO[2,3-C] QUINOLIN-6(5H)-ONE

Liquid Chromatography

Specimen (mℓ)	Extraction	Column (m × mm)	Packing (mesh)	Oven temp (°C)	Gas (mℓ/min)	Det.	Rt (min)	Internal standard (RT)	Deriv.	Other compounds (RT)	Ref.
Plasma (1)	I-1	1.5 × 2	OV-17 μ-Partisorb	170	Methane[a]	MS-CI	1.6	[^{13}C]-Chlorocitric acid	Methyl	—	1

[a] 10-Methoxy-5-(Dimethylaminoethyl)-7H-indolo [2,3-C] quinolin-6(5H)-one.

Extraction — I-1. A 250-μℓ aliquot of the solution of the internal standard (4 ng/mℓ in the mobile phase) was evaporated. The residue was heated with the sample, and 1 mℓ of 1 *M* ammonium hydroxide. The mixture was extracted with 10 mℓ of freshly distilled diethyl ether. A 9-mℓ aliquot of the ether extract was evaporated at 40°C under a stream of nitrogen. The residue was dissolved in 250 μℓ of the mobile phase, 25 μℓ of which were injected.

Elution — E-1. Hexane-tetrahydrofuran-methanol-ammonium hydroxide (75:15:9.75:0.25).

REFERENCE

1. **Strojny, N., D'Arconte, L., and deSilva, J. A. F.,** Determination of the anti-tumor agent, 10-chloro-5-(2-dimethylaminoethyl)-7H-indolo[2,3,-C] quinolin-6(5H)-one in blood or plasma by high-performance liquid chromatography, *J. Chromatogr.*, 223, 111, 1981.

(-)-*threo*-CHLOROCITRIC ACID

Gas Chromatography

Specimen (mℓ)	Extraction	Column (m × mm)	Packing (mesh)	Oven temp (°C)	Gas (mℓ/min)	Det.	Rt (min)	Internal standard (RT)	Deriv.	Other compounds (RT)	Ref.
Plasma (1)	I-1	1.5 × 2	OV-17 μ-Partisorb	170	Methane[a]	MS-CI	1.6	[^{13}C]-Chlorocitric acid	Methyl	—	1

[a] 1.5 kg/cm^2.

Extraction — I-1. To the sample were added, 50 μℓ of the internal standard solution in methanol (0.1 mg/mℓ) and 0.5 mℓ of 2 *N* HCl. The mixture was extracted with 5 mℓ of ethyl acetate containing 10% methanol. The organic layer was back extracted into 0.5 mℓ of 0.2 *M* acetate buffer pH 5. The aqueous layer was washed with 2 mℓ of the above extraction solvent, made acidic with 0.2 mℓ of 6 *N* HCl, and re-extracted with 3 mℓ of ethyl acetate (without methanol). The organic layer was evaporated, the residue treated with 0.6 mℓ of ethereal diazomethane for 10 min at room temperature. The excess reagent was removed without applying heat. The residue was dissolved in 100 μℓ of ethyl acetate and aliquots of 2 to 5 μℓ of this solution were injected.

REFERENCE

1. **Rubio, F., DeGrazia, F., Miwa, B. J., and Garland, W. A.,** Determination of (-)-*threo*-chlorocitric acid in human plasma by gas chromatography-positive chemical ionization mass spectrometry, *J. Chromatogr.*, 233, 149, 1982.

8-CHLORO-6-(2-CHLOROPHENYL)-4H-IMIDAZO-[1,5-a]BENZODIAZEPINE-3-CARBOXAMIDE

Liquid Chromatography

Specimen (mℓ)	Extraction	Column (cm × mm)	Packing (μm)	Elution	Flow (mℓ/min)	Det. (nm)	RT (min)	Internal standard (RT)	Other compounds (RT)	Ref.
Blood, plasma (1)	I-1	25 × 4.6	Partisil silica (10)	E-1	2.1	ABS (254)	5.1	a (6.4)	4-Hydroxy-metabolite (3.4)	1

[a] 8-Chloro-6-(2-chlorophenyl)-4H-imidazo-[1,5-a] [1,4]-benzo-diazepine-3-carboxamide-5-oxide.

Extraction — I-1. To the sample were added, 100 μℓ of the internal standard solution (14 μg/mℓ in the mobile phase) 0.2 mℓ of water and 2.5 mℓ of 1 *M* phosphate buffer, pH 9. The mixture was extracted with 8 mℓ of diethyl ether-dichloromethane (70:30). An aliquot of 7 mℓ of the organic layer was evaporated at 45°C under a stream of nitrogen. The residue was dissolved in 100 μℓ of the mobile phase and a 10-μℓ aliquot of the solution was injected.

Elution — E-1. Dichloromethane-methanol-ammonium hydroxide (96:3.85:0.15).

REFERENCE

1. **Puglisi, C. V. and deSilva J. A. F.,** Determination of the anxiolytic agent 8-chloro-6-(2-chlorophenyl)-4H-imidazo-[1,5-a] [1,4]-benzodiazepine-3-carboxamide in whole blood, plasma or urine by high-performance liquid chromatography, *J. Chromatogr.*, 226, 135, 1981.

1-(2-CHLOROETHYL)-1-NITROSOUREAS

Gas Chromatography

Specimen (mℓ)	Extraction	Column (m × mm)	Packing (mesh)	Oven temp (°C)	Gas (mℓ/min)	Det.	RT[a] (min)	Internal standard (RT)	Deriv.	Other compounds (RT)	Ref.
Plasma, urine (2)	I-1	1.2 × 2	Carbowax 20 M Ultrabond (80/100)	155	N_2 (NA)	MS-CI[b]	2 = 1 3 = 1.2	c (NA)	Trifluoroacetyl	—	1, 2
Urine (1)	I-2	2 × 2	20% Apiezon L Gas Chrom Q (80/100)	185	N_2 (42)	FID	3 = 2.1[d]	N,N-Dimethylaniline (2.6)	Trifluoroacetyl	—	3, 4

Liquid Chromotography

Specimen (mℓ)	Extraction	Column (cm × mm)	Packing (μm)	Elution	Flow (mℓ/min)	Det. (nm)	RT[a] (min)	Internal standard (RT)	Other compounds (RT)	Ref.
Blood, plasma (2.5)	I-3	15 × 4.6	Ultrasphere-ODS (5)[e]	E-1	1.0	ABS (237)	1 = 6	Phenytoin (9)	—	5
Plasma (0.5)	I-4	25 × 4.6	Ultrasphere-ODS (5)[f]	E-2	1.2	ABS (230)	1 = 13	Propyl paraben (17)	—	6

Note: 1 — 1,3*bis*(2-chloroethyl)-1-Nitrosourea (BCNU); 2 — 1-(2-Chloroethyl)-3-Cyclohexyl)-1-Nitrosourea (CCNU) 3 — 1-(2-Chloroethyl)-3-*trans*-4-methylcyclohexyl)-1-Nitrosourea (MeCCNO).

[a] Different nitrosoureas are referred to by numbers, see *Note*.
[b] Methane as the reagent gas.
[c] 1,3-*bis* (2-Chloropropyl)-1-nitrosourea.
[d] Retention time of trifluoroacetylurea formed by MeCCNU.
[e] Protected by an Altex guard column.
[f] Protected by a 70 × 2 mm guard column packed with Co:Pell ODS, 30 to 38 μm.

Extraction — I-1. The sample was spiked with 0.6 to 1 μg of the internal standard and extracted twice with 1-mℓ portions of 1:1 ether-hexane. The combined extracts were dried over anhydrous magnesium sulfate and evaporated under a stream of nitrogen. The residue was dissolved in 40 μℓ of acetonitrile and 30 μℓ of trifluoroacetic anhydride. The mixture was incubated at 85°C for 3 hr for CCNU and MeCCNU and for 6 hr for BCNU. The samples were cooled, evaporated to dryness under a stream of nitrogen, and redissoved in ethyl acetate.

I-2. The sample was treated with 1 mℓ of peroxyacetic acid (5 mℓ of 100-volume H_2O_2 + 95 mℓ acetic acid, refluxed on a boiling water bath for 1 hr) and allowed to stand at room temp for 20 min. The reaction mixture was extracted with 2 mℓ of dichloromethane. The organic layer was washed with 2 and 1 mℓ of 20% sodium hydroxide, dried over anhydrous sodium sulfate, and evaporated at 40°C under a stream of dry nitrogen. The residue was treated with 0.2 mℓ of trifluoroacetic anhydride. The mixture was allowed to stand at room temp for 20 min and at 45°C for a further 5 min. Excess reagent was removed under a stream of dry nitrogen and the residue was dissolved in 10 μℓ of a 50 μℓ/mℓ solution of N,N-dimethylamiline in dichloromethane.

I-3. The sample was treated with an equal volume of acetonitrile containing 200 mg of the internal standard. After centrifugation and filtration, the clear filtrates were analyzed immediately.

I-4. The sample (adjusted to pH 4 prior to storage) was spiked with 1.68 μg of the internal standard (in 5% ethanol in water) and extracted with 4 mℓ of ethanol-diethyl ether (1.25:98.75). The organic layer was evaporated at 35°C. The residue was dissolved in the mobile phase and the resulting solution was injected.

Elution — E-1. Methanol-water (1:1).

E-2. Acetonitrile-0.1% acetic acid (35:65).

REFERENCES

1. **Smith, R. G. and Cheung, L. K.,** Determination of two nitrosourea antitumor agents by chemical ionization gas chromatography-mass spectrometry, *J. Chromatogr.*, 229, 464, 1982.
2. **Smith, R. G., Blackstock, S. C., Cheung, L. K., and Loo, T. L.,** Analysis for nitrosourea antitumor agents by gas chromatography-mass spctrometry, *Anal. Chem.*, 53, 1205, 1981.
3. **Caddy, B. and Idowu, O. R.,** Gas-chromatographic determination of 1-(2-chloroethyl)-3-(*trans*-4-methylcyclohexyl)-1-nitrosourea (Methyl-CCNU), *Analyst*, 107, 556, 1982.
4. **Caddy, B. and Idowu, O. R.,** Gas-chromatographic determination of 1-(2-chloroethyl)-3-(*trans*-4-methylcyclohexyl)-1-nitrosourea (Methyl-CCNU), *Analyst*, 107, 301, 1982.
5. **Krull, I. S., Strauss, J., Hochberg, F., and Zervas, N. T.,** An improved trace analysis for N-nitrosoureas from biological media, *J. Anal. Toxicol.*, 5, 42, 1981.
6. **Yeager, R. L., Oldfield, E. H., and Chatterji, D. C.,** Quantitation of 1,3-*bis*(2-chloroethyl)-1-nitrosourea in plasma using high-performance liquid chromatography, *J. Chromatogr.*, 305, 496, 1984.

2-CHLOROPROCAINE

Gas Chromatography

Specimen (mℓ)	Extraction	Column (m × mm)	Packing (mesh)	Oven temp (°C)	Gas (mℓ/min)	Det.	Rt (min)	Internal standard (RT)	Deriv.	Other compounds (RT)	Ref.
Plasma (0.3—1)	I-1	1 × 2	3% OV-1-OV-17 Supelcoport (80/100)	T.P.[a]	He (20)	MS-EI	1.5	Procaine (1)	—	—	1

[a] Initial temp = 220°C; rate = 16°C/min; final temp = 240°C; final time = 2 min.

Extraction — I-1. To the sample were added an aqueous solution of the internal standard (25 ng), 0.5 mℓ of 2 *M* sodium carbonate saturated with sodium chloride. The mixture was extracted with 5 mℓ of diethyl ether. The ether extract was evaporated under nitrogen at room temperature. The residue was reconstituted with 30 μℓ of benzene. A 2-μℓ aliquot of the final solution was injected.

REFERENCE

1. **Kuhnert, B. R., Kuhnert, P. M., Reese, A. L. P., and Knapp, D. R.,** Measurement of 2-chloroprocaine in plasma by selected ion monitoring, *J. Chromatogr.*, 224, 488, 1981.

CHLOROPROTHIXENE

Liquid Chromatography

Specimen (mℓ)	Extraction	Column (cm × mm)	Packing (μm)	Elution	Flow (mℓ/min)	Det. (nm)	RT (min)	Internal standard (RT)	Other compounds (RT)	Ref.
Plasma (1)	I-1	25 × 4.6	Supeleo-CN (5)	E-1	2.0	ABS[a] (229)	7	Thioridazine (8)	Chlorprothixene sulfoxide (4) Desmethychlor-prothixene (6.5)	1

Thin-Layer Chromatography

Specimen (mℓ)	Extraction	Plate (Manufacturer)	Layer (mm)	Solvent	Post-separation treatment	Det. (nm)	Rf	Internal standard (RF)	Other compounds (RF)	Ref.
Blood, urine, stomach contents	I-2	20 × 20 cm (Brinkman)	Silica gel, F_{254} (0.25)	S-1	Sp: 10% ethanolic sulfuric acid; potassium iodoplatinate reagent	Visual[b]	0.53	—	Chlorprothixene sulfoxide (0.42)	2

[a] An electrochemical detector was also used.
[b] The TLC spots were scraped, eluted, and determined spectrofluorometrically.

Extraction — I-1. The sample was mixed with 50 μℓ of an aqueous solution of the internal standard (5 μg/mℓ), 2 mℓ of water, and 2 mℓ of 2 *M* sodium hydroxide. The mixture was extracted with 10 mℓ of 1% isoamyl alcohol in heptane. An aliquot of 8.5 mℓ of the organic layer was evaporated under a stream of nitrogen. The residue was dissolved in 250 μℓ of acetonitrile-water (60:40) and a 50-μℓ aliquot of this solution was injected.

Elution — E-1. Acetonitrile-0.02 *M* KH_2PO_4, pH 4.5 (60:40).

Solvent — S-1. Methanol: ammonia (100:2).

REFERENCES

1. **Brooks, M. A. and DiDonato, G.,** Determination of chlorprothixene and its sulfoxide metabolit in plasma by high-performance liquid chromatography with ultraviolet and amperometric detection, *J. Chromatogr.*, 337, 351, 1985.
2. **Poklis, A., Maginn, D., and Mackell, M. A.,** Chlorprothixene and chlorprothixene-sulfoxide in body fluids from a case of drug overdose, *J. Anal. Toxicol.*, 7, 29, 1983.

CHLOROQUINE

Gas Chromatography

Specimen (mℓ)	Extraction	Column (m × mm)	Packing (mesh)	Oven temp (°C)	Gas (mℓ/min)	Det.	RT (min)	Internal standard (RT)	Deriv.	Other compounds (RT)	Ref.
Urine (1)	I-1	1.8 × 2	3% OV-17 Supeleoport (80/100)	230	N_2 (30)	FID	2	9-Bromo-phenan-threne (1.5)	Rearrangement	—	1
Blood (1)	I-2	25 × 0.3	OV-1	T.P[a]	He (3)	NPD	7	Iodoquine, (9) N-Isopropyl-didesethyl-chloro-quine (8.1)	Pentafluro-propionyl	Desethyl-chloroquine (7.8)	2
Plasma (0.5—2)	I-3	12 × 3.2	DB-5 (0.25)[b]	255	N_2 (1)	NPD	2.5	Bromoquine (3.5)	—	Desethyl-chloroquine (2.2) Didesethyl-chloroquine (1.8)	3

Liquid Chromatography

Specimen (mℓ)	Extraction	Column (cm × mm)	Packing (µm)	Elution	Flow (mℓ/min)	Det. (nm)	RT (min)	Internal standard (RT)	Other compounds (RT)	Ref.
Blood, plasma, urine (1)	I-4	30 × 3.9	µ-Bondapack C_{18} (10)[c]	E-1	2.0	ABS (225)	4	Chlorphenir-amine (2)	Desethyl chloro-quine (8)	4
Blood (0.1)[d]	I-5	25 × 4.6	Zorbax Sil (5)	E-2	1.0	Fl (320, 380)	4	N-Isopropyl-didesethyl-chloroquine (6)	Desethyl-chloro-quine (7)	5

Plasma, urine (1)	I-6	30 × 3.9	μ-Bondapack-C_{18} (10)	E-3	1.0	ABS (340)	9	—	Didesthyl-chloroquine (8) Desethyl-chloroquine (7)	6, 7, 8
Blood (0.075)[d]	I-7	15 × 4.6	LiChrosorb Si 60 (5)	E-4	1.0	Fl (335, 370)	6.4	N,N-Dimethyl-didesethyl-chloroquine (4.8)	Desethyl-chloroquine	9, 10
Blood, plasma (1)	I-8	15 × 3.9	Novapack-C_{18} (5)	E-5	0.6	ABS (340)	6	6-Chloro-chloroquine (10)	Amodiaquine (7) Desethyl-chloroquine (4.5) Desethyl-amodiaquine (5)	11

Thin-Layer Chromatography

Specimen (mℓ)	Extraction	Plate (Manufacturer)	Layer (mm)	Solvent	Post-separation treatment	Det. (nm)	Rf	Internal standard (Rf)	Other compounds (Rf)	Ref.
Urine, blood (5)	I-9	20 × 20 cm (Merck)	Silica gel G (0.5)	S-1	Sp: dragendorff reagent	Visual[e]	0.67		Desthyl-chloroquine (0.56) Didesethyl-chloroquine (0.47) Chloroquine side-chain N-oxide (0.34) Chloroquine di-N-oxide (0.14)	12

CHLOROQUINE (continued)

[a] Initial temp = 125°C, initial time = 1 min, rate = 20°C/min; final temp = 230°C.
[b] Film thickness.
[c] Protected by a 23 × 3.9 mm guard column packed with C_{18} corasil.
[d] Blood was adsorbed on filter paper after collection.
[e] Spots corresponding to authentic compounds were scraped, eluted, and determined spectrophotometrically.

I-1. The sample was made alkaline with an equal volume of 1 *N* sodium hydroxide containing 10% sodium chloride and extracted with 10 mℓ of dichloroethane. The organic phase was evaporated at 35°C under a gentle stream of nitrogen. The residue was reconstituted in 200 μℓ of dichloromethane; 10 mg of anhydrous sodium carbonate and 10 μℓ of isobutyl chloroformate were added and the mixture was left at room temperature for 1 hr. The mixture was then vigorously shaken after the addition of 0.5 mℓ of 0.5 *M* alcoholic alkali solution. The mixture was treated with 0.5 mℓ of water. After centrifugation, 1- to 2-μℓ aliquot of the organic layer was injected.
I-2. The sample was treated with 100 μℓ of a solution of the mixture of internal standards (6.6 μg iodoquine + 5.1 μg N-isopropyldidesethyl chloroquine in 0.002 *M* HCl) and 0.5 mℓ of 5 *N* sodium hydroxide. The mixture was extracted with 3 mℓ of hexane. The organic layer was evaporated at 60°C under a stream of nitrogen. The residue was treated with 50 μℓ of pentafluoropropionic anhydride and 10 μℓ of dry pyridine and the mixture incubated at 60°C for 30 min. To the cooled mixture 2 mℓ of hexane and 2 mℓ of 1 N ammonia were added. After mixing and centrifugation, the hexane layer was evaporated at 60°C. The residue was dissolved in 25 μℓ of n-octane and a 1- to 2-μℓ aliquot was injected.
I-3. The sample was spiked with 100 μℓ of an aqueous solution of the internal standard, made alkaline with 1 mℓ of 1 N sodium hydroxide, and extracted with 4 mℓ of hexane-1-pentanol (9:1). After centrifugation, an aliquot of 3 mℓ of the organic layer was back extracted into 3 mℓ of 0.2 *M* HCl. The aqueous phase was made alkaline with 200 μℓ of 5 *M* soidum hydroxide re-extracted with 200 μℓ of chloroform and a 2 μℓ of the chloroform layer was injected.
I-4. The sample was mixed with 1 mℓ of an aqueous solution of the internal standard (0.5 μg/mℓ), 1 mℓ of 1 *N* sodium hydroxide, and the mixture was extracted with 8 mℓ of chloroform. The organic layer was evaporated under a stream of nitrogen at 40°C. The residue was reconstituted with 100 μℓ of methanol. A 10-μℓ aliquot of this solution was injected.
I-5. Dried spots of blood on the filter paper were cut and palced in the extraction vials, 50 μℓ of the internal standard solution (1.02 μg/mℓ in 0.002 *M* HCl) and 3 mℓ of 0.2 *M* HCl were added. The filter paper in each vial was then macerated using a clean glass rod. After vortexing the contents were filtered. The filtrate was made alkaline with 0.5 mℓ of 5 *M* NaOH and extracted with 3 mℓ of methyl-tert-butyl ether-hexane (1:1) and 0.5 mℓ of 5 *M* NaOH. The organic extract was evaporated on a water bath under a stream of nitrogen. The residue was reconstituted in 100 μℓ of the mobile phase and 30-μℓ aliquots of the resulting solutions were injected.
I-6. The sample was made alkaline with an equal volume of 1 *N* sodium hydroxide and extracted with 30 mℓ of *n*-heptane. An aliquot of 25 mℓ of the organic phase was evaporated at 30°C with a stream of nitrogen. The residue was dissolved in 1.0 mℓ of methanol-0.1 *M* phosphoric acid (1:1). The solution was transferred to a small vial and again evaporated. The residue was finally dissolved in 100 μℓ of methanol-phosphoric acid and aliquots of this solution were injected.
I-7. The dried blood spot on the paper was cut in pieces, put into a screw capped tube together with 1 mℓ of a 0.1% aqueous solution of diethylamine and 100 μℓ of the internal standard solution (850 nmol/ℓ). The mixture was exposed to ultrasonic treatment (10 min) and allowed to stand for 0.5 hr, made alkaline with 0.5 mℓ of 1 *M* NaOH and extracted with 6 mℓ of diethyl ether. The organic layer was evaporated under a stream of nitrogen. The residue was dissolved in 200 μℓ of the mobile phase and 50 to 100 μℓ of this solution were injected.

I-8. To the sample were added 30 μℓ of an aqueous solution of the internal standard and 250 μℓ of phosphate buffer, pH 9.5. The mixture was extracted with 7 mℓ of dichloromethane. The organic layer was washed with 1 mℓ of 1 *N* NaOH. Finally the organic phase was back extracted into 250 μℓ of 0.1 *M* HCl and 100 to 200 μℓ of the aqueous phase were injected.
I-9. The sample was made alkaline with 0.5 mℓ of ammonia and extracted twice with 5-mℓ portions of chloroform. The combined extracts were evaporated under reduced pressure. The residue was dissolved in 0.2 mℓ of methanol and aliquots of this solution were spotted on a TLC plate.

Elution — E-1. Methanol-0.18% ammonium hydroxide (70:30).
E-2. Hexane-methyl *tert*-butyl ether-methanol-diethylamine (37.25:37.25:25.0 0.5).
E-3. Acetonitrile-0.02 *M* 1-heptane sulfonic acid, pH 3.4 (66:34).
E-4. Acetonitrile-methanol-diethylamine (80:19,5:0.5).
E-5. Acetonitrile-45 m*M* phosphate buffer, pH 3 (12:88).

Solvent — S-1. Benzene-methanol-diethylamine (7.5:1.5:1).

REFERENCES

1. **Kuye, J. O., Wilson, M. J., and Walle, T.,** Gas chromatographic analysis of chloroquine after a unique reaction with chloroformates, *J. Chromatogr.*, 272, 307, 1983.
2. **Chuchill, F. C., Mount, D. L., and Schwartz, I. K.,** Determination of chloroquine and its major metabolite in blood using perfluoroacylation followed by fused-silica capillary gas chromatography with nitrogen-sensitive detection, *J. Chromatogr.*, 274, 111, 1983.
3. **Berggvist, Y. and Eckerbom, S.,** An improved gas chromatographic method for the simultaneous determination of chloroquine and two metabolites using capillary columns, *J. Chromatogr.*, 306, 147, 1984.
4. **Akintonwa, A., Meyer, M. C., and Hwang, T. R.,** Simultaneous determination of chloroquine and desethylchloroquine in blood, plasma and urine by high-performance liquid chromatography, *J. Liq. Chromatogr.*, 6, 1513, 1983.
5. **Patchen, L. C., Mount, D. L., Schwartz, I. K., and Churchill, F. C.,** Analysis of filter-paper-absorbed, finger-stick blood samples for chloroquine and its major metabolite using high-performance liquid chromatography with fluorescence detection, *J. Chromatogr.*, 278, 81, 1983.
6. **Brown, N.D., Poon, B. T., and Chulay, J. D.,** Determination of chloroquine and its de-ethylated metabolites in human plasma by ion-pair high-performance liquid chromatography, *J. Chromatogr.*, 229, 248, 1982.
7. **Brown, N. D., Stermer-Cox, M. G., Poon, B. T., and Chulay, J. D.,** Separation and identification of a plasma and urinary monoacetylated conjugate of chloroquine in man by ion-pair high-performance liquid chromatography, *J. Chromatogr.*, 309, 426, 1984.
8. **Brown, N. D., Poon, B. T., and Chulay, J. D.,** Chloroquine metabolism in man: urinary excretion of 7-chloro-4-hydroxyquinoline and 7-chloro-4-aminoquinoline metabolites, *J. Chromatogr.*, 345, 209, 1985.
9. **Lindstrom, B., Ericsson, O., Alvan, G., Rombo, L., Ekman, L., Rais, M., and Sjoqvist, F.,** Determination of chloroquine and its desethyl metabolite in whole blood: an application for samples collected in capillary tubes and dried on filter paper, *Ther. Drug Monit.*, 7, 207, 1985.
10. **Alvan, G., Ekman, L., and Lindstrom, B.,** Determination of chloroquine and its desethyl metabolite in plasma, red blood cells and urine by liquid chromatography, *J. Chromatogr.*, 229, 241, 1982.
11. **Pussard, E., Verdier, F., and Blayo, M. C.,** Simultaneous determination of chloroquine, amodiaquine and their metabolites in human plasma, red blood cells, whole blood and urine by column liquid chromatography, *J. Chromatogr.*, 374, 111, 1986.
12. **Essien, E. E. and Afamefuna, G. C.,** Chloroquine and its metabolites in human cord blood, neonatal blood, urine after maternal medication, *Clin. Chem.*, 28, 1148, 1982.

CHLORPHENIRAMINE

Gas Chromatography

Specimens (mℓ)	Extraction	Column (m × mm)	Packing (mesh)	Oven temp (°C)	Gas (mℓ/min)	Det.	RT (min)	Internal standard (RT)	Deriv.	Other compounds (RT)	Ref.
Serum (1—2)	I-1	1.8 × 2	3% OV-22 Supelcoport (80/100)	270	He (30)	MS-EI	1	[$^2H_4$]-Chlorphe-niramine	—	Desmethylchlor-pheniramene	1
Urine (5)	I-2	2 × 4	3% OV-17 Chromosorb Q (100/120)	210	N_2[a]	FID	6	Brompheniramine (11) n	—	Desmethylchlor-pheniramine (7.5) Didesmethylchlor-pheniramine (8.5)	2

Liquid Chromatography

Specimen (mℓ)	Extraction	Column (cm × mm)	Packing (μm)	Elution	Flow (mℓ/min)	Det. (nm)	RT (min)	Internal standard (RT)	Other compounds (RT)	Ref.
Urine (2)	I-3	30 × 3.9	μ-Bondapak-CN (10)	E-1	2.0	ABS (254)	14.6	Imipramine (19.5)	Didesmethyl-chlorpheniramine (10.6) Desmethyl-chlorpheniramine (12.3) Pseudo-ephedrine (7.2)	3

									Norpseudo-ephedrine (6.3)	
Plasma (2)	I-4	25 × 4.6	Spherisorb-CN (5)	E-2	3.0	ABS (229)	4.2	Prochlor-perazine (3.1)	—	4

[a] Head pressure = 1.68 kg/cm^2.

Extraction — I-1. The sample was spiked with 85 ng of the internal standard dissolved in 20 μℓ of methanol and 1 mℓ of saturated sodium borate solution was added. The mixture was extracted with 8 mℓ of hexane. The organic layer was evaporated under a stream of nitrogen to about 1 mℓ which was then back extracted into 0.5 mℓ of 0.1 HCl. The aqueous solution was brought to pH 9 with a saturated sodium borate solution and re-extracted with 1 mℓ of ethyl acetate. The organic extract was evaporated under a stream of nitrogen. The residue was dissolved in 50 μℓ of acetonitrile, the solution concentrated to 5 μℓ and the entire solution was injected.

I-2. To the sample were added 1 mℓ of an aqueous solution of the internal standard (10 μg/mℓ) and 0.5 mℓ of 20% sodium hydroxide. The mixture was extracted with 100 μℓ of chloroform. A 5-μℓ of the extract was injected.

I-3. The sample was made alkaline with 0.5 mℓ of 5% KOH and extracted with 4 mℓ of ether-dichloromethane (70:30). The organic layer was mixed with 100 μℓ of the internal standard solution (8 μg/mℓ in 0.5% phosphoric acid). After centrifugation, 20 μℓ of the aqueous phase was injected.

I-4. To the sample were added 1 mℓ of an aqueous solution of the internal standard (100 μg/mℓ), 0.5 mℓ of a saturated sodium carbonate soution, and the mixture was extracted twice with 50-mℓ portions of pentane containing 3% isopropyl alcohol. The combined organic extracts were evaporated at 65°C. The dried residue was dissolved in 200 μℓ of acetonitrile and an aliquot of 100 μℓ of this solution was injected.

Elution — E-1. Acetonitrile-methanol-0.005 *M* phosphate buffer, pH 6.6 (25:25:50).
E-2. Acetonitrile-0.05 *M* ammonium acetate (70:30).

REFERENCES

1. **Thompson, J. A. and Leffert, F. H.,** Sensitive GLC-mass spectrometric determination of chlorpheniramine in serum, *J. Pharm. Sci.*, 69, 707, 1980.
2. **Ali, H. M. and Beckett, A. H.,** Rapid method for the determination of chlorpheniramine in urine, *J. Chromatogr.*, 223, 208, 1981.
3. **Lai, C. M., Stoll, R. G., Look, Z. M., and Yacobi, A.,** *J. Pharm. Sci.*, 68, 1243, 1979.
4. **Midha, K. K., Rauw, G., McKay, G., Cooper, J. K., and McVittie, J.,** Subnanogram quantitation of chlorpheniramine in plasma by a new radioimmunoassay and comparison with a liquid chromatographic method, *J. Pharm. Sci.*, 73, 1144, 1984.

CHLORPROMAZINE

Gas Chromatography

Specimen (mℓ)	Extraction	Column (m × mm)	Packing (mesh)	Oven temp (°C)	Gas (mℓ/min)	DET.	RT (min)	Internal standard (RT)	Deriv.	Other compounds (RT)	Ref.
Plasma (2)	I-1	1.8 × 4	3% OV-17 Gas Chrom Q (100/120)	240	N_2 (30)	NPD	5	2,4-Dichloro-promazine (8)	—	Chlorpro-mazine sulf-oxide (16)	1
Plasma (2)	I-2	1.2 × 2	3% OV-1 Gas Chrom Q (100/120)	280	He (30)	MS-EI	1.2	Prochlor-perazine (3)	—	—	2
Plasma (2)	I-3	2 × 2	1% OV-17 (NA)	T.P.[a]	He (NA)	MS-EI	NA	[$^2H_2$] Chlorpro-mazine[b]	Trifluoro-acetyl	Chlorpro-mazine sulf-oxide Chlorpro-mazine N-oxide Desmethyl-chlorpro-mazine Didesme-thylchlor-pro-mazine 7-Hydroxy-chlorpro-mazine	3

Liquid Chromatography

Specimen (mℓ)	Extraction	Column (cm × mm)	Packing (μm)	Elution	Flow (mℓ/min)	Det. (nm)	RT (min)	Internal standard (RT)	Other compounds (RT)	Ref.
Dosage	—	30 × 3.9	μ-Bondapak-NH_2 (10)	E-1	1.0	Fl (280, 450)	7	—	Chlorpromazine sulfone (7.5) Chlorpromazine sulfoxide (11.5)	4
Dosage	—	25 × 3.2	LiChrosorb RP-2 (10)	E-2	2.0	ABS (254)	17	—	Chlorpromazine sulfone (8.5) Chlorpromazine sulfoxide (10) Desmethyl chlorpromazine sulfoxide (13) Desmethyl chlorpromazine (22)	5
Plasma (1—1.5)	I-4	25 × 5	Spherisorb silica (5)	E-3	1.5	ABS (250)	3.8	—	Chlorpromazine sulfoxide (5.8) 7-Hydroxy-chlorpromazine (3.2) Desmethyl-chlorpromazine (2.9)	6
Plasma (1)	I-5	15 × 4	Nucleosil-C_{18} (5)	E-4	0.7	Electrochem[c]	8	Thioridazine (9.5)	Levomepromazine (7)	7

CHLORPROMAZINE (continued)

Liquid Chromatography

Specimen (mℓ)	Extraction	Column (cm × mm)	Packing (μm)	Elution	Flow (mℓ/min)	Det. (nm)	RT (min)	Internal standard (RT)	Other compounds (RT)	Ref.
Plasma (2)	I-6	25 × 4.6	Spherisorb-CN (5)	E-5	4.0	Electro-chem[d]	2.4	Prochlorperazine (4.4)	—	8, 9, 10
Serum (0.2)	I-7	12.5 × 4.9	Spherisorb-CN (5)	E-6	2.0	Electro-chem	NA	Butaclamol	—	11

Thin-Layer Chromatography

Specimen (mℓ)	Extraction	Plate (Manufacturer)	Layer (mm)	Solvent	Post-separation treatment	Det. (nm)	Rf	Internal standard (Rf)	Other compounds (Rf)	Ref.
Microsomal incubate (2)	I-8	20 × 20 cm (Aluminum backed) (Merck)	Silica gel F_{254} (0.2)	S-1	—	Visual (254)[e]	0.59	Promazine (NA)	Desmethyl-chlorpro-mazine (0.31) Dides-methyl-chlorpro-mazine (0.76) 7-Hydroxy-chlorpro-mazine (0.55) Chlorpro-mazine sulfoxide (0.43)	12

Plasma (1)	I-9	10 × 10 cm (Merck)	Silica gel[f]	S-2[g]	E: Nitrous acid for 30 min	Reflectance (365)	0.50	Methotrimep- (0.60)	h	13

[a] Different temperature protocols were used for different compounds.
[b] A mixture of deuterium labeled chlorpromazine and a number of metabolites were used as the internal standard.
[c] Potential = 0.95 V.
[d] Potential = 0.9 V.
[e] The separated spots were cut, eluted, and evaporated. The residue was oxidized with H_2O_2 and the oxidation products determined fluorometrically.
[f] High performance TLC plate.
[g] Two stage development in two solvents. First in solvent A and then in solvent B.
[h] Rf values of a number of a number of drugs and metabolites are given.

Extraction — I-1. The sample after the addition of 1 mℓ of the internal standard solution (200 ng/mℓ in 4 *M* HCl) was washed with 7 mℓ of pentane. The aqueous phase was made alkaline and extracted with 10 mℓ of pentane. The organic layer was evaporated, the residue dissolved in 20 μℓ of methanol and aliquots of 1 to 2 μℓ of the solution were injected.
I-2. The sample was mixed with 1 mℓ of aqueous internal standard solution (100 ng/mℓ), 0.5 mℓ of saturated sodium carbonate solution added and the mixture extracted twice with 5-mℓ portions of pentane containing 3% isopropanol. The combined organic extracts were evaporated at 65°C. The residue was reconstituted with 30 μℓ of methanol and aliquots of 4 μℓ of this solution were injected.
I-3. The sample was spiked with the mixture of the internal standards. The mixture was incubated at 37°C for 30 min. The pH was adjusted to 12 with aqueous 1 *N* NaOH and the mixture was extracted at 0°C with 8 mℓ of 1.5% isoamyl alcohol in hexane. The organic layer was back extracted into 3 mℓ of 2 *N* acetic acid. The acid extract was made alkaline (pH 12) with 15% sodium hydroxide and extracted with 4 mℓ of 1.5% isoamyl alcohol in hexane. The organic layer was evaporated to about 100 μℓ. This solution was analyzed for chlorpromazine, chlorpromazine sulfoxide, and was then derivatized for the determination of nor_1 and nor_2 chlorpromazines. The aqueous phase was extracted with dichloromethane to isolate chlorpromazine-N-oxide. The aqueous layer left after this extraction was adjusted to pH 9.5 and extracted with ethyl acetate to isolate 7-hydroxy metabolite.
I-4. The sample was mixed with 1 mℓ of 1 *M* sodium hydroxide and extracted with 10 mℓ of heptane containing 1.5% amyl alcohol. An aliquot of 8 mℓ of the organic layer was evaporated at 80°C with a stream of nitrogen. The residue was dissolved in 70% methanol containing 0.1% trifluoroacetic acid and aliquots of 100 μℓ of this solution were injected. For the extraction of 7-hydroxychlorpromazine, another aliquot of the sample was extracted with ether after the addition of 1 mℓ of 1 *M* ammonium hydroxide. The ether layer was back extracted into 2 mℓ of 0.1 *M* HCl. The aqueous layer was treated with 0.5 mℓ of 1 *M* ammonium hydroxide and 0.5 mℓ of 1 *M* K_2HPO_4 and reextracted with ether. The ether layer was evaporated.
I-5. The sample was diluted with 4 mℓ of water; 1 mℓ of an aqueous solution of the internal standard (160 ng/mℓ) and 0.8 mℓ of 1 *N* sodium hydroxide were added. The mixture was extracted three times with 15-mℓ portions of *n*-heptane containing 1.5% isoamyl alcohol. The combined organic extracts were evaporated. The residue was dissolved in 10 mℓ of 0.05 *N* HCl and the solution was washed with 20 mℓ of diethyl ether. The aqueous layer was made alkaline with 1 mℓ of 5 *N* NaOH and re-extracted with 10 mℓ of heptane-isoamyl alcohol mixture. The organic layer was evaporated. The residue dissolved in 1 mℓ of acetonitrile and 50 μℓ were injected.
I-6. To the sample were added 1 mℓ of aqueous internal standard solution (100 ng/mℓ), 0.5 mℓ of saturated sodium carbonate, and the mixture was extracted twice with 5-mℓ portions of 3% isopropanol in pentane. The combined organic extracts were evaporated. The residue was reconstituted with 200 μℓ of acetonitrile and aliquots of 100 μℓ were injected.

I-7. The sample was mixed with 10 mg of sodium carbonate and extracted with 1 mℓ of hexane-isobutanol (96:4) containing the internal standard (500 ng/mℓ). The organic extract was evaporated at 50°C under nitrogen flow. The residue was dissolved in 50 μℓ of methanol and 10 μℓ were injected.
I-8. To the sample were added 100 μℓ of 5 *N* sulfuric acid, 50 μℓ of an aqueous solution (4 m*M*) of the internal standard, and the mixture extracted with 5 mℓ of 15% *n*-propanol in dichloromethane. The aqueous phase was brought to pH 12 by the addition of 500 μℓ of 2.5 *N* NaOH and was evaporated to dryness under vacuo at 55°C. the residue was dissolved in 25 μℓ of chloroform-methanol (2:1) and the entire solution was applied to the TLC plate.
I-9. To the sample were added 100 μℓ of the internal standard solution (1 μg/mℓ) in 5% isoamyl alcohol in heptane), 250 μℓ of a saturated sodium carbonate solution, and the mixture was extracted with 4.5 mℓ of pentane. The organic extract was evaporated at 40°C under a nitrogen stream. The residue was dissolved in 50 μℓ of heptane:ethanol:dodecane (75:25:0.05) and 20-μℓ aliquots were spotted.

Elution — E-1. Acetonitrile-benzene-water containing 0.01% sodium metabisulfite + 0.01% d-araboascorbic acid (16:4:1).
E-2. Acetonitrile-0.04% ammonium carbonate (1:1).
E-3. Methanol-water (70:30) containing 0.1% trifluoroacetic acid and 10 m*M* sodium heptane sulfonate.
E-4. Pyridine-tetrahydrofuran-acetonitrile-0.1 *M* acetate buffer, pH 3.5 (0.,1:1.0:68.9:30) containing 20 m*M*/ℓ of $NaClO_4$.
E-5. Acetonitrile-0.1 *M* ammonium acetate (9:1).
E-6. Acetonitrile-0.02 *M* phosphate buffer, pH 6 (65:35).

Solvent — S-1. Methanol-acetone-ammonia (50:50:1).
S-2. (A) Toluene-acetone (50:50); (B) toluene-acetone-ammonia (50:50:2.4).

REFERENCES

1. **Gupta, R. N., Bartolucci, G., and Molnar, G.,** Analysis of chlorpromazine in plasma: effect of specimen storage, *Clin. Chem. Acta*, 109, 351, 1981.
2. **McKay, G., Hall, K., Cooper, J. K., Hawes, E. M., and Midha, K. K.,** Gas chromatographic-mass spectrometric procedure for the quantitation of chlorpromazine in plasma and its comparison with a new high-performance liquid chromatographic assay with electrochemical detection, *J. Chromatogr.*, 232, 275, 1982.
3. **Gruenke, L. D., Craig, J. C., Klein, F. D., Nguyen, T. L., Hitzemann, B. A., Holaday, J. W., Loh, H. H., Braff, L., Fischer, A., Glick, I. D., Hartmann, F., and Bissell, D. M.,** Determination of chlorpromazine and its major metabolites by gas chromatography/mass spectrometry: application to biological fluids, *Biomed. Mass. Spectrom.*, 12, 707, 1985.
4. **Takahashi, D. M.,** Rapid determination of chlorpromazine hydrochloride and two oxidation products in various pharmaceutical samples using high-performance liquid chromatography and fluorescence detection, *J. Pharm. Sci.*, 69, 184, 1980.
5. **Smith, D. J.,** The separation and determination of chlorpromazine and some of its related compounds by reversed-phase high performance liquid chromatography, *J. Chromatogr. Sci.*, 19, 65, 1981.
6. **Stevenson, D. and Reid, E.,** Determination of chlorpromazine and its sulfoxide and 7-hydroxy metabolites by ion-pair high pressure liquid chromatography, *Anal. Lett.*, 14, 1785, 1981.

7. **Murakami, K., Murkami, K., Ueno, T., Hijikata, J., Shirasawa, K., and Muto, T.,** Simultaneous determination of chlorpromazine and levomepromazine in human plasma and urine by high-performance liquid chromatography using electrochemical detection, *J. Chromatogr.*, 227, 103, 1982.
8. **Cooper, J. K., McKay, G., and Midha, K. K.,** Subnanogram quantitation of chlorpromazine in plasma by high -performance liquid chromatography with electrochemical detection, *J. Pharm. Sci.*, 72, 1259, 1983.
9. **Hubbard, J. W., Cooper, J. K., Hawes, E. M., Jenden, D. J., May, P. R. A., Martin, M., McKay, G., Van Putten, T., and Midha, K. K.,** Therapeutic monitoring of chlorpromazine I: pitfalls in plasma analysis, *Ther. Drug Monit.*, 7, 222, 1985.
10. **McKay, G., Cooper, J. K., Hawes, E. M., Hubbard, J. W., Martin, M., and Midha K. K.,** Therapeutic monitoring of chlorpromazine II: pitfalls in whole blood analysis, *Ther. Drug Monit.*, 7, 472, 1985.
11. **Krska, J., Addison, G. M., and Soni, S. D.,** Determination of chlorpromazine in serum by radioreceptor assay and HPLC, *Ann. Clin. Biochem.*, 23, 340, 1986.
12. **Blanchard, D. S., Durke, M. D., and Orton, T. C.,** A TLC fluorescence derivatization assay for chlorpromazine and its non-conjugated hepatic microsomal metabolites, *J. Pharm. Biomed. Anal.*, 1, 195, 1983.
13. **Davis, C. M. and Harrington,** Quantitative determination of chlorpromazine and thioridazine by high-performance thin layer chromatography, *J. Chromatogr. Sci.*, 22, 71, 1984.

CHLORPROTHIXENE

Liquid Chromatography

Specimen (mℓ)	Extraction	Column (cm × mm)	Packing (μm)	Elution	Flow (mℓ/min)	Det. (nm)	RT (min)	Internal standard (RT)	Other compounds (RT)	Ref.
Plasma (1)	I-1	25 × 4.6	Supelco LC-PCN (5)[a]	E-1	2.0	ABS (229)	7.6	Thioridazine (8.5)	Chlorprothixene sulfoxide (4.8) Desmethylchlor-prothixene (6.9)	1

[a] Column temp = 40°C

Extraction — I-1. To the sample were added 50 μℓ of the internal standard solution (50 μg/mℓ in methanol), 2 mℓ of water and 2 mℓ of 2 *M* sodium hydroxide. The mixture was extracted with 10 mℓ of 1% isoamyl alcohol in heptane. An aliquot of 8.5 mℓ of the organic layer was evaporated under a stream of nitrogen at 60°C. The residue was dissolved in 250 μℓ of acetonitrile-water (60:40). A 50-μℓ aliquot of this solution was injected.

Elution — E-1. Acetonitrile-0.02 *M* KH_2PO_4, pH 4.5 (60:40)

CHLORPROTHIXENE (continued)

REFERENCE

1. **Brooks, M. A. and DiDonato, G.,** Determinaton of chlorprothixene and its sulfoxide metabolite in plasma by high-performance liquid chromatography with ultraviolet and amperometric detection, *J. Chromatogr.*, 337, 351, 1985.

CHLORTETRACYCLINE

Liquid Chromatography

Specimen (mℓ)	Extraction	Column (cm × mm)	Packing (μm)	Elution	Flow (mℓ/min)	Det. (nm)	RT (min)	Internal standard (RT)	Other compounds (RT)	Ref.
Dosage	—	30 × 3.9	μ-Bondapak-phenyl (10)	E-1; grad	2.6	ABS (254)	14.2	—	Tetracycline (7) 4-Epitetracycline (5) 4-Epianhydrotetracycline (17.8) Anhydrotetracycline (20)	1

Elution E-1. (A) Acetonitrile-0.2 *M* phosphate buffer, pH 2.2 (240:1760). (B) Acetonitrile-0.2 *M* phosphate buffer, pH 2.2 (440:1560). Gradient: Isocratic (A) 10 min; Isocratic (B) for 15 min.

REFERENCE

1. **Muhammad, N. and Bodnar, J. A.,** Separation and quantitation of chlortetracycline, 4-epitetracycline, 4-epianhydrotetracycline, and anhydrotetracycline in tetracycline by high-performance liquid chromatography, *J. Pharm. Sci.*, 69, 928, 1980.

CHLORTHALIDONE

Liquid Chromatography

Specimen (mℓ)	Extraction	Column (cm × mm)	Packing (μm)	Elution	Flow (mℓ/min)	Det. (nm)	RT (min)	Internal standard (RT)	Other compounds (RT)	Ref.
Plasma, blood, urine (0.2)	I-1	15 × 4.6	LiChrosorb RP-18 (5)	E-1	1.6	ABS (226)	5.5	Probenecid (6.5)	—	1
Dosage		25 × 4.6	Zorbax-ODS (5)	E-2	1.0	ABS (254)	4	Caffeine (5.5)	Clonidine (7.5)[a]	2
Blood (1)	I-2	10 × 8	Rad-Pak-C_{18} (10)	E-3	2.0	ABS (210)	8.6	Probenecid (3.5)	—	3
Blood (0.16)	I-3	30 × 3.9	μ-Bondapak-CN (10)	E-4	2.5	ABS (214)	8.7	Sulfinilamide (13.9)	—	4

[a] Different conditions are described for the assay of clonidine in tablets.

Extraction — I-1. A 0.2-mℓ aliquot of 0.067 *M* KH_2PO_4 solution containing 1 mg/ℓ of the internal standard was added to the sample and the mixture was extracted with 1 mℓ of ether. The organic extract was evaporated, the residue dissolved in 0.15 mℓ of the mobile phase, and 0.1 mℓ of the solution was injected.

I-2. The sample was treated with 0.1 mℓ of 50 μg/mℓ of the internal standard solution in 0.067 *M* phosphate buffer (pH 7.4) and then with 4 mℓ of ice cold 0.33 *N* perchloric acid. The supernatant was collected and the pellet was rinsed with 2 mℓ of water. The combined supernatants were applied to a prewashed (4 mℓ methanol, 4 mℓ water) 1-mℓ BondElut-C_{18} column. After the sample had passed through, the column was washed with 0.2 *N* HCl (2 mℓ), water (2 mℓ), 50% acetonitrile in 0.01 *M* sodium acetate (50 μℓ), and was finally eluted with 200 μℓ of 50% acetonitrile in 0.01 *M* sodium acetate. Aliquots of this eluate were injected with an autosampler.

I-3. The sample was treated with 480 μℓ of an aqueous solution of the internal standard (1.33 μg/mℓ). The mixture was vortexed and then sonicated for 15 min in an ultrasonic bath. Acetonitrile (6 mℓ) was added followed by vortexing for 30 sec and further sonication for 15 min. The supernatant obtained after centrifugation was evaporated under nitrogen at 40°C. The residue was dissolved in 80 μℓ of the mobile phase and 25 μℓ of this solution were injected.

Elution — E-1. Acetonitrile-0.01 *M* sodium acetate (100:400).

E-2. Methanol-water (50:50).

E-3. Acetonitrile-0.01 *M* sodium acetate (20:80).

E-4. Acetonitrile-tetrahydrofuran-water (0.5:2:97.5) containing 10 m*M* dibutylamine phosphate adjusted to pH 5 with 2 *M* sodium hydroxide.

CHLORTHALIDONE (continued)

REFERENCES

1. **Guelen, P. J. M., Baars, A. M., Vree, T. B., Nijkert, A. J., and Vermeer, J. M.,** Rapid and sensitive determination of chlorthalidone in blood, plasma and urine of man using high-performance liquid chromatography, *J. Chromatogr.*, 181, 497, 1980.
2. **Walters, S. M. and Stonys, D. B.,** Determination of chlorthalidone and clonidine hydrochloride in tablets by HPLC, *J. Chromatogr. Sci.*, 21, 43, 1983.
3. **MacGregor, T. R., Farina, P. R., Hagopian, M., Hay, N., Esber, H. J., and Keirns, J. J.,** Analysis of chlorthalidone in biological fluids by high-performance liquid chromatography using a rapid column cleanup procedure, *Ther. Drug Monit.*, 6, 83, 1984.
4. **Rosenberg, M. J., Lam, K. K., and Dorsey, T. E.,** Analysis of chlorthalidone in whole blood by high-performance liquid chromatography, *J. Chromatogr.*, 375, 438, 1986.

CI-923

Liquid Chromatography

Specimen (mℓ)	Extraction	Column (cm × mm)	Packing (μm)	Elution	Flow (mℓ/min)	Det. (nm)	RT (min)	Internal standard (RT)	Other compounds (RT)	Ref.
Plasma (1)	I-1	10 × 4.6	Microsorb-C_{18} (3)[a]	E-1; grad	1.0	Fl (484, 518)	14.1	8-Methoxy CI-923 (15.5)	—	1

[a] Protected by a 20 × 2 mm precolumn packed with Co:Pell-ODS.

Extraction — I-1. The sample was spiked with 0.25 mℓ of an aqueous solution of the internal standard (20 ng/mℓ) and 1 mℓ of 5% ammonia was added. The mixture was extracted with 4 mℓ of diethyl ether. The organic extract was evaporated at 55 to 60°C under a stream of nitrogen. The residue was treated with 0.25 mℓ of the derivatization reagent (2 mg of fluorescein-6-iso-thiocyanate in 10 mℓ of 0.05% pyridine in acetone containing 100 μg of triethylamine; 1.25 mℓ of this solution diluted 25 mℓ with acetone just before use) followed by 0.5- and 1.0-mℓ aliquots of acetone. Each aliquot was individually evaporated at 55 to 60°C under nitrogen. The residue was reconstituted in 1.75 mℓ of 20% aqueous acetone and a 175 μℓ aliquot was injected. The derivative was purified by an on-line isolation column (130 × 4.6 mm) packed with 5 μm silica gel.

Elution — E-1. (A) Acetonitrile; (B) 12.5 m*M* ammonium sulfate + 2.5 m*M* tetrabutylammonium hydrogen sulfate. Gradient: isocratic 30% (A) until 4.5 min; linear gradient to 33% (A) from 4.5 to 6.5 min; isocratic 33% (A) until 13.1 min; isocratic 75% (A) until 17.0 min.

REFERENCE

1. **Reynolds, D. L. and Pachla, L. A.,** Analysis of 3-(2-ethylamino)propyl)-2,2,3,4-tetrahydro-5H(1)benzopyrano(3,4-c)pyridin-5-one in plasma by liquid chromatographic column switching after derivatizing the secondary amine with fluorescein-6-isothiocyanate, *J. Pharm. Sci.*, 74, 1091, 1985.

CIANOPRAMINE

Liquid Chromatography

Specimen (mℓ)	Extraction	Column (cm × mm)	Packing (μm)	Elution	Flow (mℓ/min)	Det. (nm)	RT (min)	Internal standard (RT)	Other compounds (RT)	Ref.
Plasma (1)	I-1	25 × 4.6	Hypersil silica (5)	E-1	—	ABS[a] (235)	7.3	Nomifensine (4.6)	—	1[b]

[a] A fluorescence detector ex = 284 nm, em = 450 nm was also used. In this mode the internal standard could not be used.

[b] The authors report that the gas chromatographic procedure described in the same paper is tedious and not suitable for routine analysis. It is not clear why the authors prefered an electron capture detector to a nitrogen specific detector.

Extraction — I-1. The sample was spiked with 100 μℓ of the internal standard solution (10 μg/mℓ in methanol) and 0.6 mℓ of 1 *M* sodium hydroxide was added. The mixture was extracted twice with 2-mℓ portions of ether. The combined ether extracts were evaporated under nitrogen. The residues were dissolved in 400 μℓ of the mobile phase and 20-μℓ aliquots were injected.

Elution — E-1. Acetonitrile-isopropanol-ammonia (54:46:0.7).

REFERENCE

1. **Hojabri, H. and Glennon, J. D.,** Determination of cianopramine in human plasma by high-performance liquid chromatography and gas-liquid chromatography with ultraviolet, fluorescence and electron-capture detection, *J. Chromatogr.*, 342, 97, 1985.

CIBENZOLINE

Liquid Chromatography

Specimen (mℓ)	Extraction	Column (cm × mm)	Packing (μm)	Elution	Flow (mℓ/min)	Det. (nm)	RT (min)	Internal standard (RT)	Other compounds (RT)	Ref.
Plasma, urine (1)	I-1	25 × 4.6	Zorbax SCX (7—8)	E-1	1.5	ABS (214)	5.4	Dimethyl cibenzoline (4.4)	—	1

Extraction — I-1. To the sample were added 2 mℓ of 1 *M* phosphate buffer, pH 11, and 50 μℓ of the internal standard solution (2 μg/mℓ in acetonitrile). The mixture was extracted with 2.5 mℓ of benzene. The organic layer was evaporated at 65°C under a stream of nitrogen. The residue was reconstituted in 400 μℓ of the mobile phase and aliqouts of 50 μℓ of this solution were injected with an auto sampler.

Elution — E-1. Acetonitrile-0.015 *M* phosphate buffer, pH 6 (80:20).

REFERENCE

1. **Hackman, M. R., Lee, T. L., and Brooks, M. A.,** Determination of cibenzoline in plasma and urine by high-performance liquid chromatography, *J. Chromatogr.*, 273, 347, 1983.

CICLETANIDE

Liquid Chromatography

Specimen (mℓ)	Extraction	Column (cm × mm)	Packing (μm)	Elution	Flow (mℓ/min)	Det. (nm)	RT (min)	Internal standard (RT)	Other compounds (RT)	Ref.
Plasma, urine, saliva, red cells (1)	I-1	25 × 4.6	Nucleosil C_{18} (10)	E-1	1.0	ABS (280)	7.8	5-Methylcicletanide (10.8)	—	1

Elution — E-1. Methanol: 0.1 *M* acetic acid (50:50).

REFERENCE

1. **Cuisinaud, G., Terrier, M., Ferry, N., Proust, S., and Sassard, J.,** High-performance liquid chromatographic determination of cicletanide, a new diuretic, in plasma, red blood cells, urine and saliva, *J. Chromatogr.*, 341, 97, 1985.

CICLOPIROX

Liquid Chromatography

Specimen (mℓ)	Extraction	Column (cm × mm)	Packing (μm)	Elution	Flow (mℓ/min)	Det. (nm)	RT (min)	Internal standard (RT)	Other compounds (RT)	Ref.
Plasma (1)	I-1	12.5 × 4.6	Nucleosil-C_{18} (5)	E-1	2.0	ABS (300)	5	6-Cyclohexyl-methyl-1-hydroxy-4-methyl-2-(1H) pyridone (8.5)	—	1

Extraction — I-1. The sample was incubated with 1 mℓ of 1/15 *M* phosphate buffer (pH 5) and 10 μℓ of β-glucuronidase solution at 37°C for 24 hr; 40 μℓ of the internal standard solution (10 μg/mℓ in water) were then added. To the mixture, 0.5 mℓ of 2 *M* soidum hydroxide solution and 200 μℓ of dimethyl sulfate were added and the mixture incubated at 37°C for 15 min. Subsequently, 200 μℓ of triethylamine were added, the mixture vortex-mixed and extracted with 5 mℓ of n-hexane. The organic phase was applied to a prewashed (2 mℓ of acetonitrile) 1-mℓ BondElut CN column. After the sample had passed through, the column was washed with 1 mℓ of toluene, then eluted with 300 μℓ of the mobile phase and 100 μℓ of this eluate were injected.

Elution — E-1. Acetonitrile-water (40:60).

REFERENCE

1. **Lehr, K. H. and Damm, P.,** Quantification of ciclopirox by high-performance liquid chromatography after pre-column derivatization an example of efficient clean-up using silica-bonded cyano phases, *J. Chromatogr.*, 339, 451, 1985.

CIGLITAZONE

Liquid Chromatography

Specimen (mℓ)	Extraction	Column (cm × mm)	Packing (μm)	Elution	Flow (mℓ/min)	Det. (nm)	RT (min)	Internal standard (RT)	Other compounds (RT)	Ref.
Plasma (0.2)	I-1	15 × 4.6	Zorbax C_8 (5)[a]	E-1	2.0	ABS (229)	6	Testosterone propionate (7.5)	—	1

[a] Protected by a Brownlee 3-cm guard column packed with LiChrosorb C_{18} (10 μm).

Extraction — I-1. The sample was treated with 30 μℓ of acetonitrile containing 4 μg/mℓ of testosterone propionate. After mixing and centrifugation aliquots of 75 μℓ of the supernate were injected.

Elution — E-1. Acetonitrile-0.1% acetic acid (65:35).

REFERENCE

1. **Harrington, E. L. and Cox, S. R.,** Determination of ciglitazone in dog plasma by reversed-phase high-performance liquid chromatography, *J. Pharm. Biomed. Anal.*, 3, 483, 1985.

CILASTATIN

Liquid Chromatography

Specimen (mℓ)	Extraction	Column (cm × mm)	Packing (μm)	Elution	Flow (mℓ/min)	Det. (nm)	RT (nm)	Internal standard (RT)	Other compounds (RT)	Ref.
Urine (2)	I-1	25 × 4.6	Ultrasphere-ODS (5)	E-1	1.0	ABS (210)	4.9	N-Propionyl cilastatin (11.4)	N-Acetylcilastatin (7.5) Imipenem (2.8)	1

Extraction — I-1. To the sample were added 0.1 mℓ of an aqueous solution of the internal standard (4 mg/mℓ) and 20 μℓ of triethylamine. The mixture was applied to a prewashed (3 mℓ methanol, 3 mℓ water) 3-mℓ Bond-Elut SAX column. The column was washed with 5 mℓ of methanol followed by 1 mℓ of water. The column was eluted with 2 mℓ of 1 *M* sodium chloride. Aliquots of the eluate were injected with an autosampler.

Elution — E-1. Methanol-acetonitrile-0.85% phosphoric acid adjusted to pH 4 with triethylamine (8:2:15).

REFERENCE

1. **Hsieh, J. Y. K., Maglietto, B. K., and Bayne, W. F.,** Separation, identification, and quantification of N-acetyl cilastatin in human urine, *J. Liq. Chromatogr.*, 8, 513, 1985.

CILOSTAZOL

Liquid Chromatography

Specimen (mℓ)	Extraction	Column (cm × mm)	Packing (μm)	Elution	Flow (mℓ/min)	Det. (nm)	RT (min)	Internal standard (RT)	Other compounds (RT)	Ref.
Plasma (1)	I-1	30 × 3.9	μ-Bondapak-C_{18} (10)	E-1	1.7	ABS (254)	7.5	OPC-13012 (12)	—	1

Extraction — I-1. To the sample were added 10 μℓ of methanol containing 600 ng of the internal standard and 4 mℓ of acetonitrile. After mixing and centrifugation, the supernatant was evaporated under a stream of air. The residue was dissolved in 1 mℓ of 0.2 *M* sodium hydroxide and extracted with 5 mℓ of chloroform. The organic layer was evaporated under a stream of air. The residue was dissolved in 1 mℓ of 0.2 *M* sodium hydroxide and extracted with 5 mℓ of diethyl ether. The organic extract was again evaporated, residue dissolved in 100 μℓ of methanol, and an aliquot of 40 μℓ of the solution was injected.

Elution — E-1. Acetonitrile-water (42:58)

REFERENCE

1. **Akiyama, H., Kudo, S., Odomi, M., and Shimizu, T.,** High-performance liquid chromatographic procedure for the determination of a new antithrombotic and vasodilating agent, cilostazol, in human plasma, *J. Chromatogr.*, 338, 456, 1985.

CIMETIDINE

Liquid Chromatography

Specimen (mℓ)	Extraction	Column (cm × mm)	Packing (μm)	Elution	Flow (mℓ/min)	Det. (nm)	RT (min)	Internal standard (RT)	Other compounds (RT)	Ref.
Plasma, urine (3)	I-1	25 × 6.6	Partisil-ODS (10)[a]	E-1	2.5	ABS (228)	10.5	Metiamide (8)	—	1
Plasma (1)	I-2	10 × 8	Rad-Pak-C_{18} (10)	E-2	3.0	ABS (228)	3.8	Procainamide (2.6)	—	2, 3
Plasma (0.5)	I-3	30 × 3.9	μ-Bondapak-C_{18} (10)[b]	E-3	1.5	ABS (229)	5.8	SKF 92373 (4.3) AH 204480 (9.7)	Ranitidine (7) Cimetidine sulfoxide (3.6) Cimetidine amide (12.8) Desmethylranitidine (6.4)	4
Plasma (0.5)	I-4	10 × 8	Rad-Pak C_{18} (10)	E-4	4.0	ABS (229)	10.8	SKF 92374 (5.1)	Cimetidine sulfoxide (2.4) Hydroxymethylcimetidine (3.3) Guanyl urea cimetidine (6.9)	5, 6
Plasma (0.1)	I-5	25 × 4.9	Spherisorb silica (5)	E-5	1.0	ABS (228)	12.5	SKF 92374 (15)	Antipyrine (4)	7
Plasma (0.5)	I-6	15 × 4.6	Ultrasphere-ODS (5)[c]	E-6[d]	2.0	ABS (229)	5.8	Ornidazole (8.5)	Cimetidine sulfoxide (2)	8
Serum (0.1)	I-7	10 × 8	Rad-Pak CN (10)	E-7	2.0	ABS (220)	7.5	Ranitidine (11.5)	—	9
Plasma (0.25)	I-8	10 × 8	Rad-Pak-C_{18} (10)	E-8	3.0	ABS (228)	2.8	SKF 92374 (2)	—	10
Plasma (0.1)	I-9	25 × 4	LiChrosorb RP-8 (10)[e]	E-9	1.0	ABS (228)	9.5	Metiamide (10.5)	—	11
Plasma (1)	I-10	5 × 4.6	Sepralyte-C_{18} (3)	E-10	1.5	ABS (228)	2.7	Codeine (3.8)	—	12

[a] A 25 × 0.46-cm column packed with Porasil-C_{18} was placed between the pump and the injector to saturate the mobile phase with silica.

[b] Protected by a precolumn packed with Corasil-C_{18}.

[c] Protected by a 50 × 4.6 mm precolumn packed with Permaphase-ODS (30 μm).

[d] The mobile phase is recycled.

[e] Column temp = 40°C.

Extraction — I-1. A Sep-Pak cartridge was washed successively with methanol (4 mℓ), water (4 mℓ), methanol (4 mℓ), and water (4 mℓ). The sample was spiked with 10 μℓ of methanol containing 3 μg of the internal standard and applied to the washed cartridge. The cartridge was then washed with 5 mℓ of 1/15 *M* KH_2PO_4, and followed by 5 mℓ of 0.1 *M* Na_2CO_3. Finally, the cartridge was eluted with 3 mℓ of acetonitrile. The eluate was evaporated at 37°C under nitrogen, the residue reconstituted in 0.1 mℓ of the mobile phase and a 40-μℓ aliquot was injected.

I-2. To the sample were added 100 μℓ of the internal standard solution (20 μg/mℓ in 0.09% sodium metabisulfate solution) and 0.5 mℓ of 2 *M* sodium hydroxide. The mixture was extracted with 20 mℓ of dichloromethane. The organic layer was evaporated at 45°C under nitrogen. The residue was reconstituted in 100 μℓ of the mobile phase and 40 μℓ of this solution was injected.

I-3. To the sample were added the methanolic solution of the mixed internal standards and 50 μℓ of 2 *M* sodium hydroxide. The mixture was extracted with 4 mℓ of a mixture of ether-chloroform and isopropanol (2:1:1). The organic layer was back extracted into 100 μℓ of 2% acetic acid. A 20-μℓ aliquot of the aqueous layer was injected.

I-4. The sample was added to 2 mℓ of acetonitrile containing the internal standard. After mixing and centrifugation, the supernatant was saturated with K_2HPO_4. An aliquot of 1.8 mℓ of the upper organic phase was evaporated under a stream of nitrogen. The residue was reconstituted with 0.5 mℓ of 50 m*M* HCl. This solution was washed with 3 mℓ of water saturated isoamyl alcohol. The aqueous phase was again saturated with K_2HPO_4 and extracted with 3 mℓ of dichloromethane. The organic layer was evaporated. The residue was reconstituted in 100 μℓ of the mobile phase and 25 μℓ were injected.

I-5. To the sample were added 50 μℓ of the internal standard solution (30 mg/ℓ in acetonitrile) and 50 μℓ of 2 *M* sodium hydroxide. The mixture was extracted with 5 mℓ of dichloromethane. The organic layer was evaporated under a stream of air. The residue was reconstituted in 100 μℓ of the mobile phase and 20 μℓ were injected.

I-6. The sample was spiked with 0.5 mℓ of the internal standard solution (4 mg/ℓ in 0.2 *M* sodium phosphate buffer, pH 7) and was applied to a prewashed (1 mℓ each of methanol, water, 0.1 *M*, pH 7 phosphate buffer) 1-mℓ Bond-Elut C_{18} column. After the sample had passed through, the column was washed with 2 mℓ each of phosphate buffer and then water. The column was then eluted with 0.2 mℓ of methanol. An aliquot of 20 μℓ of this solution was injected.

I-7. The sample was spiked with 50 μℓ of the internal standard solution (20 μg/mℓ in 0.2 *M* sodium carbonate) and extracted with 5 mℓ of ethyl acetate. The organic layer was back extracted into 200 μℓ of dilute phosphoric acid (pH 3). A 100-μℓ aliquot of the aqueous phase was injected.

I-8. The sample was mixed with 150 μℓ of methanol containing 1200 ng of the internal standard, 25 μℓ of 5 *M* sodium hydroxide and 2 mℓ of acetonitrile. The supernatant was treated with 1 mℓ of 20 m *M* HCl saturated with NaCl (300 g/ℓ). After mixing and centrifugation, the aqueous layer was made alkaline with 100 μℓ of 5 *M* sodium hydroxide and extracted with 2 mℓ of acetonitrile. The organic layer was evaporated at 40°C under nitrogen. The residue was dissolved in 250 μℓ of methanol containing 1 mmol of HCl per liter, 15 μℓ of this solution was injected.

I-9. The sample was mixed with 100 μℓ of the internal standard solution (50 μg/mℓ) and 100 μℓ of 6 *M* sodium hydroxide. The mixture was extracted with 4.5 mℓ of ethyl acetate. The organic phase was evaporated, the residue dissolved in 100 μℓ of distilled water and 10-μℓ aliquots of this solution were injected.

I-10. The sample was mixed with 100 μℓ of the internal standard solution [10 μg/mℓ in methanol (1) + water (20)] and 100 μℓ of 1 *M* potassium hydroxide and applied to a prewashed (1 mℓ methanol, 1 mℓ 0.1 *M* sodium carbonate) 1-mℓ BondElut-C_2 column. The column was washed twice with 1-mℓ portions of 0.1 *M* sodium carbonate. The column was eluted with 250 μℓ of methanol and then 250 μℓ of the buffer of the mobile phase. An aliquot of 15 μℓ of the combined eluates were injected.

CIMETIDINE (continued)

Elution — E-1. Methanol-1/15 *M* dibasic sodium phosphate-1/15 monabasic potassium phosphate (185:100:815).
E-2. Acetonitrile-triethylamine-water (3:1:96) adjusted to pH 3 with phosphoric acid.
E-3. Acetonitrile-water (20:80) containing 0.005 *M* octane sulfonic acid.
E-4. Methanol-water (20:80) containing 5 m*M* n-butylamine adjusted to pH 7.1 with phosphoric acid.
E-5. Acetonitrile-water-ammonia (95:5:0.2).
E-6. Acetonitrile-0.02 *M* acetic acid-diethylamine (120:880:0.15).
E-7. Acetonitrile-water (33:67) containing 5 m*M* triethylamine adjusted to pH 3 with phosphoric acid.
E-8. Phosphate buffer (8.7 mmol KH_2PO_4 + 3.04 mmol Na_2HPO_4, pH 7.4 + 340 g of methanol per liter).
E-9. Acetonitrile-water-1/30 *M* phosphate buffer, pH 4.9 (40:39:1).
E-10. Acetonitrile-water containing 0.1 *M* sodium perchlorate and 0.01 *M* phosphoric acid (8:92).

REFERENCES

1. **Bartlett, J. M. and Segelman, A. B.,** Bioanalysis of cimetidine by high-performance liquid chromatography, *J. Chromatogr.*, 255, 239, 1983.
2. **Ching, M. S., Mihaly, G. W., Jones, D. B., and Smallwood, R. A.,** Liquid chromatographic analysis of cimetidine with procainamide as internal standard, *J. Pharm. Sci.*, 73, 1015, 1984.
3. **Mihaly, G. W., Cockbain, S., Jones, D. B., Hanson, R. G., and Smallwood, R. A.,** High-pressure liquid chromatographic determination of cimetidine in plasma and urine, *J. Pharm. Sci.*, 71, 590, 1982.
4. **Boutagy, J., More, D. G., Munro, I. A., and Shenfield, G. M.,** Simultaneous analysis of cimetidine and ranitidine in human plasma by HPLC, *J. Liq. Chromatogr.*, 7, 1651, 1984.
5. **Lloyd, C. W., Martin, W. J., Nagle, J., and Hauser, A. R.,** Determination of cimetidine and metabolites in plasma by reversed-phase high-performance liquid chromatographic radial compression technique, *J. Chromatogr.*, 339, 139, 1985.
6. **Ziemniak, J. A., Chiarmonte, D. A., and Schentag, J. J.,** Liquid-chromatographic determination of cimetidine, its known metabolites, and creatinine in serum and urine, *Clin. Chem.*, 27, 272, 1981.
7. **Adedoyin, A., Aarons, L., and Houston, J. B.,** High-performance liquid chromatographic method for the simultaneous determination of cimetidine and antipyrine in plasma, *J. Chromatogr.*, 345, 192, 1985.
8. **Lin, Q., Lensmeyer, G. L. and Larson, F. C.,** Quantitation of cimetidine and cimetidine sulfoxide in serum by solid-phase extraction and solvent-recycled liquid chromatography, *J. Anal. Toxicol.*, 9, 161, 1985.
9. **Kubo, H., Kobayashi, Y., and Tokunaga, K.,** Improved method for the determination of cimetidine in human serum by high-performance liquid chromatography, *Anal. Lett.*, 18, 245, 1985.
10. **Abdel-Rahim, M., Ezra, D., Peck, C., and Lazar, J.,** Liquid-chromatographic assay of cimetidine in plasma and gastric fluid, *Clin. Chem.*, 31, 621, 1985.
11. **Kaneniwa, N., Funaki, T., Furuta, S., and Watari, N.,** High-performance liquid chromatographic determination of cimetidine in rat plasma, urine and bile, *J. Chromatogr.*, 374, 430, 1986.
12. **Chiou, R., Stubbs, R. J., and Bayne, W. F.,** Determination of cimetidine in plasma and urine by high-performance liquid chromatography, *J. Chromatogr.*, 377, 441, 1986.

CIMOXATONE

Liquid Chromatography

Specimen (mℓ)	Extraction	Column (cm × mm)	Packing (μm)	Elution	Flow (mℓ/min)	Det. (nm)	RT (min)	Internal standard (RT)	Other compounds (RT)	Ref.
Plasma (1)	I-1	NA	Spherisorb-ODS (5)	E-1	1.0	ABS (240)	7	Ethyldesmethylcimoxatone (9)	Desmethyl-cimoxatone (4.5) Lorazepam (5.5)	1

Extraction — I-1. The sample was spiked with 100 μℓ of the internal standard solution (5 μg/mℓ in water-acetonitrile, 9:1) and was extracted with 5 mℓ of toluene. The organic layer was evaporated at 40°C under nitrogen. The residue was dissolved in 500 μℓ of *n*-heptane and the solution mixed with 500 μℓ of water-acetonitrile (9:1). After mixing and centrifugation aliquots of the aqueous phase were injected.

Elution — E-1. Methanol-0.005 *M* phosphate buffer, pH 4 (55:45).

REFERENCE

1. **Rovei, V., Rigal, M., Sanjuan, M., and Thiola, A.,** High-performance liquid chromatographic determination of cimoxatone and its O-demethyl metabolite in plasma, *J. Chromatogr.*, 277, 391, 1983.

CINNARIZINE

Gas Chromatography

Specimen (mℓ)	Extraction	Column (m × mm)	Packing (mesh)	Oven temp (°C)	Gas (mℓ/min)	Det.	RT (min)	Internal standard (RT)	Deriv.	Other compounds (RT)	Ref.
Plasma (2)	I-1	1 × 2	3% OV-17 Supelcoport (80/100)	275	N_2 (25)	NPD	2.8	R 13 415 (4.3)	—	Flunarizine (2.3)	1

CINNARIZINE (continued)

Liquid Chromatography

Specimen (mℓ)	Extraction	Column (cm × mm)	Packing (μm)	Elution	Flow (mℓ/min)	Det. (nm)	RT (min)	Internal standard (RT)	Other compounds (RT)	Ref.
Plasma (5)	I-2	30 × 4	Spherisorb-ODS (5)	E-1	1.0	ABS (285)	6.2	Chlorbenoxamine (8.7)	—	2
Plasma (1)	I-3	25 × 4	LiChrosorb RP-8 (5)[a,b]	E-2	2.0	ABS (250)	2.1	DPA 14 (3)	—	3
Plasma, blood (4)	I-4	25 × 4.6	Ultrasphere-ODS (5)[c]	E-3	1.5	ABS (254)	6.8	Meclozine (10.7)	Chlorcyclizine (5.8) Clocinizine (8.9) Buclizine (17.6)	4

[a] Protected by a 40 × 3 mm guard column packed with LiChrosorb RP-2 (30 μm).
[b] Column temp = 60°C.
[c] Protected by a 30 × 4.6 mm guard column packed with RP-18 silica (10 μm).

Extraction — I-1. The sample was spiked with 100 μℓ of the internal standard solution (1 μg/mℓ in methanol) and buffered with 2 mℓ of borate buffer (pH 8.5). The mixture was extracted twice with 4-mℓ portions of heptane-isoamyl alcohol (98:5:1.5). The combined organic extract was back extracted into 3 mℓ of 0.05 *M* sulfuric acid. The aqueous phase was made alkaline and re-extracted twice with 2-mℓ aliquots of heptane-isoamyl alcohol mixture. The combined organic layers were evaporated at 55°C under nitrogen.
I-2. To the sample were added, 100 μℓ of the internal standard solution (0.4 mg/mℓ in methanol) and 1 mℓ of 0.25 *M* sodium hydroxide. The mixture was extracted with 10 mℓ of diethyl ether. The ether layer was back extracted into 2 mℓ of 0.05 M sulfuric acid. The aqueous phase was made alkaline by adding 0.5 mℓ of 0.5 *M* sodium hydroxide and re-extracted with 5 mℓ of diethyl ether. The ether layer was dissolved in 120 μℓ of the mobile phase and 100 μℓ of this solution were injected.
I-3. The sample was adjusted to pH 3.5 to 5 with 0.15 mℓ of 0.5 *N* HCl, 5 μℓ of the internal standard solution (60 μg/mℓ in carbon tetrachloride) was added and the mixture was extracted with 0.5 mℓ of carbon tetrachloride. The organic phase was collected as much as possible and evaporated under nitrogen. The residue was dissolved in 100 μℓ of methanol and this solution was injected through a 50-μℓ loop.
I-4. To the sample were added 500 μℓ of the internal standard solution (4 mg/100 mℓ of 0.1 *N* HCl) and 10 mℓ of 0.4 *M* phosphate buffer pH 3. The mixture was extracted with 5 mℓ of chloroform-hexane (2:3). The organic layer was evaporated at 40°C under nitrogen. The residue was dissolved in 200 μℓ of methanol and 100 μℓ of this solution were injected.

Elution — E-1. Methanol-0.05 *M* ammonium dihydrogen phosphate (850:150).
E-2. Methanol-10 m*M* acetate buffer, pH 5.2 (85:15).
E-3. Methanol-0.1 *M* phosphate buffer, pH 7 (9:1).

REFERENCES

1. **Woestenborghs, R., Michielsen, L., Lorreyne, W., and Heykants, J.,** Sensitive gas chromatographic method for the determination of cinnarizine and flunarizine in biological samples, *J. Chromatogr.*, 232, 85, 1982.
2. **Hundt, H. K. L., Brown, L. W., and Clark, E. C.,** Determination of cinnarizine in plasma by high-performance liquid chromatography, *J. Chromatogr.*, 183, 378, 1980.
3. **Nitsche, V. and Mascher, H.,** Rapid high-performance liquid chromatographic assay of cinnarizine in human plasma, *J. Chromatogr.*, 227, 521, 1982.
4. **Puttemans, M., Bogaert, M., Hoogewijs, G., Dryon, L., Massart, D. L., and Vanhaelst, L.,** Determination of cinnarizine in whole blood and plasma by reversed phase HPLC and its application to a pharmacokinetic study, *J. Liq. Chromatogr.*, 7, 2237, 1984.

CINROMIDE

Liquid Chromatography

Specimen (mℓ)	Extraction	Column (cm × mm)	Packing (μm)	Elution	Flow (mℓ/min)	Det. (nm)	RT (min)	Internal standard (RT)	Other compounds (RT)	Ref.
Serum (0.2)	I-1	30 × 3.9	μ-Bondapak-C_{18} (10)	E-1	2.5	ABS (273)	NA	3-Bromo-N-isopropylcinnamide	3-Bromocinnamide 3-Bromocinnamic acid	1
Plasma (1)	I-2	25 × 3.9	Partisil silica (10)[a]	E-2	1.0	ABS (254)	12	—	3-Bromocinnamide (21)	2

Thin-Layer Chromatography

Specimen (mℓ)	Extraction	Plate (Manufacturer)	Layer (mm)	Solvent	Post-separation treatment	Det. (nm)	Rf	Internal standard (Rf)	Other compounds (Rf)	Ref.
Plasma (1)	I-2	20 × 20 cm (Merck)	Silica gel 60 (0.25)	S-1	—	Reflectance (270)	0.4	—	3-Bromocinnamide (0.26) 3-Bromocinnamic acid[b]	2

[a] Protected by a 5-cm guard column packed with Co:Pell silica.
[b] A separate developing solvent is required for the assay of this metabolite.

CINROMIDE (continued)

Extraction — I-1. The sample was mixed with 0.4 mℓ of the internal standard solution (1.9 μg/mℓ in acetonitrile). Hexane 0.4 mℓ was added, the mixture vortexed and 0.5 mℓ of saturated KCl solution was added. An aliquot of 30 μℓ of the middle phase was injected.
I-2. The sample was acidified with 0.5 mℓ of 1 *N* HCl and extracted with 6 mℓ of benzene. An aliquot of the organic layer (0.5 to 2 mℓ) was evaporated under nitrogen. The residue was dissolved in 80 μℓ of chloroform-methanol (85:15) and the entire solution was spotted on a TLC plate. For HPLC, the residue was dissolved in 100 μℓ of chloroform and 50 μℓ was injected.

Elution — E-1. Acetonitrile-1 m*M* phosphate buffer, pH 4.4 (35:65).
E-2. Chloroform-acetic acid (95:5).

Solvent — S-1. Ethyl acetate-chloroform-ammonium hydroxide (84:15:1).

REFERENCES

1. **Maddox, R. R. and Wannamaker, B. B.,** Pharmacokinetics of cinromide (BW 122U) and two metabolites (BW 432U and BW A800U), *Curr. Ther. Res.*, 32, 165, 1982.
2. **DeAngelis, R. L., Robinson, M. M., Brown, A. R., Johnson, T. E., and Welch, R. M.,** Quantitation of the anticonvulsant cinromide (3-bromo-N-ethylcinnamamide) and its major plasma metabolites by thin-layer chromatography, *J. Chromatogr.*, 221, 353, 1980.

CIPROFIBRATE

Liquid Chromatography

Specimen (mℓ)	Extraction	Column (cm × mm)	Packing (μm)	Elution	Flow (mℓ/min)	Det. (nm)	RT (min)	Internal standard (RT)	Other compounds (RT)	Ref.
Plasma (1)	I-1	30 × 3.9	μ-Bondapak-C_2-phenyl (10)[a]	E-1	2.0	ABS (232)	3.8	b (6)	—	1

[a] Protected by a 25 × 3.9 mm guard column packed with Corasilphenyl (37 to 50 μm).
[b] 2-[4-(2,2-Dichloro-3-phenylcyclopropyl)-phenoxy]-2-methyl-propanoic acid.

Extraction — I-1. To the sample were added 50 μℓ of the internal standard solution (0.1 mg/mℓ in ethyl acetate), 2 mℓ of 1 *N* HCl, and 0.2 mℓ of 60% perchloric acid, and the mixture was extracted twice with 10 mℓ of hexane. The combined organic phases were evaporated under nitrogen. The residue was dissolved in 2 mℓ of acetonitrile and the solution washed with 2 mℓ of hexane. The acetonitrile solution was then evaporated under nitrogen. The residue was dissolved in 0.5 mℓ of acetonitrile-tetrahydrofuran (10:1) and 0.5 mℓ of 0.1 *M* K_2HPO_4 buffer, pH 4 was added. After vigorous shaking a 100-μℓ aliquot of the solution was injected.

Elution — E-1. Acetonitrile-tetrahydrofuran-0.1 *M* K_2HPO_4 buffer, pH 4 (96:10:104).

REFERENCE

1. **Park, G. B., Biddlecome, C. E., Koblantz, C., and Edelson, J.,** Determination of ciprofibrate in human plasma by high-performance liquid chromatography, *J. Chromatogr.*, 227, 534, 1982.

CIPROFLOXACIN

Liquid Chromatography

Specimen (mℓ)	Extraction	Column (cm × mm)	Packing (μm)	Elution	Flow (mℓ/min)	Det. (nm)	RT (min)	Internal Standard (RT)	Other compounds (RT)	Ref.
Serum, urine (0.5)	I-1	30 × 3.9	μ-Bondapak-C_{18} (10)	E-1	1.5	Fl (270, 440)	8	A-56619 (12.8)	—	1
Plasma, urine (1)	I-2	25 × 4	Spherisorb ODS II (5)	E-2	2.0	Fl (277, 445)	5	—	—	2
Serum	I-3	30 × 3.9	μ-Bondapak-C_{18} (10)	E-3	2.0	Fl (254, 425)	3.6	—	—	3
Serum, urine (0.5)	I-4	15 × 4.6	Ultrasphere-C_{18} (5)	E-4	2.0	ABS (254)	2.4	—	—	4
Serum, urine (0.3)	I-5	25 × 4.6	Spherisorb-ODS (5)	E-5	2.0	Fl (278, 456)	4	—	—	5
Serum (0.05)	I-6	10 × 8	Rad-Pak-C_{18} (10)	E-6[a]	2.0	Fl[b] (277, 445)	6.4	—	—	6

[a] A different mobile phase was used for a column of a different lot number.
[b] An absorbance detector (277 nm) was also used.

Extraction — I-1. The sample was treated with 50 μℓ of the internal standard solution (30 μg/mℓ) in water). While vortex mixing 25 μℓ of perchloric acid was added dropwise. After centrifugation, a 25-μℓ aliquot was injected.
I-2. The sample is diluted with 2 mℓ of 0.16 *N* HCl and filtered through a 0.22-μm membrane filter. An aliquot (10 μℓ) of the filtrate is injected.

CIPROFLOXACIN (continued)

I-3. The sample was treated with 2 volumes of methanol. After mixing and centrifugation, aliquots of the supernate were injected.
I-4. The sample was extracted with 3.5 mℓ of dichloromethane. An aliquot of 3 mℓ of the organic phase was back extracted into 200 μℓ of dilute phosphoric acid (pH 2). After centrifugation, 20 μℓ of the upper aqueous phase were injected.
I-5. The sample was treated with 6% aqueous trichloroacetic acid. After mixing and centrifugation, aliquots of the supernate were injected.
I-6. The sample was mixed with an equal volume of acetonitrile. After mixing and centrifugation, aliquots of 20 μℓ of the clear supernatant were injected.

Elution — E-1. Acetonitrile-methanol-20 m*M* phosphate buffer, pH 3, containing 5 m*M* tetrabutylammonium hydroxide (5:14:81).
E-2. Acetonitrile-0.25 *M* tetrabutylammonium phosphate, pH 3 (5:95).
E-3. Acetonitrile-30 m*M* tetrabutylammonium phosphate, pH 3 (12:88).
E-4. Acetonitrile-0.005 *M* tetrabutylammonium phosphate, pH 2 (10:90).
E-5. Acetonitrile-0.025 *M* tetrabutylammonium phosphate, pH 3 (11:89).
E-6. Methanol-67 m*M* phosphate buffer, pH 3.5 (35:65).

REFERENCES

1. **Nix, D. E., De Vito, J. M., and Schentag, J. J.,** Liquid-chromatographic determination of ciprofloxacin in serum and urine, *Clin. Chem.*, 31, 684, 1985.
2. **Gau, W., Ploschke, H. J., Schmidt, K., and Weber, B.,** Determination of ciprofloxacin (bay o 9867) in biological fluids by high-performance liquid chromatography, *J. Liq. Chromatogr.*, 8, 485, 1985.
3. **Fasching, C. E. and Peterson, L. R.,** High pressure liquid chromatography of (Bay o 9867) ciprofloxacin in serum samples, *J. Liq. Chromatogr.*, 8, 555, 1985.
4. **Jehl, F., Gallion, C., Debs, J., Brogard, J. M., Monteil, H., and Minck, R.,** High-performance liquid chromatographic method for determination of ciprofloxacin in biological fluids, *J. Chromatogr.*, 339, 347, 1985.
5. **Joos, B., Ledergerber, B., Flepp, M., Bettex, J. D., Luthy, R., and Siegenthaler, W.,** Comparison of high-pressure liquid chromatography and bioassay for determination of ciprofloxacin in serum and urine, *Antimicrob. Agents Chemother.*, 27, 353, 1985.
6. **Weber, A., Chaffin, D., Smith, A., and Opheim, K. E.,** Quantitation of ciprofloxacin in body fluids by high-pressure liquid chromatography, *Antimicrob. Agents Chemother.*, 27, 531, 1985.

CIRAMADOL

Gas Chromatography

Specimen (mℓ)	Extraction	Column (m × mm)	Packing (mesh)	Oven temp (°C)	Gas (mℓ/min)	Det.	RT (min)	Internal standard (RT)	Deriv.	Other compounds (RT)	Ref.
Plasma (1)	I-1	1.8 × 2	2% OV-101 Chromosorb W (80/100)	230	N_2 (NA)	ECD	5.8	WY 15623 (7.6)	Pentafluoro-benzoyl	—	1

Liquid Chromatography

Specimen (mℓ)	Extraction	Column (cm × mm)	Packing (μm)	Elution	Flow (mℓ/min)	Det. (nm)	RT (min)	Internal standard (RT)	Other compounds (RT)	Ref.
Plasma, serum (0.5)	I-2	12 × 3.9	μ-Bondapak-C_{18} (10)	E-1	0.8	Electrochem[a]	4.1	WY 15623 (9.7)	Dezocine[b]	2

[a] ESA Coulometric detector. Electrode 1 = + 0.58 V; electrode 2 = + 0.90 V.
[b] A different mobile phase and a different internal standard are required for the assay of this compound.

Extraction — I-1. To the sample were added 100 μℓ of an aqueous solution of the internal standard (1 μg/mℓ), 1 mℓ of 5 *N* NH_4OH and the mixture was extracted twice with 5-mℓ portions of diethyl ether. The combined ether extract was back extracted into 1 mℓ of 5 *N* acetic acid. The aqueous layer was washed with 5 mℓ of ethyl acetate, saturated with NaCl (800 mg), made basic with 1 mℓ of 5 *N* NH_4OH, and extracted with 1 mℓ of toluene. An aliquot of 0.7 μℓ of toluene extract was treated with 100 μℓ of pyridine and 50 μℓ of pentafluorobenzoyl chloride at 25°C for 30 min. The mixture was washed with 1 mℓ of 10% Na_2CO_3, 1 mℓ of 0.2 *N* H_2SO_4, 1 mℓ of 10% Na_2CO_3 and 1 mℓ of water. Aliquots of the toluene layer were injected.
I-2. To the sample were added 50 μℓ of an aqueous solution of the internal standard (1 μg/mℓ), 1 mℓ of 5 *N* NH_4OH and the mixture was extracted with 5 mℓ of ethyl acetate. The organic layer was evaporated at 40°C under vacuum. The residue was reconstituted with 200 μℓ of the mobile phase. A 20 to 50-μℓ of this solution was injected with an autosampler.

Elution — E-1. Water-methanol-acetonitrile-butanol-phosphoric acid (750:90:10:10:1).

CIRAMADOL (continued)

REFERENCES

1. **Sisenwine, S. F., Kimmel, H. B., Tio, C. O., Liu, A. L., and Ruelius, H. W.,** Determination of ciramadol in plasma by gas-liquid chromatography, *J. Pharm. Sci.*, 72, 85, 1983.
2. **Locniskar, A. and Greenblatt, D. J.,** Determination of ciramadol and dezocine, two new analgesics, by high-performance liquid chromatography using electrochemical detection, *J. Chromatogr.*, 374, 215, 1986.

CISPLATIN[a]

Liquid Chromatography

Specimen (mℓ)	Extraction[b]	Column (cm × mm)	Packing (μm)	Elution	Flow (mℓ/min)	Det. (nm)	RT (min)	Internal standard (RT)	Other compounds (RT)	Ref.
Urine	—	15 × 4.6	Technicon-C_8 (5)	E-1	2.0	Electrochem[c]	4	—	—	3
Plasma, urine	—	25 × 5	Spherisorb-ODS (NA)[d]	E-2; grad	0.5	ABS (225)[e]	10	—	f	4
Plasma	I-1	25 × 4.6	Spheri-CN (5)	E-3[g]	1.5	ABS (254)	5.3	Palladium chloride (8.1)	—	5
Urine	—	10 × 4.6	Hypersil-ODS (5)	E-4	1.1	h	4.1	—	Carboplatin (2.5)	6
Plasma, urine	I-2	30 × 3.9	μ-Bondapak-CN (10)	E-5	4.0	ABS (254)	3.9	Nickel chloride (2.4)	—	7
Dosage	—	11 × 4.6	Spherisorb ODS-2 (5)[i]	E-6	1.0	Electrochem[b]	—	—	Spiroplatin[j]	8
Plasma	I-3	10 × 8	Rad-Pak-C_{18} (10)	E-7	1.5	ABS (254)	7.8	Nickel chloride (9.3)	—	9

[a] Electrochemical detection of platinum complexes used for cancer chemotherapy has been evaluated.[1,2]
[b] Platinum complexes are not amenable to extraction. Protein free filtrates of plasma or filtered urine samples are injected directly.
[c] Polarographic analyzer with a static mercury drop electrode.
[d] Protected by a precolumn of the same material.

[e] Fractions were collected and Pt determined by flameless atomic absorption spectrometry.
[f] A number of metabolites containing Pt have been isolated from plasma and urine.
[g] The mobile phase was heated to 40°C prior to passage through the column.
[h] Quenched phosphorescence, (ex = 415 nm, em = 520 nm).
[i] Protected by a 40 × 4.6 mm precolumn packed with Permaphase ODS.
[j] Separation of spiroplatin and a number of its possible derivatives in aqueous solution is shown.

Extraction — I-1. To the plasma ultrafiltrate were added 100 μℓ of an aqueous solution of the internal standard, 100 μℓ of a 10% solution of diethyldithiocarbamate in 0.1 *M* sodium hydroxide, and 200 μℓ of a saturated solution of sodium nitrite. The mixture was allowed to stand at room temperature for 1 hr and extracted with 3 mℓ of chloroform. The organic layer was dried over anhydrous sodium sulfate and then evaporated under nitrogen at 40°C. The residue was reconstituted in 25 μℓ of chloroform and a 10-μℓ aliquot of this solution was injected.
I-2. To the plasma filtrate or urine were added 100 μℓ of an aqueous solution of the internal standard (1 g/ℓ), 200 μℓ of a 2% solution of sodium diethyldithiocarbamate. The mixture was incubated at 37°C for 1 hr and then extracted with 1 mℓ of chloroform. An aliquot of 10 μℓ of the chloroform extract was injected.
I-3. To the plasma filtrate were added 5 μℓ of an aqueous solution of the internal standard (109 μg/mℓ), 50 μℓ of 10% diethyldithiocarbamate in 0.1 *N* soidum hydroxide. The mixture was incubated at 37°C for 30 min and after cooling was extracted with 0.2 mℓ of chloroform. A 20-μℓ aliquot of the chloroform layer was injected.

Elution — E-1. 10 m*M* citrate-0.1 *M* hexadecyltrimethyl ammonium bromide, pH 7.3.
E-2. (A) 5 m*M* sodium dodecylsulfate; (B) 90% Acetonitrile. Gradient: isocratic A for 10 min; linear gradient from 0 to 100% (A) over 10 to 80 min.
E-3. Heptane-isopropanol (90:10).
E-4. 1 m*M* Citrate buffer, pH 5-methanol (99:1) containing 10 m*M* biacetyl and 0.02 m*M* hexadecyltrimethylammonium bromide.
E-5. Heptane-isopropanol (90:10).
E-6. Methanol-0.05 *M* sodium sulfate, pH 3 (20:80).
E-7. Methanol-water (4:1).

REFERENCES

1. **Richmond, W. N. and Baldwin, R. P.,** Chloride-assisted electrochemical detection of Cis-dichlorodiammineplatinum(II) after liquid chromatography, *Anal. Chim. Acta*, 154, 133, 1983.
2. **Ding, X. D. and Krull, I. S.,** Dual electrode liquid chromatography-electrochemical detection (LCEC) for platinum-derived cancer chemotherapy agents, *J. Liq. Chromatogr.*, 6, 2173, 1983.
3. **Bannister, S. J., Sternson, L. A., and Repta, A. J.,** Evaluation of reductive amperometric detection in the liquid chromatographic determination of antineoplastic platinum complexes, *J. Chromatogr.*, 273, 301, 1983.
4. **Daley-Yate, P. T. and McBrien, D. C. H.,** Cisplatin metabolites: a method for their separation and for measurement of their renal clearance *In Vivo, Biochem. Pharmacol.*, 32, 181, 1983.
5. **Reece, P. A.,** Sensitive high-performance liquid chromatographic assay for platinum in plasma ultrafiltrate, *J. Chromatogr.*, 306, 417, 1984.
6. **Gooijer, C., Veltkamp, A. C., Baumann, R. A., Velthorst, N. H., Frei, R. W., and Van Der Vijgh, W. J. F.,** Analysis of platinum complexes by liquid chromatography with quenched phosphorescence detection, *J. Chromatogr.*, 312, 337, 1984.

CISPLATIN[a] (continued)

7. **Drummer, O. H., Proudfoot, A., Howes, L., and Louis, W. J.,** High-performance liquid chromatographic determination of platinum(II) in plasma ultrafiltrate and urine: comparison with a flameless atomic absorption spectrometric method, *Clin. Chim. Acta*, 136, 65, 1984.
8. **Elferink, F., Van Der Vijgh, W. J. F., and Pinedo, H. M.,** Analysis of antitumour [1,1-*bis*(aminomethyl)cyclohexane] platinum(II) complexes derived from spiroplatin by high-performance liquid chromatography with differential pulse amperometric detection, *J. Chromatogr.*, 320, 379, 1985.
9. **Andrews, P. A., Wung, W. E., and Howell, S. B.,** A high-performance liquid chromatographic assay with improved selectivity for cisplatin and active platinum (II) complexes in plasma ultrafiltrate, *Anal. Biochem.*, 143, 46, 1984.

CITALOPRAM

Liquid Chromatography

Specimen (mℓ)	Extraction	Column (cm × mm)	Packing (μm)	Elution	Flow (mℓ/min)	Det. (nm)	RT (min)	Internal standard (RT)	Other compounds (RT)	Ref.
Plasma, urine (1)	I-1	25 × 3	Spherisorb-ODS (5)[a]	E-1	1.3	Fl (140, 296)	6.3	LU 10-202[b] (8.4)	Lu 11-161[c] (3.7) LU 11-305 (5.2) LU 11-109 (4.6)	1, 2
Plasma (1—2)	I-2	30 × 3.9	μ-Bondapack-C_{18} (10)	E-2	0.5	ABS (254)	11.3	Desipramine (13.8)	LU 11-109 (10)[d]	3

Thin-Layer Chromatography

Specimen (mℓ)	Extraction	Plate (Manufacturer)	Layer (mm)	Solvent	Post-separation treatment	Det. (nm)	Rf	Internal standard (Rf)	Other compounds (Rf)	Ref.
Plasma (2)	I-3	10 × 20 cm (Merck)	Silica gel (HPTLC)	S-1	—	Fl. Reflectance (240,295)	0.43	—	LU 11-109 (0.51) LU 11-161 (0.58)	4

[a] Protected by a 30 × 4.6 mm MPLC RP-18 guard column.
[b] Chloro analog.
[c] Metabolites of citalopram; LU 11-109 = monodesmethyl; LU 11-161 = didesmethyl, and LU 11-30 = N-oxide.
[d] Conditions for the assay of other antidepressant drugs are described.

Extraction — I-1. To the sample were added 75 μℓ of an aqueous solution of the internal standard (1 μg/mℓ) and 50 μℓ of 1 *N* NaOH. The mixture was extracted twice with 3-mℓ portions of diethyl ether. The combined ether extracts were evaporated in the presence of 50 μℓ of 0.1 *N* HCl with a stream of nitrogen at 40°C. The aqueous phase was washed with 0.5 mℓ of ether and 15 to 20 μℓ of the residual extract was injected.
I-2. To the sample were added 100 μℓ of the internal standard solution (1 μg/mℓ in methanol) and 1 mℓ of 2 *M* sodium hydroxide. The mixture was extracted with 10 mℓ of ether. The organic phase was back extracted into 2 mℓ of 0.5 *M* sulfuric acid. The aqueous layer was made alkaline with 3 mℓ of 2 *M* sodium hydroxide and re-extracted with 10 mℓ of diethyl ether. The organic phase was evaporated under nitrogen. The residue was dissolved in 100 μℓ of the mobile phase and 20 to 50 μℓ were injected.
I-3. The sample was adjusted to pH 10 with 100 μℓ of 1 *N* NaOH and extracted twice with 6-mℓ portions *n*-hexane containing 1% triethylamine. The combined organic extracts were evaporated in 50 μℓ of chloroform. The entire solution together with a 50-μℓ rinse was spotted with an autospotter.

Elution — E-1. Acetonitrile-0.6% KH_2PO_4, pH 3 (55:45).
E-2. Acetonitile-0.025 *M* KH_2PO_4-water (45:50:5).

Solvent — S-1. Dichloroethane-ethyl acetate-ethanol-acetic acid-water (15:26:12:8:7.5).

REFERENCES

1. **Oyehaug, E., Ostensen, E. T., and Salvesen, B.,** Determination of the antidepressant agent citalopram and metabolites in plasma by liquid chromatography with fluorescence detection, *J. Chromatogr.*, 227, 129, 1982.
2. **Oyehaug, E., Ostensen, E. T., and Salvesen, B.,** High-performance liquid chromatographic determination of citalopram and four of its metabolites in plasma and urine samples from psychiatric patients, *J. Chromatogr.*, 308, 199, 1984.
3. **Viala, A., Durand, A., and Conquy, T.,** Determination of citalopram, amitriptyline and clomipramine in plasma by reversed-phase high-performance liquid chromatography, *J. Chromatogr.*, 338, 171, 1985.
4. **Overo, K. F.,** Fluorescence assay of citalopram and its metabolites in plasma by scanning densitometry of thin-layer chromatograms, *J. Chromatogr.*, 224, 526, 1981.

CIROVORUM FACTOR

Liquid Chromatography

Specimen (mℓ)	Extraction	Column (cm × mm)	Packing (μm)	Elution	Flow (mℓ/min)	Det. (nm)	RT (min)	Internal standard (RT)	Other compounds (RT)	Ref.
Plasma, urine (0.025)	I-1	25 × 4.6	LiChrosorb (10)[a]	E-1	1.4	ABS (254)	19	—	5-Methyltetrahydrofoalte (13)	1

[a] Protected by a precolumn filled with Co:Pell-ODS

Extraction — I-1. The sample was diluted with 25 μℓ of water containing 2.5 μg each of citrovorum factor acid, 5-methyltetrahydrofoalte, as carriers, diluted sample was injected directly. Urine samples were injected without the addition of carriers.

REFERENCE

1. **Straw, J. A., Covey, J. M., and Szapary, D.,** Differences in the pharmacokinetics of the diastereoisomers of citrovorum factor in dogs, *Cancer Res.*, 41, 3936, 1981.

CLAVULANIC ACID

Liquid Chromatography

Specimen (mℓ)	Extraction	Column (cm × mm)	Packing (μm)	Elution	Flow (mℓ/min)	Det. (nm)	RT (min)	Internal standard (RT)	Other compounds (RT)	Ref.
Fermentation broth	—	15 × 4.1	Hamilton PRP-1 (10)	E-1	1.0	ABS (220)	8.5	—	—	1
Serum, urine (0.5)	I-1	25 × 4.6	Sperisorb-ODS (5)	E-2	1.5	ABS (313)	7	—	Ticarcillin[a]	2
Plasma, urine (0.15—0.2)	I-2	15 × 4.6	Develosil-ODS (5)[b]	E-3[c]	0.8	Fl (386, 460)	4	—	Amoxicillin	3

[a] Different conditions for the assay of this drug are described.
[b] Protected by a 30 × 4.6 mm packed with the same material as in the analytical column.
[c] A different mobile phase is used when amoxicillin is also present.

Extraction — I-1. The sample was diluted with an equal volume of 0.1 *M* phosphate buffer, pH 7 and filtered through an Amicon MPS-1 ultrafiltration apparatus. An aliquot of 100 μℓ of the ultrafiltrate was treated with 100 μℓ of imidazole reagent. Aliquots of this mixture were injected. (See Ref. 1 under amoxicillin for the preparation of imidazole reagent).
I-2. The sample was ultrafiltered. To a 50-μℓ aliquot of the ultrafiltrate, 150 μℓ of 1 *M* phosphate buffer (pH 3.8) and 20 μℓ of 2% benzadehyde solution were added. The mixture was incubated at 100°C for 20 min and was then immediately cooled to room temperature. A 20- to 25-μℓ aliquot of this solution was injected.

Elution — E-1. Acetonitrile-1 m*M* phosphate buffer, pH 6 containing 3 m*M* tetrabutylammonium bromide (1:9).
E-2. Methanol-0.1 *M* potassium dihydrogen phosphate (10:90) containing 0.05 *M* pentane sulfonic acid and 0.1 *M* ethanolamine.
E-3. Water-methanol (1:2).

REFERENCES

1. **Salto, F. and Alemany, M. T.,** Ion interaction chromatography of clavulanic acid on a poly(styrene-divinylbenzene) adsorbent in the presence of tetrabutylammonium salts, *J. Liq. Chromatogr.*, 7, 1477, 1984.
2. **Watson, I. D.,** Clavulanate-potentiated ticarcillin: high-performance liquid chromatographic assays for clavulanic acid and ticarcillin isomers in serum and urine, *J. Chromatogr.*, 337, 301, 1985.
3. **Haginaka, J., Yasuda, H., and Uno, T.,** High-performance liquid chromatographic assay of clavulanate in human plasma and urine by fluorimetric detection, *J. Chromatogr.*, 377, 269, 1986.

CLENBUTEROL

Liquid Chromatography

Specimen (mℓ)	Extraction	Column (cm × mm)	Packing (μm)	Elution	Flow (mℓ/min)	Det. (nm)	RT (min)	Internal standard (RT)	Other compounds (RT)	Ref.
Dosage	I-1	30 × 4	μ-Bondapak-CN (10)	E-1	1.5	ABS (254)	9	—	—	1
Urine (18)	I-2	30 × 4	μ-Bondapak-CN (10)	E-2	2.0	ABS (222)	9	—	—	2

CLENBUTEROL (continued)

Extraction — I-1. The sample was acidified with 1 *N* HCl and washed with ether. The ether layer was back-washed with 1 *N* HCl. The combined aqueous layer was diluted to a volume and aliquots of this solution were injected.
I-2. The sample (18 mℓ) was diluted to 25 mℓ and pH adjusted to 11 with NH_4OH. The sample was applied to ClinElut and allowed to stand for 10 min. The column was eluted three times with 20-mℓ portions of hexane. The combined eluate was back extracted with 200 μℓ of 1 *N* HCl. A 20-μℓ aliquot of the aqueous layer was injected.

Elution — E-1. Methanol-isopropanol-0.15% sodium heptanesulfonate, pH 3 (18:2:80).
E-2. Isopropanol-0.4% dodecyl sulfate (30:70), pH, 4.85.

REFERENCES

1. **Hamann, J. A., Johnson, K., and Jeter, D. T.,** HPLC determination of clenbuterol in pharmaceutical gel formulations, *J. Chromatogr. Sci.*, 23, 34, 1985.
2. **Eddins, C., Hamann, J., and Johnson, K.,** HPLC analysis of clenbuterol, a beta-adrenergic drug, in equine urine, *J. Chromatogr. Sci.*, 23, 308, 1985.

CLINDAMYCIN

Liquid Chromatography

Specimen (mℓ)	Extraction	Column (cm × mm)	Packing (μm)	Elution	Flow (mℓ/min)	Det. (nm)	RT (min)	Internal standard (RT)	Other compounds (RT)	Ref.
Dosage	—	30 × 3.9	μ-Bondapak-C_{18} (10)	E-1	1.0	Refractive index	11.3	Phenethyl alcohol (5.9)	Lincomycin B (4.4) Lincomycin (4.5) Clindomycin B (7.7) 7-Epiclindamycin (9.2)	1
Dosage	—	25 × 4.6	Zorbax-C (10)	E-2	NA	ABS (210)	8[a]	*p*-Hydroxyacetophen-one (10.2)	Lincomycin B HCl (2.6) Lincomycin-2-phos-phate (2.8) Lincomycin HCl (3.2)	2

Clindamycin B-2-phosphate (4.4)
Clindamycin HCl (13.5)

[a] Retention time of clindamycin-2-phosphate.

Elution — E-1. Methanol-water (60:40), 0.035 *M* acetic acid and 0.005 *M* camphor sulfonate, pH 6.0.
E-2. Acetonitrile-0.1 *M* phosphate buffer, pH 2.5 (225:775).

REFERENCES

1. **Landis, J. B., Grant, M. E., and Nelson, S. A.,** Determination of clindamycin in pharmaceuticals by high-performance liquid chromatography using ion-pair formation, *J. Chromatogr.*, 202, 99, 1980.
2. **Munson, J. W. and Kubiak, E. J.,** A high-performance liquid chromatographic assay for clindamycin phosphate and its principal degradation product in bulk drug and formulations, *J. Pharm. Biomed. Anal.*, 3, 523, 1985.

CLOBAZAM

Gas Chromatography

Specimen (mℓ)	Extraction	Column (m × mm)	Packing (mesh)	Oven temp (°C)	Gas (mℓ/min)	Det.	RT (min)	Internal standard (RT)	Deriv.	Other compounds (RT)	Ref.
Plasma (1)	I-1	1.2 × 4	10% OV-101 Chromosorb W (80/100)	265	Ar: 95-Methane:5 (25)	ECD	7.4	Diazepam (5.5)	—	N–Desmethyl-clobazam	1
Serum (1)	I-2	1.2 × 4	2% OV-101 Chromosorb W (120/150)[a]	240	He (40)	NPD	3.5	Diazepam (2.2)	—	N-Desmethyl-clobazam	2
Plasma (1)	I-3	0.5 × 3	2% SP-2510-DA (100/120)	245	N_2 (80)	ECD	3.5	Methyl-clonazepam (5.5)	—	b	3

CLOBAZAM (continued)

Liquid Chromatography

Specimen (mℓ)	Extraction	Column (cm × mm)	Packing (μm)	Elution	Flow (mℓ/min)	Det. (nm)	RT (min)	Internal standard (RT)	Other compounds (RT)	Ref.
Serum (1)	I-4	25 × 4.6	Spherisorb-ODS (5)	E-1	1.8	ABS (228)	4.9	Prazepam (6.1)	N-Desmethylclobazam (3.5)	—
Plasma (0.5)	I-5	30 × 3.9	μ-Bondapak-C_{18} (10)	E-2	1.0	ABS (230)	8.7	Diazepam (11.4)	N-Desmethylclobazam (6.5) 4′-Hydroxyclobazam[c] 4′-Hydroxy-N-desmethyl clobazam[c]	5

[a] Particle size in micrometers.
[b] The metabolic N-desmethylclobazam cannot be determined with this liquid phase.
[c] Chromatographed at a different flow rate separately with nitrazapam as the internal standard.

Extraction — I-1. The sample was extracted after the addition of the internal standard with 5 mℓ of benzene containing 1.5% isopentanol. The organic layer was evaporated at 40°C under nitrogen. The residue was dissolved in 100 to 200 μℓ of toluene containing 15% isopentanol, of which 1 to 6 μℓ was injected.
I-2. The sample was mixed with 0.1 mℓ of the internal standard solution (1 mg/ℓ in water) and extracted with 5 mℓ of diethyl ether. The organic layer was evaporated at 40°C under nitrogen flow. The residue was dissolved in 50 μℓ of ethyl acetate and 1 to 2 μℓ of this solution were injected.
I-3. To the sample were added 100 μℓ of the internal standard solution (40 μg/mℓ in methanol) and 0.5 mℓ of 1 *N* HCl. The mixture was extracted with 4 mℓ of benzene. The organic phase was evaporated at 45°C in a vacuum evaporator. The residue was dissolved in 100 μℓ of acetone and 1 to 2 μℓ of this solution were injected.
I-4. To the sample were added 0.1 mℓ of the internal standard solution (1 mg/mℓ in methanol) and 0.5 mℓ of saturated trisodium orthophosphate. The mixture was extracted with 5 mℓ of dichloromethane. The organic layer was evaporated at 40°C under a stream of nitrogen. The residue was dissolved in 50 μℓ of acetonitrile and 10-μℓ aliquots were injected.
I-5. A suitable volume of methanolic soution of the internal standard was evaporated. The residue was treated with 0.5 mℓ of plasma and 1 mℓ of saturated sodium phosphate and the mixture was extracted twice with 5-mℓ portions of diethyl ether. The combined organic extracts were evaporated. The residue was dissolved in 100 μℓ of methanol. For the extraction of hydroxy metabolites, the aqueous layer left after the extraction with ether was adjusted to pH 9 and re-extracted with ether after the addition of the internal standard.

Elution — E-1. Acetonitrile-water (53:47).
E-2. Acetonitrole-water (47:53).

REFERENCES

1. **Greenblatt, D. J.,** Electron-capture GLC determination of clobazam and desmethylclobazam in plasma, *J. Pharm. Sci.*, 69, 1351, 1980.
2. **Hajdu, P., Uihlein, M., and Damm, D.,** Quantitation determination of clobazam in serum and urine by gas chromatography, thin layer chromatography and fluorometry, *J. Clin. Chem. Clin. Biochem.*, 18, 209, 1980.
3. **Riva, R., Tedeschi, G., Albani, F., and Baruzzi, A.,** Quantitative determination of clobazam in the plasma of epileptic patients by gas-liquid chromatography with electron-capture detection, *J. Chromatogr.*, 225, 219, 1981.
4. **Ratnaraj, N., Goldberg, V., and Lascelles, P.,** Determination of clobazam and desmethylclobazam in serum using high-performance liquid chromatography, *Analyst*, 109, 813, 1984.
5. **Tomasini, J. L., Coassolo, P., Aubert, C., and Cano, J. P.,** Determination of clobazam, N-desmethylclobazam and their hydroxy metabolites in plasma and urine by high-performance liquid chromatography, *J. Chromatogr.*, 343, 369, 1985.

CLOCAPRAMINE

Liquid Chromatography

Specimen (mℓ)	Extraction	Column (cm × mm)	Packing (μm)	Elution	Flow (mℓ/min)	Det. (nm)	RT (min)	Internal standard (RT)	Other compounds (RT)	Ref.
Plasma, feed (1)	I-1	15 × 4.6	LiChrosorb SI-60 (5)	E-1	0.9	ABS (254)	3.5	Opipramol (4.7)	—	1

Extraction — I-1. The sample was mixed with 0.2 mℓ of 1 *N* NaOH and extracted with 5 mℓ of chloroform containing 0.5 μg/mℓ of the internal standard. The organic layer was evaporated at 50°C under nitrogen. The residue was dissolved in 0.5 mℓ of the mobile phase and aliquots of this solution (20 μℓ) were injected.

Elution — E-1. Dichloromethane-methanol-ammonia (100:10:25).

REFERENCE

1. **Geahchan, A., Chambon, P., and Genoux, P.,** High performance liquid chromatographic determination of clocapramine in feed and plasma, *J. Assoc. Off. Anal. Chem.*, 65, 706, 1982.

CLOFAZIMINE

Liquid Chromatography

Specimen (mℓ)	Extraction	Column (cm × mm)	Packing (μm)	Elution	Flow (mℓ/min)	Det. (nm)	RT (min)	Internal standard (RT)	Other compounds (RT)	Ref.
Plasma (1)	I-1	25 × 4.6	Ultrasphere Octyl (5)[a]	E-1	1.5	ABS (285)	9.6	—	—	1

Thin-Layer Chromatography

Specimen (mℓ)	Extraction	Plate (Manufacturer)	Layer (mm)	Solvent	Post-separation treatment	Det. (nm)	Rf	Internal standard (Rf)	Other compounds (Rf)	Ref.
Plasma (1—3)	I-2	20 × 10 cm (Merck)	Silica gel 60 (HPTLC) (0.25)	S-1	—	Reflectance (545)	0.36	—	—	2

[a] Column temp = 40°C

Extraction — I-1. The sample was mixed with 1 mℓ of phosphate-citrate buffer, pH 6 and extracted with 14 mℓ of chloroform-methanol (4:1). An aliquot of 10 mℓ of the organic layer was evaporated under nitrogen. The residue was reconstituted in 150 μℓ of the mobile phase. The solution was washed with 0.5 mℓ of hexane. Aliquots of the aqueous phase were injected.
I-2. The sample was mixed with 2 mℓ of 1 *M* acetate buffer, pH 5 and extracted with 6 mℓ of toluene. An aliquot of 5 mℓ of the organic phase was evaporated under nitrogen at 40°C. The residue was dissolved in 100 μℓ of toluene and aliquots of this solution were applied on the TLC plate.

Elution — E-1. Methanol-0.0425 *M* phosphoric acid, pH 2.4 (82:19)

Solvent — S-1. Toluene-acetic acid-water (50:50:4)

REFERENCES

1. **Peters, J. H., Hamme, K. J., and Gordon, G. R.,** Determination of clofazimine in plasma by high-performance liquid chromatography, *J. Chromatogr.*, 229, 503, 1982.
2. **Lanyi, Z. and Dubois, J. P.,** Determination of clofazimine in human plasma by thin-layer chromatography, *J. Chromatogr.*, 232, 219, 1982.

CLOFIBRIC ACID

Gas Chromatography

Specimen (mℓ)	Extraction	Column (m × mm)	Packing (mesh)	Oven temp (°C)	Gas (mℓ/min)	Det.	RT (min)	Internal standard (RT)	Deriv.	Other compounds (RT)	Ref.
Plasma (1)	I-1	1 × 2	10% Silar 10C Gas Chrom Q (100/120)	220	N_2 (24)	FID	2.5	Banzocaine (6.2)	—	Clofibrate (1.8)	1
Serum (0.5)	I-2	2 × 1.8 (Steel)	3% OV-17[a] Chromosorb W (80/100)	125	N_2 (27)	ABS[b] (280)	NA	*p*-Chloro-phenoxy acetic acid	Methyl	—	2

Liquid Chromatography

Specimen (mℓ)	Extraction	Column (cm × mm)	Packing (μm)	Elution	Flow (mℓmin)	Det. (nm)	RT (min)	Internal standard (RT)	Other compounds (RT)	Ref.
Plasma, urine (0.1)	I-3	30 × 3.9	μ-Bondapak-C_{18} (10)	E-1	2.0	ABS (235)	4.6	Flurbiprofen (8.2)	Probencid (5.8)	3
Plasma, dosage (0.5)	I-4	30 × 3.9	μ-Bondapak-alkyl phenyl (10)	E-2	1.5	ABS (225)	3.4	Morphine (4.7)	Clofibrate (10.7)	4
Plasma (0.5)	I-5	30 × 4.6	μ-Bondapak-C_{18} (10)	E-3	1.3	ABS (230)	5.5	*p*-Chlorophenoxy acetic acid (4)	Fenofibric acid (12)	5

[a] An alternative column (1.2 m × 2.1 mm) packed with 10% EGA was also used.
[b] The effluent of the column were scrubbed into a stream of 2-propanol, at a flow of 0.5 mℓ/min. This was then debubbled and a portion was drawn through the 20 μℓ UV flow cell. A FID was also used.

Extraction — I-1. The sample was treated with 550 mg of ammonium sulfate, mixed, and centrifuged. To the supernate (0.5 mℓ) 100 μg of benzocaine was added and the mixture extracted with 1 mℓ of diethyl ether. Aliquots of 1 μℓ of the ether layer were injected.

I-2. The sample was spiked with the internal standard solution, made acidic with 500 μℓ of 1 *N* sulfuric acid, and extracted twice with 4% isopropanol in benzene. The combined organic layers were evaporated. The residue was treated with 1 mℓ of sulfuric acid-isooctane-methanol (5:25:93.5). The mixture was incubated at 70°C

CLOFIBRIC ACID (continued)

overnight. After cooling, the mixture was diluted with 1 mℓ of water and extracted twice with 1.5 mℓ of light petroleum. The combined organic layers were evaporated at room temperature under a stream of nitrogen. The residue was dissolved in 25 μℓ of isooctane of which 5 μℓ was injected.
I-3. The sample was treated with 0.25 mℓ of the internal standard solution (30 mg/ℓ in acetonitrile). After mixing and centrifugation 20- to 50-μℓ aliquots of the supernatant was injected.
I-4. The sample was extracted with 4 mℓ of chloroform. An aliquot of 2 mℓ of organic extract was evaporated and the residue reconstituted in 100 μℓ of the mobile phase containing morphine sulfate as the internal standard.
I-5. The sample was mixed with 30 μℓ of the internal standard solution (100 μg/mℓ in mobile phase), 100 μℓ of 0.017 *M* acetic acid, and the mixture was extracted with 7 mℓ of diethyl ether. The organic layer was evaporated. The residue was dissolved in 200 μℓ of the mobile phase, a 50-μℓ aliquot was injected.

Elution — E-1. Acetonitrile-acetic acid-water (450:5:545).
E-2. Acetonitrile-0.1 *N* acetate buffer, pH 3.8 (50:50).
E-3. Methanol-water (45:55) containing 1% acetic acid.

REFERENCES

1. **Wolf, M. S. and Zimmerman, J. J.,** Simultaneous GLC determination of clofibrate and clofibric acid in human plasma, *J. Pharm. Sci.*, 69, 92, 1980.
2. **Karmen, A., Pritchett, T., and Lam, S.,** Therapeutic drug assays with gas-liquid chromatography and optical detection, *J. Chromatogr.*, 217, 247, 1981.
3. **Veenendaal, J. R. and Meffin, P. J.,** The simultaneous analysis of clofibric acid and probenecid and the direct analysis of clofibric acid glucuronide by high-performance liquid chromatography, *J. Chromatogr.*, 223, 147, 1981.
4. **Garrett, E. R. and Gardner, M. R.,** Prediction of stability in pharmaceutical preparations. XIX. Stability evaluation and bioanalysis of clofibric acid esters by high-pressure liquid chromatography, *J. Pharm. Sci.*, 71, 14, 1982.
5. **Paillet, M. and Doucet, D.,** Rapic determination of clofibric acid in human plasma by high-performance liquid chromatography, *J. Chromatogr.*, 375, 179, 1986.

CLOFILIUM

Gas Chromatography

Specimen (mℓ)	Extraction	Column (m × mm)	Packing (mesh)	Oven temp (°C)	Gas (mℓ/min)	Det.	RT (min)	Internal standard (RT)	Deriv.	Other compounds (RT)	Ref.
Plasma (0.2)	I-1	0.6 × 2	1% SP-2100 Supelcoport (100/120)	195	He (30)	MS-EI	2	4-Chloro-N,N-dimethyl-N-heptylbenzene butanaminium bromide (2.0)	Pyrolysis	—	1

Extraction — I-1. To the sample were added an aqueous solution of the internal standard (100 ng), 1.5 mℓ of 0.1 *M* sodium bromide and the mixture was extracted twice with 5-mℓ portions of dichloromethane. The combined extracts were evaporated at room temperature under vacuum. The residue was taken up in 10 μℓ of methanol containing 0.01 *M* KOH. An aliquot of 3 mℓ of this solution was injected.

REFERENCE

1. **Lindstrom, T. D. and Wolen, R. L.,** Determination of clofilium, a new antifibrillatory agent, in plasma by gas chromatography-mass spectrometry, *J. Chromatogr.*, 233, 175, 1982.

CLOMIPHENE

Liquid Chromatography

Specimen (mℓ)	Extraction	Column (cm × mm)	Packing (μm)	Elution	Flow (mℓ/min)	Det. (nm)	RT (min)	Internal Standard (RT)	Other compounds (RT)	Ref.
Plasma (3)	I-1	25 × 4.6	Zorbax-silica (6)	E-1	0.8	Fl[a] (257, 367)	*cis* 6.5 *trans* 7.4	—	Metabolites (11), (13)[b]	1
Plasma (3)	I-2	15 × 4.6	Supelco LC-8 (5)	E-2	1.5	Fl[a] (255, 378)	*cis* 24 *trans* 21	—	—	2

[a] The column eluate passes through a Teflon coil wound around the silica window of a mercury lamp prior to detection.
[b] Unidentified metaboites.

Extraction — I-1. The sample was mixed with 1 mℓ of borate buffer, pH 9 and extracted with 9 mℓ of redistilled diethyl ether. The ether extract was dried over anhydrous sodium sulfate and evaporated under a stream of air. The residue was dissolved in 75 μℓ of the mobile phase and 10 μℓ of the resulting solution was injected.

I-2. The sample was extracted with 12 mℓ of *tert*-butyl methyl ether. The extract was evaporated under nitrogen. The residue was dissolved in 0.5 mℓ of methanol-water (80:20). Aliquots of 100 μℓ of this solution were injected.

Elution — E-1. Chloroform-methanol (80:20).

E-2. Methanol-water (80:20) containing 2.3 mℓ of phosphoric acid and 10 μℓ of triethylamine per liter.

CLOMIPHENE (continued)

REFERENCES

1. **Harman, P. J., Blackman, G. L., and Phillipou, G.,** High-performance liquid chromatographic determination of clomiphene using post-column on-line photolysis and fluorescence detection, *J. Chromatogr.*, 225, 131, 1981.
2. **Baustian, C. L. and Mikkelson, T. J.,** Analysis of clomiphene isomers in human plasma and detection of metabolites using reversed-phase chromatography and fluorescence detection, *J. Pharm. Biomed. Anal.*, 4, 237, 1986.

CLOMIPRAMINE

Gas Chromatography

Specimen (mℓ)	Extraction	Column (m × mm)	Packing (mesh)	Oven temp (°C)	Gas (mℓ/min)	Det.	RT (min)	Internal standard (RT)	Deriv.	Other compounds (RT)	Ref.
Plasma (0.5)	I-1	2 × 2	1% OV-17 (NA)	230	NA	MS-Ei	4	N-Ethyldesmethylclomipramine (6)	—	—	1

Liquid Chromatography

Specimen (mℓ)	Extraction	Column (cm × mm)	Packing (μm)	Elution	Flow (mℓ/min)	Det. (nm)	RT (min)	Internal standard (RT)	Other compounds (RT)	Ref.
Plasma (1)	I-2	12.5 × 4.6	LiChrosorb Si 60 (5)	E-1	1.5	ABS (254)	3.3	Imipramine + Desipramine (4.1) (15.3)	Desmethylclomipramine (11.3)	2
Plasma (2)	I-3	10 × 5	Xo A 800 silica (5)	E-2	1.5	ABS (222)	131	Imipramine (4.1)	Desmethylclomipramine (14)	3

Extraction — I-1. The sample was mixed with 50 μℓ of drug free plasma containing 1 μg/mℓ of the internal standard and 0.5 mℓ of dilute ammonia. The mixture was extracted twice with 10 mℓ portions of petroleum ether. The combined extract was evaporated under nitrogen, residue dissolved in 0.5 mℓ of methanol. The solution was applied to a 3 × 0.5 cm sulfo ethyl-sephadex-LH 20 column and the basic fraction was eluted with ammoniacal methanol. The eluate was evaporated under nitrogen.

I-2. To the sample were added 100 μℓ of the solution of the internal standards, 1 mℓ borate buffer (0.2 *M*, pH 10), and the mixture was extracted with 5 mℓ of *n*-heptane-isoamyl alcohol (99:1). The organic layer was back extracted into 1 mℓ of 0.1 *N* H_2SO_4. The aqueous phase was made alkaline with 0.5 mℓ of 2 *N* NaOH and extracted with 5 mℓ of *n*-heptane-isoamyl alcohol. The organic extract was evaporated at 60°C under nitrogen and the residue was dissolved in 300 μℓ of mobile phase.
I-3. To the sample were added 50 μℓ of the internal standard solution (5 μg/mℓ in 0.01 *M* HCl), 0.5 mℓ of 2 *M* sodium carbonate, and the mixture was extracted twice with 9-mℓ portions of hexane. The combined organic extracts were evaporated at 60°C under nitrogen. The residue was dissolved in 100 μℓ of the mobile phase and an aliquot of 20 μℓ was injected.

Elution — E-1. Ethanol-hexane-dichloromethane-diethylamine (30:62:8:0.005).
E-2. Diethylamine-water-acetonitrile-ethanol (0.5:1.5:8:990).

REFERENCES

1. **Gaskell, S. J.,** Gas chromatography/high-resolution mass spectrometry as a reference method for clomipramine determination, *Postgrad. Med. J.*, 56, 90, 1980.
2. **Godbillon, J. and Gauron, S.,** Determination of clomipramine or imipramine and their mono-demethylated metabolites in human blood or plasma by high-performance liquid chromatography, *J. Chromatogr.*, 204, 303, 1981.
3. **Diquet, B., Gaudel, G., Colin, J. N., and Singlas, E.,** Dosage plasmatique de la clomipramine et de la demethylclomipramine par chromatographie liquide a haute performance, *Ann. Biol. Clin.*, 40, 321, 1982.

CLONAZEPAM

Gas Chromatography

Specimen (mℓ)	Extraction	Column (m × mm)	Packing (mesh)	Oven temp (°C)	Gas (mℓ/min)	Det.	RT (min)	Internal standard (RT)	Deriv.	Other compounds (RT)	Ref.
Plasma (0.5)	I-1	1.5 × 4	3% OV-17 Gas Chrom Q (100/120)	255	N_2 (120)	ECD	3[a]	Desmethylfluni-traze-pam (2)[a]	Hydrolysis	—	1
Plasma (1)	I-2	1.8 × 2	3% SE-30 Gas Chrom Q (80/100)	215	He (30)	NPD	8[a]	b (10)	Hydrolysis	—	2
Plasma (0.1)	I-3	0.9 × 2	2% SP-2510 DA Supelcoport (100/120)	260	N_2 (40)	ECD	8.3	Methylclonaze-pam	—	—	3
Plasma (0.2—0.5)	I-4	1.8 × 2	3% SP-2250 Supelcoport (100/120)	T.P.[c]	N_2 (35)	ECD	3.6	Desmethyldiazepam (2.5)	—	7-Aminoclonazepam (4)	4

CLONAZEPAM (continued)

Liquid Chromatography

Specimen (mℓ)	Extraction	Column (cm × mm)	Packing (μm)	Elution	Flow (mℓ/min)	Det. (nm)	RT (min)	Internal standard (RT)	Other compounds (RT)	Ref.
Plasma (1)	I-5	30 × 3.9	μ-Bondapak-C_{18} (10)	E-1	1.5	ABS (254)	8.3	Chlordiazepoxide (9.2)	—	5
Liver microsomal incubation	I-6	25 × 4.6	Zorbax-C_8 (10)[d]	E-2, gradient	2.0	ABS (254)	21	Flunitrazepam (23)	3-Hydroxyaminoclonazepam (5) 3-Hydroxyacetamidoclonazepam (6.5) Aminoclonazepam (7.5) Acetaminoclonazepam (9) 3-Hydroxyclonazepam (16.5)	6
Serum (2)	I-7	15 × 4.6	Ultrasphere-C_{18} (5)	E-3	1.5	ABS (308)	3.2[e]	Desalkylflurazepam (4.1)[e]	f	7
Serum (2)	I-8	10 × 8	Rad-Pak-C_{18} (10)	E-4	NA	ABS (313)	4.4	Nordiazepam (6.5)	f	8
Serum (1)	I-9	15 × 4.6	Utrasphere-C_{18} (5)[g]	E-5	2.0	ABS (254)	6	Methylclonazepam (11)	f	9

[a] Retention time of hydrolysis product.
[b] 2-Amino (diethylamino ethyl) 5-chloro-2'-fluorobenzophenone (hydrolysis product of flurazepam).
[c] Initial temp = 270°C, initial time = 1 min; rate = 50°C/min; final temp = 330°C; final time = 5 min.
[d] Column temp = 45°C.
[e] Retention times of methyl derivatives.
[f] Retention times of a number of drugs are given.
[g] Column temp = 50°C.

Extraction — I-1. To the sample were added 50 μℓ of the internal standard solution (700 ng/mℓ in acetone-heptane 1:4) 1 mℓ of borate buffer, pH 9 and the mixture was extracted with 2 mℓ of diethyl ether-heptane (40:60). The organic layer was back extracted into 0.5 mℓ of concentrated HCl. To the aqueous phase, 1.5 mℓ of methanol was added and the mixture heated in a boiling water bath for 10 min. After cooling 3 mℓ of 2 *M* sodium hydroxide was added and the mixture extracted with 2 mℓ of ether. The organic layer was evaporated, residue dissolved in 50 μℓ of toluene, and 1 to 2 μℓ of the solution was injected.
I-2. The sample was mixed with 50 ng of the internal standard (50 ng/mℓ in water), 2 mℓ of 1 *M* borate-KCl buffer (pH 9), and extracted with 10 mℓ of ether. The ether layer was back extracted into 2 mℓ 6 *N* HCl-6 *N* H_2SO_4 (95:5). The aqueous layer was washed with 5 mℓ of ether and placed in boiling water bath for 50 min. After cooling, acid was neutralized with 6 *N* NaOH and the mixture extracted with 4 mℓ of diethyl ether. The organic phase was evaporated at 50°C under nitrogen, the residue dissolved in 25 μℓ of ethyl acetate, and 5 to 7 μℓ of this solution were injected.
I-3. The sample was extracted with 0.5 mℓ of ethyl acetate-cyclohexane (4:1) containing 20 μmol/ℓ of the internal standard. The organic phase was evaporated at 40°C under nitrogen to about 50 μℓ, aliquots of 5 μℓ of the resulting solution were injected.
I-4. The sample was mixed with 20 μℓ of acetonitrile containing 5 ng of the internal standard about 200 μℓ of 1 *M* borate buffer, pH 9 and the mixture was extracted with 100 μℓ of toluene. Aliquots of 4 μℓ of the organic extract were injected.
I-5. The sample was spiked with 50 μℓ of an aqueous solution of the internal standard, pH adjusted to 9.5 with 0.5 *N* sodium hydroxide, and extracted twice with 5-mℓ portions of ethyl ether. The combined ethereal extracts were evaporated under nitrogen at 35°C. The residue was dissolved in 200 μℓ of 0.2 *M* HCl. The solution was washed with 200 μℓ of hexane and 100 μℓ of the aqueous phase was injected.
I-6. Incubations were adjusted to pH 9.5 and extracted three times with 10-mℓ portions of chloroform-ethyl acetate (1:1). The combined extracts were evaporated, residue dissolved in 200 μℓ of methanol, and an aliquot of 20 μℓ was injected.
I-7. The sample was mixed with 50 μℓ of the internal standard solution (20 mg/ℓ in isopropanol), 1 mℓ of ammonia buffer, pH 9.7, and the mixture extracted with 15 mℓ of benzene. The organic layer was back extracted into 2 mℓ of 2 *M* HCl. To the aqueous layer were added 1.5 mℓ of 25% ammonium hydroxide containing 0.2% diethylamine, 2 mℓ of buffer, and the mixture extracted with 6 mℓ of benzene. The organic layer was evaporated at 56°C in air stream. The residue was treated with 100 μℓ of acetone, 20 μℓ of iodomethane, and 20 μℓ of 0.02 *M* tetramethylammonium hydroxide in methanol. The mixture was incubated at 56°C for 20 min and then the reagents were evaporated. The residue was dissolved in 100 μℓ of the mobile phase and 30 μℓ were injected.
I-8. The sample was mixed with 100 μℓ of an aqueous solution of the internal standard (10 μg/mℓ) and 2 mℓ of 0.3 *M* phosphate buffer. The sample was then loaded onto the type W extraction column and the extraction was carried out with Prep I automated sample processor. Alternatively, the buffered sample was extracted with 6 mℓ of chloroform. The organic layer was evaporated at 60°C under nitrogen. The residue was dissolved in 100 μℓ of the mobile phase and an aliquot of 30 μℓ was injected.
I-9. A 1-mℓ BonElut-C_{18} column was washed with 2 column volumes of methanol followed by 2 column volumes of water. One hundred microliters of the internal standard solution (1 μg/mℓ in 1 *M* glycine buffer, pH 10.5) was placed on the washed column followed by 1 mℓ of the sample. When the sample had passed through, the column was washed with 2 mℓ of water and 50 μℓ of methanol. Finally, the column was eluted with 200 μℓ of methanol. The eluate was evaporated at 45°C under a stream of nitrogen. The residue was dissolved in 40 μℓ of methanol and the entire solution was injected.

CLONAZEPAM (continued)

Elution — E-1. Acetonitrile-water (40:60).
E-2. (A) Methanol-tetrahydrofuran-10 m*M* phosphate buffer, pH 7.0 (53:15:212); (B) Methanol-acetonitrile-10 m*M* phosphate buffer, pH 7.0 (50:35:125). Isocratic A from 0 to 9 min; Isocratic B from 9 to 24 min.
E-3. Acetonitrile-methanol-water-diethylamine-phosphoric acid (1100:800:1200:4:1) containing 1.5 g sodium pentane solfonate, pH 6.5.
E-4. Acetonitrile-methanol-0.05 *M* acetate buffer, pH 5.4 (235:265:450).
E-5. Acetonitrile-0.02 *M* phosphate buffer, pH 3.8 (30:70).

REFERENCES

1. **Larking, P.,** Gas chromatographic determination of clonazepam, *J. Chromatogr.*, 221, 399, 1980.
2. **Dhar, A. K. and Kutt, H.,** Improved gas chromatographic procedure for the determination of clonazepam levels in plasma using a nitrogen-sensitive detector, *J. Chromatogr.*, 222, 203, 1981.
3. **Badcock, N. R. and Pollard, A. C.,** Micro-determination of clonazepam in plasma or serum by electron-capture gas-liquid chromatography, *J. Chromatogr.*, 230, 353, 1982.
4. **Loscher, W. and Al-Tahan, F. J. O.,** Rapid gas chromatographic assay of underivatized clonazepam in plasma, *Ther. Drug Monit.*, 5, 229, 1983.
5. **Bouquet, S., Aucouturier, P., Brisson, A. M., Courtois, P., and Fourtillan, J. B.,** High-performance liquid chromatographic determination in human plasma of an anticonvulsant benzodiazepine: clonazepam.
6. **Rammel, R. P. and Elmer, G. W.,** Separation of clonazepam and five metabolites by reverse phase HPLC and quantitation from rat liver microsomal incubations, *J. Liq. Chromatogr.*, 6, 585, 1983.
7. **Shaw, W., Long, G., and McHan, J.,** An HPLC method for analysis of clonazepam in serum, *J. Anal. Toxicol.*, 7, 119, 1983.
8. **Taylor, E. H., Sloniewsky, D., and Gadsden, R. H.,** Automated extraction and high-performance liquid chromatographic determination of serum clonazepam, *Ther. Drug Monit.*, 6, 474, 1984.
9. **Kabra, P. M. and Nzekwe, E. U.,** Liquid chromatographic analysis of clonazepam in human serum with solid-phase (Bond-Elut) extraction, *J. Chromatogr.*, 341, 383, 1985.

CLONIDINE

Gas Chromatography

Specimen (mℓ)	Extraction	Column (m × mm)	Packing (mesh)	Oven temp (°C)	Gas (mℓ/min)	Det.	RT (min)	Internal standard (RT)	Deriv.	Other compounds (RT)	Ref.
Plasma (4)	I-1	1.8 × 2	3% OV-17 Chromosorb W (100/120)	175	N_2 (25)	ECD	5	4-Methylclonidine (6.5)	Heptafluorobutyryl	—	1

Serum (5)	I-2	25 × 0.25[a]	OV-17	230	N_2[b]	ECD	2.0	2-(2,4-Dichlorophenylamino)-20imidazoline	Pentafluorobenzyl (28)	—	2
Plasma (4)	I-3	1.8 × 2	3% OV-17 Gas Chrom Q (100/120)	260	He (15)	MS-CI[c]	4.8	[$^2H_4$]Clonidine	3,5-*bis*-(Trifluoromethyl) benzoyl	—	3

[a] A number of alternative columns were used.
[b] Linear velocity = 35 cm/sec.
[c] Ammonia as a reagent gas (ion source pressure = 0.3 torr).

Extraction — I-1. The sample was spiked with the internal standard (4 ng), made alkaline with 2 mℓ of 1 *M* carbonate buffer (pH 9.75), and extracted with 12 mℓ of 10% ethyl acetate in dichloromethane. The organic layer was back extracted into 3 mℓ of 0.1 *N* sulfuric acid. The aqueous phase was adjusted to pH 10 with a saturated solution of sodium carbonate and extracted with 3 mℓ of benzene. The benzene extract was evaporated at 40°C under a stream of nitrogen. The residue was dissolved in 100 μℓ of ethyl acetate and 4 μℓ of heptafluorobutyric anhydride were added. The mixture was incubated at 45°C for 15 min and then evaporated under a stream of nitrogen. The residue was dissolved in 400 μℓ of hexane. The hexane solution was washed with 1 mℓ of 0.1 *M* carbonate buffer, pH 9.2 and applied to a prewashed (twice with benzene, twice with hexane) minicolumn packed with silica. After the sample has passed through, the column was washed with 1 to 2 mℓ of hexane. Finally, the column was eluted with 2 mℓ of 25% benzene in hexane. The eluate was evaporated at 40°C under nitrogen, the residue dissolved in 25 to 50 μℓof 20% ethyl acetate in hexane. Aliquots of 2 to 5 μℓ of this solution were injected.
I-2. The sample was mixed with 0.2 mℓ of an aqueous solution of the internal standard (50 ng/mℓ) and 15 mℓ of 0.02 *M* ammonia. The mixture was applied to an Extrelut column. After 10 min, the column was eluted with 30 mℓ of cyclohexane-butanol (9:1). The eluate was back extracted with 1 mℓ of 0.1 *M* sulfuric acid. The aqueous phase was washed with 3 mℓ of cyclohexane-butanol, made alkaline with 1 mℓ of 13 *M* ammonia, and extracted with 10 mℓ of cyclohexane-butanol. The organic phase was evaporated under nitrogen at 60°C and the residue treated with 1 mℓ of 1% pentafluorobenzylbromide solution in acetone and 5 to 25 mg of potassium carbonate. The mixture was refluxed for 45 min and evaporated under a stream of nitrogen. The residue was dissolved in 1 mℓ of 1 *M* sulfuric acid which was washed twice with 1-mℓ portions of heptane, made alkaline with 0.25 mℓ of 13 *M* ammonia, and extracted with 3 mℓ of cyclohexane-butanol. Finally, the organic layer was evaporated, the residue dissolved in 100 μℓ of ethyl acetate and 1 to 3 μℓ of this solution were injected.
I-3. To the sample were added 100 μℓ of a methanolic solution of the internal standard (20 μg/mℓ) and 1 mℓ of 1 *M* sodium carbonate. The mixture was extracted with 20 mℓ of ethyl acetate. The organic extract was back extracted into 1 mℓ of 0.1 *N* HCl. The aqueous layer was made alkaline with 1 mℓ of 1 *M* sodium carbonate and extracted with 10 mℓ of diethyl ether. The ether extract was evaporated under nitrogen. The residue was dissolved in 100 μℓ of ethyl acetate, treated with 5 μℓ of 3,5-*bis*-(trifluoromethyl)-benzoyl chloride and the mixture incubated at 60°C for 2 hr. The reaction mixture was evaporated under nitrogen, residue dissolved in 10 μℓ ethylacetate, and aliquots of 1 μℓ were injected.

REFERENCES

1. **Chu, L. C., Bayne, W. F., Tao, F. T., Schmitt, L. G., and Shaw, J. E.,** Determination of submicrogram quantities of clonidine in biological fluids, *J. Pharm. Sci.*, 68, 72, 1979.

CLONIDINE (continued)

2. **Edlund, P. O.,** Determination of clonidine in human plasma by glass capillary gas chromatography with electron-capture detection, *J. Chromatogr.*, 187, 161, 1980.
3. **Murray, S. and Davies, D. S.,** *Bis* (trifluoromethyl)aryl derivatives for drug analysis by gas chromatography electron capture negative ion chemical ionization mass spectrometry. Application to the measurement of low levels of clonidine in plasma, *Biomed. Mass Spectrom.*, 11, 435, 1984.

CLOPENTHIXOL

Liquid Chromatography

Specimen (mℓ)	Extraction	Column (cm × mm)	Packing (μm)	Elution	Flow (mℓ/min)	Det. (nm)	RT (min)	Internal standard (RT)	Other compounds (RT)	Ref.
Serum (2)	I-1	25 × 4.6	Spherisorb silica (5)	E-1	1.0	ABS (254)	*cis* (Z) = 6.5	Lu 9-215 (6)	*cis* (Z)-N-Dealkyl clopenthixol (9.5) *trans* (E)-N-Dealkyl clopenthixol (12.5)	1
Plasma (2)	I-1	25 × 4.6	Spherisorb silica (5)	E-2	1.0	ABS (254)	*cis* (Z) = 6.0 *trans* (E) = 7.5	Lu 9-215 (6.8)	*cis* (Z)-N-Dealkyl clopenthixol (7.2) *trans* (E)-N-Dealkyl clopenthixol (11.6) Clopenthixol sulfoxide (24.5)	2

Extraction — I-1. To the sample were added 25 μℓ of the internal standard solution (1 μg/mℓ in ethanol), 300 μℓ of ethanol, and 100 μℓ of 7 *N* sodium hydroxide. The mixture was extracted with 8 mℓ of hexane containing 0.1% isopropylamine. The hexane phase was back extracted into 2 mℓ of 0.1 *N* HCl. The aqueous phase was made alkaline with 200 μℓ of 7 *N* soidum hydroxide and re-extracted with 4 mℓ of hexane. The organic phase was evaporated at 30°C under a stream of air. The residue was dissolved in 100 μℓ of hexane containing 0.1% isopropylamine. A 7-μℓ aliquot of this solution was injected.

Elution — E-1. n-Heptane-2-propanol-ammonia-water (85:15:0.4:0.2).
E-2. n-Heptane-2-propanol-ammonia-water (85:15:0.4:0.2).

REFERENCES

1. **Aaes-Jorgensen, T.,** Specific high-performance liquid chromatographic method for estimation of the *cis* (Z)- and *trans* (E)-isomers of clopenthixol and a N-dealkyl metabolite, *J. Chromatogr.*, 183, 239, 1980.
2. **Viala, A,. Hou, H., Durand, A., Ba, B., Aaes-Jorgensen, T., and Jorgensen, A.,** Dosage du *cis*(Z)-clopenthixol et de la fluphenazine dans le sang total et le plasma par chromatographie en phase liquide A performance avec etalonnage interne, *J. Pharm. Belg.*, 38, 6, 299, 1983.

CLORAZEPATE

Gas Chromatography

Specimen (mℓ)	Extraction	Column (m × mm)	Packing (mesh)	Oven temp (°C)	Gas (mℓ/min)	Det.	Rt (min)	Internal standard (RT)	Deriv.	Other compounds (RT)	Ref.
Plasma (0.1—1)	I-1	1.2 × 4	3% OV-17 Gas Chrom Q (60/80)	240	Ar:90 Methane:10 (40)	ECD	—	Methyl-nitrazepam (6)	—	Desmethyl-diazepam[a] (4)	1

Liquid Chromatography

Specimen (mℓ)	Extraction	Column (cm × mm)	Packing (μm)	Elution	Flow (mℓ/min)	Det. (nm)	RT (min)	Internal standard (RT)	Other compounds (RT)	Ref.
Dosage	—	30 × 4	μ-Bondapak-C_{18} (10)	E-1	1.8	ABS (230)	4.6	2,6-Dimethylaniline (7.1)	Desmethyl-diazepam (14)	
Plasma (1)	I-2	25 × 4	LiChrosorb RP-18 (10)	E-2	1.5	(225)	2.1	Diazepam (9.2)	Desmethyl-diazepam (6.2)	3

[a] Clorazapate spontaneously decarboxylates to desmethyldiazepam.

Extraction — I-1. The sample was mixed with an equal volume of saturated potassium chloride solution and extracted with 1 mℓ of benzene containing 25 μg of the internal standard. Aliquots of 10 μℓ of the benzene extract were injected.

CLORAZEPATE (continued)

I-2. The sample was mixed with 0.5 mℓ of 2 *M* glycine buffer, pH 9 and extracted with 10 mℓ of hexane-ethyl acetate (70:30) containing 6 ng/mℓ of the internal standard. The organic layer was evaporated at 38°C under vacuum. The residue was dissolved in 100 μℓ of the mobile phase and 50 μℓ were injected.

Elution — E-1. Acetonitrile + 0.005 *M* tetrabutylammonium phosphate, pH 7.5 (300:700).
E-2. Acetonitrile-0.05 Acetate buffer, pH 5 (45:55).

REFERENCES

1. **Haidukewych, D., Rodin, E. A., and Davenport, R.,** Monitoring clorazepate dipotassium as desmethyldiazepam in plasma by electron-capture gas-liquid chromatography, *Clin. Chem.*, 26, 142, 1980.
2. **Elrod, L., Shada, D. M., and Taylor, V. E.,** High-perfromance liquid chromatographic analysis of clorazepate dipotassium and monopotassium in solid dosage forms, *J. Pharm. Sci.*, 70, 793, 1981.
3. **Colin, P., Sirois, G., and Lelorier, J.,** High-performance liquid chromatography determination of dipotassium clorazepate and its major metabolite nordiazepam in plasma, *J. Chromatogr.*, 273, 367, 1983.

CLOTIAZEPAM

Gas Chromatography

Specimen (mℓ)	Extraction	Column (m × mm)	Packing (mesh)	Oven temp (°C)	Gas (mℓ/min)	Det.	RT (min)	Internal standard (RT)	Deriv.	Other compounds (RT)	Ref.
Plasma (1)	I-1	1.8 × 2	3% SP-2250 Supelcoport (80/100)	260	Ar:95 Methane:5 (30)	ECD	6.4	Diazepam	—	Hydroxy-clotiazepam (7.1); (12.6)[a] Desmethyl-clotiazepam (8.7)	1

Thin-Layer Chromatography

Specimen (mℓ)	Extraction	Plate (Manufacturer)	Layer (mm)	Solvent	Post-separation treatment	Det. (nm)	Rf	Internal standard (Rf)	Other compounds (Rf)	Ref.
Plasma (1)	I-2	20 × 10 cm (Merck)	Silica gel 60 (HPTLC) (0.25)	S-1	—	Fl[b] (313,460)	0.38	—	—	2

[a] Possibly a decomposition product of hydroxyclotiazepam.
[b] For the determination of the drug in tabletes UV-reflectance (243 nm) was used.

Extraction — I-1. The sample was added to a tube containing the residue after evaporation of 0.5 mℓ of the internal standard solution (0.1 μg/mℓ in benzene) and extracted with 2 mℓ of benzene containing 1.5% isoamyl alcohol. The organic extract was evaporated at 40 to 50°C under reduced pressure. The residue was dissolved in 0.2 mℓ of toluene (containing 15% isoamyl alcohol), of which 6 μℓ was injected.
I-2. The sample was extracted twice with 2-mℓ portions of *n*-hexane-propanol-2 (98.5:1.5). The combined organic extracts were evaporated at 37°C under a stream of nitrogen. The residue was dissolved in 100 μℓ of chloroform, aliquots of which were spotted with an auto spotter.

Solvent — S-1. Toluene-methanol (90:10).

REFERENCES

1. **Arendt, R., Ochs, H. R., and Greenblatt, D. J.,** Electron capture GLC analysis of the thienodiazepine clotiazepam, *Arzneim. Forsch.*, 32, 453, 1982.
2. **Busch, V. M., Ritter, W., and Mohrle, H.,** Dunnschicht-densitometrische Bestimmung von Clotiazepam in Tabletten und in Blutplasma, *Arzneim. Forsch.*, 35, 547, 1985.

CLOVOXAMINE

Gas Chromatography

Specimen (mℓ)	Extraction	Column (m × mm)	Packing (mesh)	Oven temp (°C)	Gas (mℓ/min)	Det.	RT (min)	Internal standard (RT)	Deriv.	Other compounds (RT)	Ref.
Plasma (2)	I-1	1.8 × 4	5% SP-2100 Supelcoport (100/120)	200	Ar:95 Methane:5 (40)	ECD	3.2[a]	Fluvoxamine (3.2)	Hydrolysis	—	1

CLOVOXAMINE (continued)

Liquid Chromatography

Specimen (mℓ)	Extraction	Column (cm × mm)	Packing (μm)	Elution	Flow (mℓ/min)	Det. (nm)	RT (min)	Internal standard (RT)	Other compounds (RT)	Ref.
Plasma (1)	I-2	15 × 4.6	LiChrosorb RP-8 (7)[b]	E-1	1.0	Fl (380, 470)	17.5	—	—	2

[a] Retention time of ketone obtained after hydrolysis
[b] A precolumn (50 × 4.6 mm) packed with LiChroprep RP-2 (32 μ) was used.

Extraction — I-1. To the sample were added 40 μℓ of the internal standard solution (1 μg/mℓ in methanol) and 0.2 mℓ of saturated solution of sodium carbonate. The mixture was extracted 3 times with 6 mℓ portions of ethyl acetate. The combined organic extracts were back extracted with 2 mℓ of 0.74 *M* phosphoric acid and the aqueous phase was incubated at 90°C for 1 hr. The cooled reaction mixture was extracted with 100 μℓ of hexane, of which 1 to 3 μℓ were injected.
I-2. The sample was treated with 1 mℓ of 0.01 *M* phosphate buffer, pH 7 and 1 mℓ of flurescamine solution (1 mg/mℓ in acetone). After mixing and centrifugation, 200 μℓ of the supernatant were injected.

Elution — E-1. Methanol-0.01 *M* phosphate buffer, pH 7 (62:38)

REFERENCES

1. **Hurst, H. E., Jones, D. R., Jarboe, C. H., and deBree, H.,** Determination of clovoxamine concentration in human plasma by electron capture gas chromatography, *Clin. Chem.*, 27, 1210, 1981.
2. **DeJong, G. J.,** The use of a pre-column for the direct high-performance liquid chromatographic determination of the antiliquid chromatographic determination of the anti-depressants clovoxamine and fluvoxamine in plasma, *J. Chromatogr.*, 183, 203, 1980.

CLOXACILLIN

Liquid Chromatography

Specimen (mℓ)	Extraction	Column (cm × mm)	Packing (μm)	Elution	Flow (mℓ/min)	Det. (nm)	RT (min)	Internal standard (RT)	Other compounds (RT)	Ref.
Plasma (0.1)	I-1	10 × 8	Rad-Pak-C_{18} (10)	E-1	2.0	ABS (195)	3.5	5-p-Hydroxy-phenyl-5-phenyl-hydantoin (8.5)	Nafcillin (4.5)	1
Serum, urine (0.5)	I-2	10 × 4.6	Brownlee-RP-8 (10)	E-2	1.6	ABS (210)	5.3	Nafcillin (6.9)	—	2

Extraction — I-1. The sample was mixed with 20 μℓ of acetic acid and extracted with 3 mℓ of chloroform containing 0.48 μg of the internal standard. After centrifugation, 2 mℓ of the organic layer was evaporated under vacuum. The residue was dissolved in 100 μℓ of the mobile phase and 20 μℓ were injected. I-2. To the sample were added 50 μℓ of an aqueous solution of the internal standard (0.24 mg/mℓ) and 50 μℓ of 1 *M* sulfuric acid. The mixture was extracted with 2 mℓ of dichloromethane. The organic layer was back extracted into 1 mℓ of 0.04 *M* NaH_2PO_4, pH 6.8 and 10 μℓ of the aqueous phase was injected.

Elution — E-1. Acetonitrile-10 m*M* phosphate buffer, pH 7 (24:76)
E-2. Acetonitrile-40 m*M* phosphate buffer, pH 4.5 (6.2:20)

REFERENCES

1. **Soldin, S. J., Tesoro, A. M., and MacLeod, S. M.,** A rapid high-performance liquid chromatographic procedure for the analysis of cloxacillin and/or nafcillin in serum, *Ther. Drug Monit.*, 2, 417, 1980.
2. **Teare, F. W., Kwan, R. H., Spino, M., and MacLeod, S. M.,** High-pressure liquid chromatographic assay of cloxacillin in serum and urine, *J. Pharm. Sci.*, 71, 938, 1982.

COCAINE[a]

Gas Chromatography

Specimen (mℓ)	Extraction	Column (m × mm)	Packing (mesh)	Oven temp (°C)	Gas (mℓ/min)	Det.	RT (min)	Internal standard (RT)	Deriv.	Other compounds (RT)	Ref.
Plasma, urine (2)	I-1	1.2 × 2	3% OV-1 Gas Chrom Q (100/120)	205	Methane (20)	MS-CI	NA[b]	[$^2H_3$]-Cocaine + [$^2H_3$]-Benzoyl-ecgonine	*n*-Propyl	Norcocaine (0.93) Benzoylecgonine (1.57)[b]	2
Blood (2)	I-2	1.2 × 4	3% OV-1 Chromosorb-W (80/100)	205	He (30)	NPD	3.5	Propylben-zoylecgonine (5.5)	Ethyl	Benzoylecgonine (4)	3
Urine (0.5—2)	I-3	0.7 × 2	2% OV-101 Gas Chrom Q (80/100)	T.P.[c]	He (20)	MS-EI	5	Phencyclidine (3)	—	Ecgonine methyl	4
Blood, bile, urine (5)	I-4	1.8 × 2	3% OV-1 Gas Chrom Q	230	He (20)	MS-EI[d]	3.2	Proadifin[e] (4.5)	Ethyl	Benzoylecgonine (3.5)	5
Plasma (1)	I-5	1.8 × 2	3% OV-101 + 0.1% KOH Chromosorb W (100/120)	200	N_2 (30)	NPD	2.6	*m*-Tolylecgonine methyl ester (3.5)	—	—	6
Plasma, urine (1.5, 10)	I-6	2 × 3	3% OV-17 Gas Chrom Q (80/100)	270	He (40)	MS-CI	4.3	[$^2H_5$]-Cocaine + [$^2H_5$]-Ben-zoylecgo-nine	Propyl	Benzeylecgonine (5.5) Ecgonine methyl ester (2.5)[f]	7
Urine	I-7	12 × NA	DB-1	220	N_2 (2)	FID	—	Codeine (5)	Butyl	Benzoyolecgon-ine (5.5)	8

Liquid Chromatography

Specimen (mℓ)	Extraction	Column (cm × mm)	Packing (μm)	Elution	Flow (mℓ/min)	Det. (nm)	RT (min)	Internal standard (RT)	Other compounds (RT)	Ref.
Plasma, tissue (0.5)	I-8	30 × 4	μ-Bondapak-C_{18} (10)	E-1	2.0	ABS (235)	9.7	Lidocaine (4.2)	Benzoylocgonine (2.9) Norcocaine (11.1)	9
Plasma (1)	I-9	25 × 2.6	Perkin-ElmerODS-HC (5)[g]	E-2[b]	0.8	ABS (232)	6.2	Tetracaine (8.5)	—	10
Dosage	—	16 × 5	Hypersil-ODS (5)	E-3	2.0	ABS (230)	3.5	—	Benzoylecgonine (8)[i]	11
Urine (3)	I-10	15 × 3.9	Nova-Pak-C_{18} (5)	E-4	1.5	ABS (234)	—	—	Benzoylecgonine (7)	12

Thin-Layer Chromatography

Specimen (mℓ)	Extraction	Plate (Manufacturer)	Layer (mm)	Solvent	Post-separation treatment	Det. (nm)	Rf	Internal standard (Rf)	Other compounds (Rf)	Ref.
Urine (15)	I-11	20 × 20 cm (Merck)	Silica gel (0.25)	S-1	sp: Acidified iodoplatinate reagent	Visual	0.74[j]	—	Nicotine (0.67) Phencyclidine (0.94) Norpropoxyphene (0.47) Propoxyphen (0.88)	13
Urine (5)	I-12	10 × 20 cm (Schleicher & Schüll)	Silica gel	S-2	sp: i. 0.05 *M* Sulfuric acid ii. Iodoplatinate	Visual	0.70, 0.93[k]	—	Benzoylecgonine (0.51, 0.83)[k]	14

[a] Analytical methods for cocaine have been reviewed.[1]
[b] Retention times relative to that of cocaine (=1) are given.
[c] Initial temp = 140°C; Rate = 15°C; Final temp = 240°C.
[d] A NPD was also used.
[e] External standard, added just prior to injection.
[f] Different temperature conditions are required.

COCAINE[a] (continued)

[g] Protected by a Brownlee RP-18 guard column.
[h] Temp = 40°C.
[i] Capacity factors of a number of local anaesthetics used to adulterate street cocaine are given.
[j] Benzoylecgonine is converted to cocaine prior to extraction.
[k] Rf values in two different solvents A and B, respectively.

Extraction — I-1. The sample was mixed with 200 μℓ of an aqueous solution of cocaine-d_3 (1 μg/mℓ) and 2 mℓ of 50% dibasic potassium phosphate. The mixture was extracted with 200 μℓ of toluene-heptane-isoamyl alcohol (70:20:10). Aliquots of 5 to 10 μℓ of the organic phase were injected.
I-2. The sample was mixed with 1 mℓ of water, 100 μℓ of the internal standard solution (10 μg/mℓ), and 100 μℓ pH 10 ammonia buffer. The mixture was applied to a modified (containing only one fourth of the packing) JETUBE and allowed to stand for 1 min. The tube was eluted with 50 mℓ of dichloromethane-ethanol (9:1). The eluate was evaporated to dryness and the residue was treated with 200 μℓ of N,N-dimethylacetamide, 25 μℓ of 0.025 *M* tetrabutylammonium hydrogen sulfate in 0.2 *M* trimethylamilinium hydroxide and 25 μℓ of ethyl iodide. The mixture was incubated at 100°C for 8 min. After cooling, the reaction mixture was extracted with 3 mℓ of hexane after the addition of 0.5 mℓ of 0.1 *M* KOH. The hexane layer was back extracted into 1 mℓ of 0.5 *M* sulphuric acid. The aqueous layer was washed with 3 mℓ of ether. To the aqueous layer 400 μℓ of ammonia buffer was added and the solution extracted as quickly as possible with 3 mℓ of ether. The organic layer was evaporated, residue dissolved in 100 μℓ of ethanol and 1 to 2 μℓ of this solution were injected.
I-3. The sample was adjusted to pH 8.5 to 9 by the addition of sodium borate and spiked with the internal standard at a concentration of 10 μg/mℓ of urine. The sample was extracted with 20 mℓ of dichloromethane-isopropanol (3:1). The organic phase was evaporated at 50°C under a stream of air. The residue was dissolved in ethanol and aliquots of this solution were injected.
I-4. The sample was extracted with 25 mℓ of 20% ethanol/chloroform. The organic layer was dried (Na_2SO_4) and evaporated at 55°C under a stream of air. The residue was treated with 0.6 mℓ of ethanol/sulfuric acid (2:1) and the mixture incubated at 85°C for 10 min. After cooling the reaction mixture was washed twice with 10-mℓ portions of ethyl ether, was made alkaline with 2.5 mℓ of 3.6 *N* sodium carbonate and extracted with 0.2 mℓ of chloroform containing 1 μg/mℓ of the external standard. After centrifugation, aliquots of the organic extract were injected.
I-5. To the sample were added 100 μℓ of the internal standard (1 μg/mℓ in 0.01 *M* sulfuric acid), 0.5 mℓ of 1 *M* carbonate buffer, pH 9.5, and the mixture was extracted with 2 mℓ of toluene-*tert*-amyl alcohol (9:1). The organic layer was back extracted into 0.5 mℓ of 0.1 *M* sulfuric acid. The aqueous layer was made alkaline with 0.5 mℓ of the carbonate buffer and extracted with 0.5 mℓ of butyl acetate. Aliquots of the organic layer were injected with the autosampler.
I-6. To the sample were added 400 ng each of the internal standard, pH adjusted to 7 to 9, and water to make the volume 10 mℓ. The mixture was applied to an Extrelut column which was eluted with 40 mℓ of chloroform-isopropanol (9:1). The eluate was evaporated. The residue was derivatized with N,N-dimethylformamide-di-*n*-propyl acetal mixture, adjusted to pH 9 and extracted with toluene-heptane-isoamyl alcohol (76:20:4). The organic layer was evaporated under nitrogen. The residue dissolved in 100 μℓ of chloroform for injection. A separate extraction procedure was used for ecgonine methyl ester.
I-7. The sample was spiked with the internal standard, pH adjusted to 9.2 with 1 *M* phosphate buffer and extracted with chloroform-ethanol. The residue of the extract was butylated with butyliodide in tetrahexylammonium hydroxide/trimethylammonium hydroxide (*Anal. Chem.*, 48, 34, 1976; *Anal. Chem.*, 49, 1974, 1977).
I-8. The sample was spiked with 100 μℓ of the internal standard solution (100 μg/mℓ in water), adjusted to pH 9 with 0.4 mℓ of 0.2 *M* carbonate buffer, and extracted with 7 mℓ of chloroform-isopropanol (3:2). The organic phase was evaporated at 40°C under a stream of nitrogen and the residue was dissolved in 250 μℓ of water for injection.

I-9. The sample was spiked with 100 μℓ of the internal standard solution (10 μg/mℓ in 50% methanol), made alkaline with 50 μℓ of saturated sodium carbonate solution, and extracted with 5 mℓ of ether. The organic layer was back extracted into 2 mℓ of 0.1 *M* acetic acid. The aqueous layer was made alkaline with 60 μℓ of saturated carbonate solution and reextracted with hexane. The organic layer was back extracted into 2 mℓ of 0.1 *M* acetic acid. The aqueous layer was made alkaline with 60 μℓ of saturated carbonate solution and reextracted with hexane. The organic layer was evaporated under nitrogen at 40°C. The residue was dissolved in 100 μℓ of the mobile phase and 20-μℓ aliquots were injected.
I-10. The sample was mixed with 3 mℓ of 0.5 *M* ammonium sulfate adjusted with ammonia to pH 9.3 and passed through a prewashed Sep-pak-C_{18} cartridge (5 mℓ methanol, 5 mℓ water). The cartridge was then washed with 20 mℓ of a 2% acetonitrile solution in 10 m*M* sodium hydroxide and 10 mℓ of a 2% acetonitrile solution in 10 m*M* sodium dihydrogen phosphate buffer, pH 2.1. Finally, the cartridge was eluted with 5 mℓ of 15% acetonitrile in 10 m*M* sodium dihydrogen phosphate buffer, pH 6.1. The eluate was mixed with 3 mℓ of the 0.5 *M* ammonium buffer and treated in a second Sep-pak cartridge in the same way as in the first one. An aliquot of 200 μℓ of the eluate was injected.
I-11. The sample was treated with 2 mℓ of 2 M phosphate buffer, pH 5 and 0.2 mℓ of dimethyl sulfate. The mixture was heated on a 70°C water bath for 15 min when the dimethyl sulfate layer disappeared. The cooled reaction mixture was treated with 3 mℓ of tris buffer and extracted with 20 mℓ of chloroform-isopropanol (95:5). The organic layer was evaporated, the residue dissolved in 0.1 mℓ of methanol and spotted in a TLC plate.
I-12. The sample was adjusted to pH 9 with 1 *M* sodium hydroxide and extracted with 250 μℓ of chloroform-isopropanol (9:1). Aliquots of 50 μℓ were spotted on two TLC plates which were in 2 different solvents (A and B).

Elution — E-1. Acetonitrile-methanol-water (1:1:8) containing 1% acetic acid and 0.3 *M* EDTA.
E-2. Methanol 0.05 *M* phosphate buffer, pH 6.6 (75:25).
E-3.Methanol-water-1% phosphoric acid-*n*-hexylamine (30:70:100:1.4), pH 2.5.
E-4.Acetonitrile-10 m*M* phosphate buffer, pH 2.1 containing 1 m*M* dodecyl sulfate (34:66).

Solvent — S-1. Hexane-chloroform-diethylamine (80:10:10).
S-2. (A) Methanol-ammonia (100:1.5); (B) Ethyl acetate-methanol-ammonia (85:10:5)

REFERENCES

1. **Lindgren, J. E.,** Guide to the analysis of cocaine and its metabolites in biological material, *J. Ethnopharmacol.*, 3, 337, 1981.
2. **Chinn, D. M., Crouch, D. J., Peat, M. A., Finkle, B. S., and Jennison, T. A.,** Gas chromatography-chemical ionization mass spectrometry of cocaine and its metabolites in biological fluids, *J. Anal. Toxicol.*, 4, 37, 1980.
3. **McCurdy, H. H.,** Quantitation of cocaine and benzoylecognine after JETUBE extraction and derivitization, *J. Anal. Toxicol.*, 4, 82, 1980.
4. **Ambre, J. J., Ruo, T. I., Smith, G. L., Backes, D., and Smith, C. M**, Ecgonine methyl ester, a major metabolite of cocaine, *J. Anal. Toxicol.*, 6, 26, 1982.
5. **Griesemer, E. C., Liu, Y., Budd, R. D., Raftogianis, L., and Noguchi, T. T.,** The determination of cocaine and its major metabolite, benzoylecgonine, in postmortem fluids and tissues by computerized gas chromatography/mass spectrometry, *J. Forensic Sci.*, 28, 894, 1983.
6. **Jacob, P., Elias-Baker, B. A., Jones, R. T., and Enowitz, N. L.,** Determination of cocaine in plasma by automated gas chromatography, *J. Chromatogr.*, 306, 173, 1984.

COCAINE[a] (continued)

7. **Matsubara, K., Maseda, C., and Fukui, Y.,** Quantitation of cocaine, benzoylecgonine and ecgonine methyl ester by GC-CI-SIM after extrelut extraction, *Forensic Sci. Int.*, 26, 181, 1984.
8. **Falk, P. M and Harrison, B. C.,** Use of DB-1 capillary columns in the GC/FID analysis of benzoylecgonine, *J. Anal. Toxicol.*, 9, 273, 1985.
9. **Evans, M. A. and Morarity, T.,** Analysis of cocaine and cocaine metabolites by high pressure liquid chromatography, *J. Anal. Toxicol.*, 4, 19, 1980.
10. **Masoud, A. N. and Krupski, D. M.,** High-performance liquid chromatographic analysis of cocaine in human plasma, *J. Anal. Toxicol.*, 4, 305, 1980.
11. **Gill, R., Abbott, R. W., and Moffat, A. C.,** High-performance liquid chromatography with applicability to the analysis of illicit cocaine samples, *J. Chromatogr.*, 301, 155, 1984.
12. **Svensson, J. O.,** Determination of benzoylecgonine in urine from drug abusers using ion pair high performing liquid chromatography, *J. Anal. Toxicol.*, 10, 122, 1986.
13. **Budd, R. D., Mathis, D. F., and Yang, F. C.,** TLC analysis of urine for benzoylecgonine and norpropoxyphene, *Clin. Toxicol.*, 16, 1, 1980.
14. **Hsu, L. S. F., Sharrard, J. I., Love, C., and Marrs, T. C.,** A rapid method for screening urine samples in suspected abuse of cocaine, *Ann. Clin. Biochem.*, 18, 368, 1981.

CODEINE

Gas Chromatography

Specimen (mℓ)	Extraction	Column (m × mm)	Packing (mesh)	Oven temp (°C)	Gas (mℓ/min)	Det.	RT (min)	Internal standard (RT)	Deriv.	Other Compounds (RT)	Ref.
Plasma (2)	I-1	0.9 × 2	1% OV-17 Chromosorb W (100/120)	235	He (30)	NPD	2.3	Oxycodone (3.8)	—	—	1
Blood, urine (3)	I-2	25 × 0.32	CP-Sil 8 (0.61 μm)[a]	T.P.[b]	He[c]	NPD	13.3	Hydro-codone (14.4)	—	Ethylmorphine (14.4)	2
Urine (10)	I-3	15 × 0.25	DB-5	240	He (1.6)	mS-EI	3.3	Nalorphine	Acetyl (6)	Morphine (4.4)	3

Liquid Chromatography

Specimen (mℓ)	Extraction	Column (m × mm)	Packing (μm)	Elution	Flow (mℓ/min)	Det. (nm)	RT (min)	Internal standard (RT)	Other Compounds (RT)	Ref.
Plasma (2)	I-4	30 × 3.9	μ-Bondapak-C_{18} (10)	E-1	2.0	Fl (213,320)	4	N-Isopropyl-codeine (7)	Morphine (2.2) Norcodeine (3.3)	4
Plasma (1)	I-5	30 × 3.9	μ-Porasil (10)[c]	E-2	1.5	ABS (254)	6.3	Methadone (7.2)	—	5
Blood, serum (2)	I-6	25 × 4.2	Spherisorb-CN (5)[d]	E-3	2.0	ABS (210)	8.3	Nalorphine (4.4)	Normorphine (4.8) Norcodeine (5.9) Morphine (7) Heroin (7.3) Hydromorphone (10.4) Hydrocodone (12.1)	6
Plasma (0.8)	I-7	25 × 4	Polygosil-C_{18} (7.5)[e]	E-4	2.0	ABS (220)	3.1	Diazepam (2.6)	Norcodeine (1.6) Morphine (1.7)	7
Plasma (2)	I-8	30 × 4	Micropak-CN (10)	E-5	1.8	Fl (213)[f]	4	N-Isopropyl-codeine (6)	Ibuprofen (8)	8
Plasma, urine (0.5, 0.1)	I-9	5 × 4.6	Speralyte-C_{18} (3)[g]	E-6	1.5	Fl (220,355)	1.5	N-Allylnor-codeine (2.8)	—	9

[a] Film thickness.
[b] Head pressure = 0.8 bar (11.6 psi).
[c] Protected by a 10 × 10.21 cm precolumn packed with Vydac 101 SC.
[d] Column temp = 40°C.
[e] Column temp = 45°C.
[f] No emission filter was used.
[g] Column temp = 50°C.

Extraction — I-1. The sample was made alkaline with 1 mℓ of 1 *N* sodium hydroxide and extracted with 10 mℓ of chloroform-butanol (9:1) containing 30 ng of the internal standard. The organic layer was back extracted into 5.5 mℓ of 0.1 *N* sulphuric acid. The aqueous layer was made alkaline and extracted with chloroform-butanol mixture. The organic layer was evaporated at 40°C under nitrogen. The residue was reconstituted in 20 μℓ of methanol and 5 μℓ of the solution was injected.
I-2. The sample was spiked with 12 μg of the internal standard solution (in methanol) and applied to a ClinElut column. The column was eluted with two 6-mℓ aliquots of chloroform. The eluate was evaporated under reduced pressure. The residue dissolved in 100 μℓ of methanol and 0.1 μℓ of this solution was injected.

CODEINE (continued)

I-3. To the sample were added 250 μℓ of the internal standard solution (40 μg/mℓ) in methanol and 1 mℓ of concentrated hydrochloric acid. The mixture was hydrolyzed at 121°C for 15 min. The cooled mixture was made alkaline with 1 mℓ of 12 *N* sodium hydroxide and 1.5 mℓ of 7.3 *M* ammonium chloride. The pH of the mixture was adjusted to 9 to 9.3 and extracted with 10 mℓ of dichloromethane-isobutanol (9:1). The organic layer was washed with 5 mℓ of 0.05 *M* phosphate buffer, pH 9, and then extracted with 3 mℓ of 0.2 *M* acetate buffer, pH 1. The aqueous layer was made alkaline with a solution of sodium carbonate (3.5 mℓ, 0.75 *M*, pH 9) and extracted with 5 mℓ of dichloromethane-isobutanol. The organic layer was evaporated at 50°C under a stream of nitrogen. The residue was treated with 0.2 mℓ of acetic anhydride and 0.2 mℓ of dry pyridine and the mixture incubated at 50°C for 15 min. The excess reagents were removed at 50°C under a stream of nitrogen. The residue was dissolved in 50 μℓ of chloroform for injection.
I-4. The sample was spiked with 200 μℓ of the internal standard solution (1 μg/mℓ in 50% methanol), made alkaline with 2 mℓ of 50 m*M* phosphate buffer solution, pH 8, and extracted twice with 6-mℓ portions of hexane-dichloromethane (2:1). The combined organic extracts were washed with 1 mℓ of 50 m*M* sodium hydroxide and evaporated under a stream of nitrogen at 50°C. The residue was dissolved in 200 μℓ of the mobile phase and an aliquot of 50 μℓ was injected.
I-5. The sample was mixed with 1 mℓ of 10% ammonia and extracted with 4 mℓ of cichloromethane containing 80 ng/mℓ of the internal standard. The organic layer was reduced to 200 to 300 μℓ and the entire residue was injected.
I-6. To the sample were added, 50 μℓ of the internal standard solution (50 mg/ℓ in methanol), 50 μℓ of 3.5 *M* sodium hydroxide and 2 mℓ of 40% K_2HPO_4, pH 9.2. The mixture was extracted with 10 mℓ of chloroform-isopropanol-heptane (50:17:33). The organic layer was back extracted into 5 mℓ of 0.2 *M* HCl. The aqueous layer was washed with 2 mℓ of heptane, buffered with 2 mℓ of 40% K_2HPO_4 and 0.5 mℓ of concentrated ammonium hydroxide, and extracted with 8 mℓ of chloroform. The organic layer was evaporated at 56°C under a stream of air. The residue was dissolved in 100 μℓ of the mobile phase and 10- to 15-μℓ aliquots were injected.
I-7. The sample was mixed with 0.3 mℓ of 0.1 *M* borate buffer, pH 8.9 and applied to a Baker 1-mℓ C_{18} extraction column. The column was washed with 1 mℓ of 0.1 *M* HCl, 1 mℓ of methanol-0.1 *M* ammonium hydroxide (20:80), and 0.3 mℓ of 0.1 *M* HCl. Finally, the column was eluted with 0.3 mℓ of methanol-0.1 *M* HCl. Ammonium hydroxide (30 μℓ of 1 *M*) and 5 μℓ of the internal standard solution (10 μg/mℓ) were added to the eluate prior to injection.
I-8. The sample was spiked with 20 μℓ of an aqueous solution of the internal standards (30 mg/ℓ). The mixture was made alkaline with 2 mℓ of 50 m*M* phosphate buffer (pH 8) and extracted twice with 6-mℓ portions of hexane-dichloromethane (2:1). The combined organic extracts were washed with 1 mℓ of 0.05 *M* sodium hydroxide for 2 min. The organic layer was evaporated under a gentle stream of nitrogen at 30°C. (Ibuprofen and its internal standard were extracted with hexane at acidic pH and the extract combined with the residue of codeine extract). The residue was dissolved in 100 μℓ of acetonitrile-water (1:1) and 20 μℓ of this solution was subjected to chromatography.
I-9. The sample was spiked with 50 μℓ of the internal standard solution (1 mg/mℓ in 0.1 *M* phosphoric acid) and mixed with 500 μℓ of 0.1 *M* sodium carbonate. The mixture was applied to a prewashed (methanol, 1 mℓ; 0.1 *M* sodium carbonate, 1 mℓ) 1-mℓ BondElut C-2 column. The column was washed with 0.1 *M* sodium carbonate (1 mℓ) and a 30:70 mixture of methanol and 0.1 *M* sodium carbonate. The column was then eluted with 200 μℓ of methanol-sodium carbonate (80:20). Perchloric acid (0.2 *M*, 400 μℓ) was added to each tube and an aliquot of 50 μℓ was injected.

Elution — E-1. Methanol-water (21:79) containing 1.5 g of phosphoric acid.
E-2. Dichloromethane-methanol-33% ammonia (90:10:0.1).
E-3. Methanol-0.1 *M* phosphate buffer, pH 6.8.
E-4. Methanol-0.1 *M* ammonium carbonate (70:30).
E-5. Acetonitrile-0.1 *M* KH_2PO_4 + 0.05 *M* sodium octanesulfonate, pH 3.7 (22:78).
E-6. Methanol-0.2 *M* sodium perchlorate + 0.1 *M* phosphosphoric acid (16:84).

REFERENCES

1. **Renzi, N. L., Jr., Stellar, S. M., and Ng, K. T.,** Gas chromatographic assay of codeine in human plasma utilizing nitrogen-selective detection, *J. Chromatogr.*, 278, 179, 1983.
2. **Demedts, P., De Waele, M., Van der Verren, J., and Heyndrickx, A.,** Application of the combined use of fused silica capillary columns and NPD for the toxicological determination of codeine and ethylmorphine in a human overdose case, *J. Anal. Toxicol.*, 7, 113, 1983.
3. **Paul, B. D., Mell, L. D., Jr., Mitchell, J. M., Irving, J., and Novak, A. J.,** Simultaneous identification and quantitation of codeine and morphine in urine by capillary gas chromatography and mass spectroscopy, *J. Anal. Toxicol.*, 9, 222, 1985.
4. **Tsina, I. W., Fass, M., Debban, J. A., and Matin, S. B.,** Liquid chromatography of codeine in plasma with fluorescence detection, *Clin. Chem.*, 28, 1137, 1982.
5. **Visser, J., Grasmeijer, G., and Moolenaar, F.,** Determination of therapeutic concentrations of codeine by high-performance liquid chromatography, *J. Chromatogr.*, 274, 372, 1983.
6. **Posey, B. L. and Kimble, S. N.,** Simultaneous determination of codeine and morphine in urine and blood by HPLC, *J. Anal. Toxicol.*, 7, 241, 1983.
7. **Nitsche, V. and Mascher, H.,** Determination of codeine in human plasma by reverse-phase high-performance liquid chromatography, *J. Pharm. Sci.*, 73, 1556, 1984.
8. **Ginman, R., Karnes, H. T., and Perrin, J.,** Simultaneous determination of codeine and ibuprofen in plasma by high-performance liquid chromatography, *J. Pharm. Biomed. Anal.*, 3, 439, 1985.
9. **Stubbs, R. J., Chiou, R., and Bayne, W. F.,** Determination of codeine in plasma and urine by reversed-phase high-performance liquid chromatography, *J. Chromatogr.*, 377, 447, 1986.

COLCHICINE

Liquid Chromatography

Specimen (mℓ)	Extraction	Column (cm × mm)	Packing (μm)	Elution	Flow (mℓ/min)	Det. (nm)	RT (min)	Internal standard (RT)	Other compounds (RT)	Ref.
Pure compounds	—	30 × 3.9	μ-Bondapak-C_{18} (10)	E-1[a]	2.0	ABS (350)	24.4	—	N-Desacetylcolchicine (14.4) Demecolcine (20.3) 3-Demethylcolchicine (8.7)	1
Blood (2)	I-1	30 × 3.9	μ-Bondapak-C_{18} (10)[b]	E-2	1.5	ABS (254)	4.5	Quinidine (6.8)	—	2

COLCHICINE (continued)

Liquid Chromatography

Specimen (mℓ)	Extraction	Column (cm × mm)	Packing (μm)	Elution	Flow (mℓ/min)	Det. (nm)	RT (min)	Internal standard (RT)	Other compounds (RT)	Ref.
Serum (1)	I-2	NA	LiChrosorb C_8 (7)	E-3	1.5	ABS (350)	4.2	—	—	3
Plasma, urine (1)	I-3	30 × 4	MicroPak MCH (10)	E-4	2.0	ABS (245)	3.5	Morpholino-propylcolchic-amide (4.7)	—	4

[a] A number of alternative mobile phases are described.
[b] Column temp = 40°C.

Extraction — I-1. To the sample were added 0.5 mℓ of 1 *M* sodium bicarbonate solution and 0.2 mℓ of the internal standard solution (10 mg/ℓ in water). The mixture was extracted with 30 mℓ of dichloromethane. The organic extract was evaporated at room temperature in the dark. The residue was dissolved in 200 μℓ of 0.1 *N* HCl.
I-2. The sample was adjusted to pH 9.5 with borate buffer and applied to an Extrelut column. The column was eluted with dichloromethane-isopropanol (85:15). The organic phase was evaporated and the residue dissolved in 200 μℓ of methanol for injection.
I-3. The sample was spiked with 100 μℓ of the internal standard solution (1 μg/mℓ in methanol) and made alkaline with 1 mℓ of 8 *M* ammonium hydroxide. The mixture was extracted twice with 15-mℓ portions of dichloromethane. The combined organic layers were mixed with 10 mℓ of ethanol and centrifuged. The supernatant was evaporated under nitrogen at 50°C. The residue was dissolved in 100 μℓ of the mobile phase, and 50 μℓ were injected.

Elution — E-1. Acetonitrile-methanol-0.05 *M* phosphate buffer, pH 6 (16:5:79).
E-2. Methanol-acetonitrile-0.1 *M* phosphate buffer, pH 7.6 (41:15:44) containing 5 m*M* pentanesulfonic aicd, final pH 6.45 with acetic acid.
E-3. Methanol-20 m*M* phosphate buffer, pH 3 (60:40).
E-4. Acetonitrile-water (50:50).

REFERENCES

1. **Davis, P. J. and Klein, A. E.,** High-performance liquid chromatographic separation of colchicine and its phenolic and N-desacetylated derivatives, *J. Chromatogr.*, 188, 280, 1980.
2. **Caplan, Y. H., Orloff, K. G., and Thompson, B. C.,** A fatal overdose with colchicine, *J. Anal. Toxicol.*, 4, 153, 1980.

3. **Harzer, K.,** Todliche Vergiftung mit Colchicin, *Z. Rechtsmed,* 93, 181, 1984.
4. **Lhermitte, M., Bernier, J. L., Mathieu, D., Mathieu-Nolf, M., Erb, F., and Roussel, P.,** Colchicine quantitation by high-performance liquid chromatography in human plasma and urine, *J. Chromatogr.*, 342, 416, 1985.

COLTEROL

Liquid Chromatography

Specimen (mℓ)	Extraction	Column (cm × mm)	Packing (μm)	Elution	Flow (mℓ/min)	Det. (nm)	RT (min)	Internal Standard (RT)	Other compounds (RT)	Ref.
Plasma, urine (1)	I-1	25 × 4.6	Ultransphere-ODS (5)	E-1	1.0	Electrochem[a]	8.5	1-(3,4-Dihydroxy phenyl)-2-cyclopentyl-aminoethanol (13.5)	—	1

[a] Potential = +0.6 V.

Extraction — I-1. To the plasma sample were added 100 μℓ of the internal standard solution (50 ng/mℓ in 0.05 *N* sulfuric acid), 2 mℓ of 0.2 *M* phosphate buffer (pH 6.9), and 10 mℓ of 1.5% di-(2-ethylhexyl)phosphoric acid in benzene. After mixing and centrifugation, the organic layer was back extracted into 130 μℓ of 0.2 *N* sulfuric acid. A 100-μℓ aliquot of the aqueous layer was injected. Urine after the addition of the internal standard was extracted with a cation exchange column (Bio-Rex 70). The eluate obtained wtih 2% boric acid solution was extracted as for plasma.

Elution — E-1. Methanol-0.1 *M* sodium sulfate + 100 mg/ℓ EDTA, pH 3 (12:88).

REFERENCE

1. **Park, G. B., Koss, R. F., Utter, J., and Edelson, J.,** Determination of colterol in human plasma and urine by reversed-phase chromatography with amperometric detection, *J. Chromatogr.*, 273, 481, 1983.

COPOVITHANE

Liquid Chromatography

Specimen (mℓ)	Extraction	Column (cm × mm)	Packing (μm)	Elution	Flow (mℓ/min)	Det. (nm)	RT (min)	Internal standard (RT)	Other compounds (RT)	Ref.
Plasma, urine (2)	I-1	25 × 4	LiChrosorb RP-18 (10)	E-1	2.0	ABS (340)	4.5	—	—	1
Plasma (2)	I-2	30 × 3.1	μ-Bondapak-C_{18} (10)	E-2	2.0	ABS (340)	16	—	—	2

Extraction — I-1. The sample was treated with 8 mℓ of acetone over a period of 2 min with constant mixing. After centrifugation, the supernatant was shaken with 12 mℓ of dichloromethane. The aqueous phase (1 mℓ) was diluted with 1 mℓ of water and applied to a prewashed (20 mℓ water, 20 mℓ methanol) Sep-Pak C_{18} cartridge. The cartridge was washed with 2 mℓ of water and eluted with 10 mℓ of 80% methanol. The eluate was evaporated at 70°C under a stream of nitrogen. The residue was dissolved in 1 mℓ of 5 *N* HCl and heated at 160°C for 16 hr. The cooled hydrolysate was adjusted to pH 8 with 0.5 mℓ of 10 *N* sodium and 2 mℓ of standard buffer, pH 8, and treated with 0.75 mℓ of 0.5% 2,4,6-trinitrobenzene sulfonic acid solution. The mixture was allowed to stand at room temperature for 2.5 hr in complete darkness and then extracted wtih 6 mℓ of toluene. An aliquot of 3 mℓ of toluene extract was evaporated at 50°C under nitrogen. The residue was taken up in 250 μℓ of the mobile phase for injection.
I-2. The sample was treated with 200 μℓ of 10 *N* perchloric acid. The supernatant was treated with 200 μℓ of 10 *N* KOH, cooled, and centrifuged. The supernatant was treated with 2 mℓ of hot (85°C) saturated sodium chloride solution and extracted three times with 3-mℓ portions of chloroform. The combined chloroform extracts were evaporated under a stream of nitrogen. The residue was treated wtih 1 mℓ of 5 *N* HCl and heated at 160°C for 16 hr. After cooling, the pH of the reaction mixture was adjusted to 8 with 1 mℓ of 5 *N* NaOH and 3 mℓ of 1 *M* sodium bicarbonate buffer, treated wtih 0.75 mℓ of a 0.5% solution of trinitrobenzene sulphonic acid in acetone, allowed to stand in the dark for 150 min, and extracted three times with 3-mℓ portions of ethyl acetate. The combined extracts were evaporated under nitrogen and the residue reconstituted in 250 μℓ of 0.2 M Na_2HPO_4 in acetonitrile for injection.

Elution — E-1. Acetonitrile-water (45:55).
E-2. Acetonitrile-water (30:70).

REFERENCES

1. **Wingender, W.,** High-performance liquid chromatographic method for the quantitative analysis of a synthetic copolymer with antitumor activity (copovithane) and methylamine in human blood plasma and urine, *J. Chromatogr.*, 273, 319, 1983.
2. **Rosenblum, M. G., Hortobagyi, G. N., Wingender, W., and Hersh, E. M.,** Analysis of the antitumor agent bay i 7433 (copovithane) in plasma and urine by high performance liquid chromatography, *J. Liq. Chromatogr.*, 7, 159, 1984.

COUMERMYCIN A_1

Liquid Chromatography

Specimen (mℓ)	Extraction	Column (cm × mm)	Packing (μm)	Elution	Flow (mℓ/min)	Det. (nm)	RT (min)	Internal standard (RT)	Other compounds (RT)	Ref.
Plasma (0.1)	I-1	10 × 8	Radial-Pak-C_{18} (5)	E-1	1.8	ABS (330)	7	Novobiocin (3)	a	1

[a] Separation of homologues designated as B, C, and D is shown.

Extraction — I-1. The sample was mixed with the residue after evaporation of 100 μℓ of the internal standard solution (25 μg/mℓ in ethanol), and 1 mℓ of 1 *M* phosphate buffer, pH 6.5. The mixture was extracted twice with 4.5-mℓ portions of methyl-*tert*-butyl ether-2-propanol (97.5:2.5). The combined organic extracts were evaporated at 20 to 25°C under a stream of nitrogen. The residue was dissolved in 2.5 mℓ of the mobile phase and aliquots of 50 μℓ were injected with an autosampler.

Elution — E-1. Methanol-2-methoxyethanol-water (800:50:150) containing 4.33 g sodium laurylsulfate, 2 mℓ of 1 *M* orthophosphoric acid, pH 2.8.

REFERENCE

1. **Strojny, N., Conzentino, P., and de Silva, J. A. F.,** Determination of coumermycin A_1 in plasma by reversed-phase high-performance liquid chromatographic analysis, *J. Chromatogr.*, 342, 145, 1985.

CROMOGLYCATE SODIUM

Liquid Chromatography

Specimen (mℓ)	Extraction	Column (cm × mm)	Packing (μm)	Elution	Flow (mℓ/min)	Det. (nm)	RT (min)	Internal standard (RT)	Other compounds (RT)	Ref.
Urine (10)	E-1	25 × 4.6	Partisil SAX (10)	E-1	3.6	ABS (325)	4.5	—	—	1

CROMOGLYCATE SODIUM (continued)

Extraction — I-1. To the sample were added 5 g of sodium chloride, 1 mℓ of water, and 1 mℓ of concentrated hydrochloric acid. The mixture was extracted twice with 10-mℓ portions of diethyl ether. The combined extracts were back extracted into 1 mℓ of 1 *M* glycine-HCl buffer (pH 3.5). Aliquots of the aqueous phase were injected.

Elution — E-1. 0.9 *M* Phosphate buffer, pH 2.3

REFERENCE

1. **Gardner, J. J.,** Determination of sodium cromoglycate in human urine by high-performance liquid chromatography on an anion-exchange column, *J. Chromatogr.*, 305, 228, 1984.

CYANAMIDE

Gas Chromatography

Specimen (mℓ)	Extraction	Column (m × mm)	Packing (mesh)	Oven temp (°C)	Gas (mℓ/min)	Det.	RT (min)	Internal standard (RT)	Deriv.	Other compounds (RT)	Ref.
Plasma (1)	I-1	1.8 × 2	3% OV-1 Chromosorb W (80/100)	185	Argon: 95 Methane:5 (20)	ECD	1.8	—	Heptafluorobutyryl	—	1

Liquid Chromatography

Specimen (mℓ)	Extraction	Column (cm × mm)	Packing (μm)	Elution	Flow (mℓ/min)	Det. (nm)	RT (min)	Internal standard (RT)	Other compounds (RT)	Ref.
Plasma (0.5)	I-2	10 × 8	Rad-Pak-C_{18} (10)[a]	E-1; grad	4	Fl (360, 495)	5.2	—	—	2

[a] Protected by a guard column packed with μ-Bondapak-C_{18}.

Extraction — I-1. The sample was treated with 0.1 *M* sodium hydroxide to adjust the pH to 10, saturated with 0.5 g of sodium chloride, and extracted twice with 2-mℓ portions of ethyl acetate. The combined organic extract was evaporated with a stream of nitrogen at 50°C. The residue was dissolved in 100 mℓ of acetonitrile and treated with 5 mℓ of heptafluorobutyric anhydride. The mixture was allowed to stand at room temp for 30 min, then evaporated to dryness at 45°C with a stream of nitrogen. The residue was redissolved in 100 mℓ of benzene and a 1-mℓ aliquot was injected.
I-2. The sample was extracted twice with 2-mℓ aliquots of ethyl acetate. The combined extracts were evaporated under a stream of nitrogen at 40°C. The residue was dissolved in 100 mℓ of 0.2 *M* carbonate buffer (pH 9), the solution treated with 100 μℓ of dansyl chloride solution (1 mg/mℓ in acetone), and the mixture incubated at 40°C for 1 hr. Aliquots of 50 mℓ of this solution were injected.

Elution — E-1. (A) 10 m*M* potassium phosphate, pH 7; (B) Acetonitrile-10 m*M* potassium phosphate, pH 7 (55:45). Gradient: 0 min; 70% (A); 7 min, 40% (A) (Curve 6); 10 min, 0% (A) (Curve 1); 12 min, 70% (A) (Curve 1).

REFERENCES

1. **Loomis, C. W. and Brien, J. F.,** Determination of carbimide in plasma by gas-liquid chromatography, *J. Chromatogr.*, 222, 421, 1981.
2. **Prunonosa, J., Obach, R., and Valles, J. M.,** Determination of cyanamide in plasma by high-performance liquid chromatography, *J. Chromatogr.*, 377, 253, 1986.

CYCLANDELATE

Gas Chromatography

Specimen (mℓ)	Extraction	Column (m × mm)	Packing (mesh)	Oven temp (°C)	Gas (mℓ/min)	Det.	RT (min)	Internal standard (RT)	Deriv.	Other compounds (RT)	Ref.
Plasma (1)	I-1	25 × 0.3	SE-30	T.P.[a]	N_2 (2)	FID	31.3	Ethylmandelate (7.1)	Trimethylsilyl	Mandelic acid (8.2)	1

[a] Initial temp = 125°C; initial time = 13 min; rate = 3°C/min; final temp = 180°C; final time = 1 min.

Extraction — I-1. The sample was spiked with 20 μℓ of the internal standard solution (20 μg/mℓ in ethyl acetate) and extracted with 10 mℓ of diethyl ether. The organic phase was dried over anhydrous sodium sulfate. The aqueous layer was made acidic with 30% HCl and re-extracted wtih 10 mℓ of ether. The extracts were combined and evaporated at 20 to 30°C under vacuum. The residue was treated with 50 μℓ of pyridine and 25 μℓ of N-*bis*(trimethylsilyl)fluoroacetamide-trimethylchlorosilane (99:1). The mixture was incubated at 60°C for 5 min. Aliquots of 1 μℓ of this solution were injected.

CYCLANDELATE (continued)

REFERENCE

1. **Andermann, G. and Dietz, M.,** Simultaneous determination of cyclandelate and its metabolite in human plasma by capillary column gas-liquid chromatography, *J. Chromatogr.*, 223, 365, 1981.

CYCLIZINE

Gas Chromatography

Specimen (mℓ)	Extraction	Column (m × mm)	Packing (mesh)	Oven temp (°C)	Gas (mℓ/min)	Det.	RT (min)	Internal standard (RT)	Deriv.	Other compounds (RT)	Ref.
Plasma (1)	I-1	1.8 × 4	5% OV-17 Chromosorb W (100/120)	246	He (50)	NPD	3.1	Chlorcyclizine (5.5)		Norcyclizine (4)	1

Extraction — I-1. To the sample were added 50 μℓ of the internal standard solution (10 μg/mℓ in water) and 2 mℓ of 2 *M* sodium hydroxide. The mixture was extracted with 10 mℓ of cyclohexane. The organic layer was back extracted into 2 mℓ of 2 *M* HCl. The aqueous layer was made alkaline with 2 mℓ of 4 *M* sodium hydroxide and re-extracted twice with 4-mℓ portions of cyclohexane. The combined extracts were evaporated in the presence of 20 μℓ of Dorotherm A at room temp under a stream of nitrogen. Aliquots of 4 mℓ of the resulting solution were injected.

REFERENCE

1. **Land, G., Dean, K., and Bye, A.,** Determination of cyclizine and norcyclizine in plasma and urine using gas-liquid chromatography with nitrogen selective detection, *J. Chromatogr.*, 222, 135, 1981.

CYCLOBENZAPRINE

Gas Chromatography

Specimen (mℓ)	Extraction	Column (m × mm)	Packing (mesh)	Oven temp (°C)	Gas (mℓ/min)	Det.	RT (min)	Internal standard (RT)	Deriv.	Other compounds (RT)	Ref.
Plasma (1)	I-1	40 × NA	DB-5 (0.25 μm)[a]	T.P.[b]	He[c]	NPD	11	5-(2-Dimethylaminoethylidene) dibenzo [a,e] cyclohepatriene (9.5)	—	—	1

Liquid Chromatography

Specimen (mℓ)	Extraction	Column (cm × mm)	Packing (μm)	Elution	Flow (mℓ/min)	Det. (nm)	RT (min)	Internal standard (RT)	Other compounds (RT)	Ref.
Dosage	—	25 × 4.6	Zorbax-C_8 (10)	E-1	1.5	ABS (254)	7.5	Naphazoline (5)	—	2

[a] Film thickness.
[b] Initial temp = 140°C; rate = 50°C/min to 230°C; 4°C/min to 245°C; 50°C/min to 300°C; final time = 3 min.
[c] Column head pressure = 30 psi.

Extraction — I-1. To the sample were added 100 μℓ (10 ng) of the internal standard solution and 1 mℓ of 0.2 *M* carbonate buffer, pH 9.8. The mixture was extracted with 5 mℓ of hexane. The organic layer was evaporated at 50°C under nitrogen. The residue was reconstituted in 20 μℓ of 0.01% triethylamine in hexane and an aliquot of 5 μℓ of this solution was injected.

Elution — E-1. Acetonitrile-0.6% phosphate buffer, pH 3 (75:25).

REFERENCES

1. **Constanzer, M. L., Vincek, W. C., and Bayne, W. F.,** Determination of cyclobenzaprine in plasma and urine using capillary gas chromatography with nitrogen-selective detection, *J. Chromatogr.*, 339, 414, 1985.
2. **Heinitz, M. L.,** Determination of cyclobenzaprine in tablets by high-performance liquid chromatography, *J. Pharm. Sci.*, 71, 656, 1982.

β-CYCLODEXTRIN

Liquid Chromatography

Specimen (mℓ)	Extraction	Column (cm × mm)	Packing (μm)	Elution	Flow (mℓ/min)	Det. (nm)	RT (min)	Internal standard (RT)	Other compounds (RT)	Ref.
Plasma (0.5)	I-1	25 × 4	LiChrosorb RP-18 (5)	E-1	0.6	Refractive index	9	—	—	1

Extraction — I-1. A protein free filtrate was obtained by centrifugal ultrafiltration using a MPS.1 for 15 min at 1760 g.

Elution — E-1. Methanol-water (16:84).

REFERENCE

1. **Koizumi, K., Kubota, Y., Okada, Y., and Utamura, T.,** Microanalyses of β-cyclodextrin in plasma by high-performance liquid chromatography, *J. Chromatogr.*, 341, 31, 1985.

CYCLOPHOSPHAMIDE

Gas Chromatography

Specimen (mℓ)	Extraction	Column (m × mm)	Packing (mesh)	Oven temp (°C)	Gas (mℓ/min)	Det.	RT (min)	Internal standard (RT)	Deriv.	Other compounds (RT)	Ref.
Plasma (0.1)	I-1	1.8 × 2	3% SE-30 Chromosorb W	195	He (35)	NPD	8	Isophosphamide (6.5)	Heptafluorobutyryl	—	1
Plasma (0.2)	I-2	0.75 × 1.2	3% SE-30 Gas Chrom Q (100/120)[a]	195	N_2 (6)	NPD1	1.6	Isophosphamide (4.4)	Trifluoroacetyl	—	2
Plasma (0.05)	I-3	10 × 0.31	OV-275	205	NA	NPD	3	5-Chlorouracil (9)	—	4-Ketocyclophosphamide (4.5)	3

										5-Fluorouracil (5.5) 5-Fluorodihydro-uracil (2.5) Carboxy-phosphamide (1)	
Plasma (0.2)	I-4	25 × 0.2	SP-2100 (0.2 μm)[b]	240	He (NA)	MS-EI		Isophosphamide	NA	—	4
Plasma (1)	I-5	25 × 0.33	SE-30	T.P.[c]	N_2 (1.5)	NPD	12.5	N-Nitrosodipheny-lamine (8)	—	—	5

Liquid Chromatography

Specimen (mℓ)	Extraction	Column (cm × mm)	Packing (μm)	Elution	Flow (mℓ/min)	Det. (nm)	RT (min)	Internal standard (RT)	Other compounds (RT)	Ref.
Urine (1)	I-6	10 × 4	LiChrosorb C_{18} (5)	E-1	1.2	ABS[d] (210)	2.7	[$^2H_{10}$] Cyclophos-phamide	Carboxycyclo-phosphamide (1.2) 4-Ketocyclophos-phamide (1.6)	6
Serum (1)	I-7	30 × 4	μ-Bondapak-C_{18} (10)	E-2	1.5	ABS (195)	7	5-Ethyl-5-*p*-tolylbar-bituric acid (12)	—	7

Thin-Layer Chromatography

Specimen (mℓ)	Extraction	Plate (Manufacturer)	Layer (mm)	Solvent	Post-separation treatment	Det. (nm)	Rf	Internal standard (Rf)	Other compounds (Rf)	Ref.
Plasma (1)	I-8	NA (Merck)	Silica gel G-60-F_{254} (0.25)[e]	S-1	Heating at 230—260°C for 10 min	Reflectance (254)	0.52	—	—	8

[a] Use of alternative columns have been described.
[b] Film thickness.
[c] Initial temp = 115°C; initial time = 1.5 min; rate = 10°C/min; final temp = 200°C; final time = 5 min.

CYCLOPHOSPHAMIDE (continued)

[d] Not suitable for the analysis of biological samples. Fractions corresponding to the retention times of a standard mixture are collected to be analyzed by field desorption was spectrometry.
[e] The plates were washed with methanol prior to use.

Extraction — I.1. To the sample were added an aliquot of the aqueous solution of the internal standard and 0.9 mℓ of 0.1 *N* sodium hydroxide. The solution was extracted twice with 5-mℓ portions of ether. The combined ether extract was evaporated under a stream of nitrogen and the residue was heated at 70°C for 30 min with 100 μℓ of ethyl acetate and 50 μℓ of heptafluorobutyric anhydride. The reaction mixture was evaporated under nitrogen and reconstituted with 100 μℓ of ethyl acetate for injection.
I-2. To the sample were added an aliquot of the aqueous solution of the internal standard and 0.1 mℓ of 0.6 *N* sodium hydroxide. The solution was extracted three times with 1-mℓ volumes of ethyl acetate. The combined extracts were evaporated under nitrogen. The residue was dissolved in 0.5 mℓ of methanol-water (9:1), the solution was washed three times with 1 mℓ portions of hexane and then evaporated. The residue was dissolved in 50 μℓ of trifluoroacetic anhydride and allowed to stand at room temperature for 30 min. The residue was dissolved in 100 μℓ of ethyl acetate for injection.
I-3. The sample was treated with 500 μℓ of a mixture of 2-propanol-diethyl ether (22:77) and an aliquot of the internal standard solution. While vortex mixing, 50 μℓ of acetone was gradually added. The supernatant was evaporated under a stream of nitrogen. The residue was dissolved in 100 μℓ of 2-propanol-diethyl ether (22:77) and 10 μℓ of this solution were transferred to the stainless needle of the solid-sample injection.
I-4. The sample was spiked with an aliquot of an aqueous internal standard solution (20 μg/mℓ), 1 mℓ of 0.6 *M* sodium hydroxide added, and extracted three times with 1-mℓ portions of ethyl acetate. The combined organic extracts were evaporated under a stream of nitrogen at room temperature. The residue was dissolved in 100 μℓ of ethyl acetate and 50 μℓ of trifluoroacetic anhydride. The mixture was incubated at 60°C for 30 min. The solution was evaporated under nitrogen. The residue dissolved in 200 μℓ of ethyl acetate and 1 μℓ was injected.
I-5. The sample was spiked with a methanolic solution of the internal standard and applied to a prewashed (10 mℓ ethyl acetate) Sep-Pak cartridge. The cartridge was eluted with 6 mℓ of ethyl acetate. The eluate was dried over anhydrous calcium chloride and evaporated in a stream of nitrogen. The residue was reconstituted in 125 μℓ of ethyl acetate and 1 μℓ of the solution was injected.
I-6. The sample was spiked with the deuterated analog, pH adjusted to 3 to 4 with 0.1 M H_2SO_4, and extracted with 5 mℓ of ethyl acetate. The organic layer was evaporated, residue dissolved in 1 mℓ of mobile phase, and 50 μℓ injected.
I-7. The sample was spiked with the internal standard and passed through a prewashed (10 mℓ methanol, 10 mℓ water) Sep-Pak-C_{18} cartridge. The cartridge was washed with 20 mℓ water and 10 mℓ air was sucked. Finally, the cartridge was eluted with 2 mℓ of methanol. The eluate was evaporated under nitrogen at 40°C. The residue was dissolved in 100 μℓ of water, the solution washed with 40 μℓ of toluene, and extracted with 2.5 mℓ of chloroform. A 2-mℓ aliquot of chloroform extract was evaporated under nitrogen. The residue was dissolved in 100 μℓ of water and 90 μℓ of this solution was injected.
I-8. The sample was diluted with 1 mℓ of borate buffer. The mixture was passed through a prewashed (3 mℓ methanol, 5 mℓ borate buffer) Sep-Pak-C_{18} cartridge. The cartridge was washed with 1.5 mℓ of buffer and 1.5 mℓ of water and eluted with 1.5 mℓ of methanol. The eluate was evaporated at 40°C under vacuum and the residue was dissolved in 100 μℓ of methanol. Aliquots of 1 μℓ of this solution were spotted.

Elution — E-1. Acetonitrile-water (28:72).
E-2. Acetonitrile-2 m*M* phosphate buffer, pH 4 (29:71).

Solvent — S-1. Methanol-dichloromethane (1:7.5.).

REFERENCES

1. **Nayar, M. S. B., Lin, L. Y., Wan, S. H., and Chan, K. K.,** Gas chromatographic analysis, of cyclophosphamide in plasma and tissues using nitrogen-phosphorus detection, *Anal. Lett.*, 12, 905, 1979.
2. **Van Den Bosch, N. and DeVos, D.,** Some aspects of the gas-liquid chromatographic analysis of cyclophosphamide in plasma, *J. Chromatogr.*, 183, 49, 1980.
3. **DeBruin, E. A., Tjaden, U. R., Van Oosterom, A. T., Leefland, P., and Leclercq, P. A.,** Determination of the underivatized antineoplastic drugs cyclophosphamide and 5-fluorouracil and some of their metabolites by capillary gas chromatogrpahy combined with electron-capture and nitrogen-phosphorus selective detection, *J. Chromatogr.*, 279, 603, 1983.
4. **Lartigue-Mattei, C., Chabard, J. L., Touzet, C., Bargnoux, H., Petit, J., and Berger, J. A.,** Plasma cyclophosphamide assay by selective ion monitoring, *J. Chromatogr.*, 310, 407, 1984.
5. **El-Yazigi, A. and Martin, C. R.,** Improved analysis of cyclophosphamide by capillary gas chromatography with thermionic (nitrogen-phosphorus) specific detection and silica sample purification, *J. Chromatogr.*, 374, 177, 1986.
6. **Bahr, U. and Schulten, H. R.,** Isolation, identification and determination of cyclophosphamide and two of its metabolites in urine of a multiple sclerosis patient by high pressure liquid chromatography and field desorption mass spectrometry, *Biomed. Mass Spectrom.*, 8, 553, 1981.
7. **Hardy, R. W., Erlichman, C., and Soldin, S. J.,** High-performance liquid chromatographic measurement of cyclophosphamide in serum, *Ther. Drug Monit.*, 6, 313, 1984.
8. **Gattavecchia, E., Tonelli, D., Ghini, S., and Breccia, A.,** Thin-layer chromatographic determination of ^{14}C-labelled and unlabelled cyclophosphamide, *Anal. Lett.*, 16, 57, 1983.

CYCLOSPORINE[a]

Liquid chromatography

Specimen (mℓ)	Extraction	Column (cm × mm)	Packing (μm)	Elution	Flow (mℓ/min)	Det. (nm)	RT (min)	Internal standard (RT)	Other compounds (RT)	Ref.
Plasma (1)	I-1	27 × 4	Bondapak-C_{18} (5)[b]	E-1	1.0	ABS (205)	4.2	—	—	3
Blood, plasma (2)	I-2	15 × 4.6	Supelco LC-18 (5)[c]	E-2	1.4	ABS (202)	5.8	Cyclosporin D (7.8)	—	4
Serum (2)	I-3	25 × 4.6	Ultrasphere-ODS (5)[d]	E-3; grad	1.0	ABS (215)	14.1	Cyclosporin D (15.7)	—	5
Blood, plasma (0.5)	I-4	15 × 4.6	LiChrosorb RP-18 (5)[d]	E-4[e]	1.7	ABS (210)	13	Cyclosporin D (15)	—	6

CYCLOSPORINE[a] (continued)

Liquid chromatography

Specimen (mℓ)	Extraction	Column (cm × mm)	Packing (μm)	Elution	Flow (mℓ/min)	Det. (nm)	RT (min)	Internal standard (RT)	Other compounds (RT)	Ref.
Plasma (1)	I-5	25 × 4.6	Ultrasphere-Octyl (5)[f]	E-5	1.5	ABS (210)	20.5	Cyclosporin D (27) + Cyclosporin C (18)	—	7
Blood, plasma (0.5)	I-6	15 × 4.6	Supelcosil LC-8 (5)[c]	E-6A	3.0	ABS (202)	—	—	—	8, 9
		15 × 4.6	Supelsocil LC-18 (5)[c]	E-6B	1.0		10		—	
Plasma (1)	I-7	25 × 4	LiChrosorb-RP-18	E-7	1.0	ABS	10.5	Cyclosporin D (13.9)	—	10
Blood, plasma (0.5-1)	I-8	25 × 4	LiChrocart-RP-18 (4)[c]	E-8	1.2	ABS (210)	6	Cyclosporin D (7.2)	—	11
Blood (0.5)	I-4	7.5 × 4.6	Ultrasphere-ODS (3)[g]	E-9	1.0	ABS (214)	14[h]	—	—	12
Blood, serum (1)	I-9	10 × 4.6	RP-8 MLPC (10)[d]	E-10	0.6	ABS (215)	9.2	Cyclosporin D (11.8)	—	13
Blood, plasma, serum (1)	I-10	25 × 4.6	Zorbax-cyano-propyl (5)[g]	E-11	1.5	ABS (214)	7.5	Cyclosporin D (9)	—	14
Blood (0.5)	I-11	15 × 4.6	Zorbax-CN (5)[b]	E-12	1.3	ABS (210)	NA	Cyclosporin D	—	15
Blood (1)	I-12	25 × 4.6	LiChrosorb CN (10)[i]	E-13	1.0	ABS (212)	13.2	—	—	16
Blood, plasma (1)	I-13	15 × 4.6	Ultrasphere CN (5)[j]	E-14	1.0	ABS (210)	11	Cyclosporin D (12)	—	17
Blood (1)	I-14	15 × 3.9	μ-Bonkapak-C_{18} (5)[c]	E-15	2.0	ABS (206)	4	Cyclosporin D (15)	—	18
Blood (1)	I-15	25 × 4.6	Alltech-C_{18} (10)[d]	E-16	0.8	ABS (200)	24	—	—	19
Blood (1)	I-16	7.5 × 4.6	Supelcosil-C_8 (3)[f]	E-17	1.9	ABS (214)	4.6	Cyclosporin D (6.1)	—	20

[a] The concentration of drug in plasma depends upon the temperature at which plasma is separated from the cells.[1,2]
[b] Column temp. = 55°C.
[c] Column temp. = 75°C.
[d] Column temp. = 70°C.
[e] Different mobile phases[5] are used for different operation of this automated procedure. E-4 is used for the separation on the analytical column.
[f] Column temp. = 72°C.
[g] Column temp. = 60°C.
[h] Estimated; retention time for only first column is shown.
[i] Column temp. = 40°C.
[j] Column temp. = 50°C.

Extraction — I-1. Plasma sample (1 mℓ) is shaken with 3 mℓ of methanol for 30 min and centrifuged. Supernatant (2 mℓ) diluted with 1 mℓ of water, passed through Sep-Pak C_{18} cartridge, washed with water (5 mℓ), 75% methanol (5 mℓ) and methanol (0.5 mℓ), and eluted with methanol (1 mℓ). Only glass syringes were used to avoid contamination from plastics; 10 μℓ injected.
I-2. Sample (2 mℓ) diluted with water (2 mℓ), extracted with ether (14 mℓ), ether collected, and evaporated. Residue treated with 0.025 *N* HCl (1 mℓ), methanol (2 mℓ) and washed with hexane twice; 0.025 *N* NaOH (1 mℓ) added; extracted with ether (7 mℓ). Residue dissolved in 100 μℓ of mobile phase, 90 μℓ injected.
I-3. Disposable column (Baker CN, 3 mℓ) washed with acetonitrile (3 mℓ); water (3 mℓ). Sample + internal standard applied, washed twice with water (3 mℓ); methanol/water (40:60) (1 mℓ). Eluted with methanol twice (1 mℓ). Evaporation with nitrogen. Residue dissolved twice in 55 μℓ of mobile phase, 100 μℓ injected.
I-4. The blood sample is treated with 1.5 mℓ of the internal standard solution (0.5 μg/mℓ in 65% methanol). Clean up of the supernatant is carried out on a precolumn automatically on the analytical instrument equipped with column and solvent switching valves.
I-5. Sample (1 mℓ) + 0.18 *N* HCl (1 mℓ) extracted with ether (10 mℓ). Ether washed with 0.95 *N* NaOH (2 mℓ) evaporated at 37°C. Residue dissolved in 250 μℓ of ammonium sulfate solution (75.8 m*M* in acetonitrile, methanol, water = 20:20:60); 100 μℓ injected.
I-6. The sample is treated with 1.2 mℓ of acetonitrile-water (97.5:2.5). The supernatant is automatically washed with hexane, separated, and injected on a C-8 column. The segment containing cyclosporin is automatically diverted to a IInd C-18 column.
I-7. The sample (1 mℓ) is incubated with 200 μℓ of proteinase K (5 mg/ℓ) at 37°C for 3 hr. The digest is diluted with methanol (1 mℓ) and ethyl acetate (1 mℓ), vortexed, and centrifuged. Supernatant (2 mℓ) applied to Extrelut column. Extraction repeated twice with 2 mℓ of ethyl acetate. Column eluted with ethyl acetate (6 mℓ). Pooled effluent evaporated. Residue dissolved in 75% acetonitrile (2 mℓ), filtered through Millex-SR filter, and filter rinsed. Combined filtrate (6 mℓ) evaporated and residue dissolved in 75% acetonitrile (1 mℓ).
I-8. Blood samples are lysed by rapid freezing and thawing. The samples are diluted with an equal volume of buffer and extracted twice with 7 mℓ portions of ether. Ether evaporated under nitrogen. Residue dissolved in 200 μℓ of methanol-0.1 *N* HCl (1:2), washed with hexane, and 30 μℓ injected.
I-9. The sample after the addition of internal standard and 3 mℓ 0.1 *M*. Tris buffer (pH 9.8) is extracted with ether. Ether is evaporated in the presence of 200 μℓ of 75% methanol and the residue applied to washed (twice with methanol; twice with water) CN Baker extraction columns (3 mℓ). Column is washed: 25% acetonitrile (3 mℓ); hexane (6 mℓ); dried, and eluted with 3 × 200 μℓ of methanol. Eluate evaporated and dissolved in 200 μℓ of mobile phase; 165 μℓ injected.
I-10. Sample (1 mℓ) is treated with 2 mℓ of diluent (acetonitrile/water-30:70) containing internal standard. The supernatant is applied to washed once with acetonitrile; twice with 20% acetonitrile BondElut cyanopropyl-1 mℓ column. The column is washed twice with 0.5 *N* acetic acid in 20% acetonitrile and with 0.25 mℓ of 0.5 *N* acetic acid in 40% acetonitirile. The column is eluted with 0.4 mℓ of acetonitrile. The eluate is evaporated, reconstituted with 150 μℓ of mobile phase, and 50 μℓ injected.

CYCLOSPORINE[a] (continued)

I-11. The sample was mixed with 1 mℓ of 10% isopropanol in acetonitrile containing 750 ng of the internal standard. The supernatant was mixed with 1.5 mℓ of water and applied to a prewashed (2 mℓ of acetonitrile, 2.5 mℓ of 70% methanol) 1 mℓ Baker C_{-18} column. After the sample had passed through, the column was washed with 5 mℓ of 70% methanol and eluted with 1.3 mℓ of acetonitrile. The eluate was evaporated at 75°C under a stream of air. The residue was reconstituted with 100 μℓ of 50% acetonitrile.
I-12. The sample was extracted with 5 mℓ of ether. The ether extract was evaporated under nitrogen at 50°C. The residue was dissolved in 4 mℓ of methanol-water (7:3) and extracted with 5 mℓ of carbon tetrachloride. The lower layer was then washed with 4 mℓ of methanol-2 *N* NaOH (7:3) and 4 mℓ of methanol-2 *N* HCl (7:3). Finally, the carbon tetrachloride extract was evaporated under nitrogen. The residue was reconstituted with 200 μℓ of the mobile phase for injection.
I-13. The sample was mixed with 2 mℓ of the internal standard solution [200 μg/ℓ in acetonitrile (96) + dimethylsulfoxide (4)]. The supernatant was diluted with water and applied to a prewashed (5 mℓ acetonitrile-water, 1:3) 1-mℓ Bond-Elut-C_{18} column. After the sample had passed through, the column was washed with the above-mentioned wash solution and then eluted with 300 μℓ of ethanol-tetrahydrofuran (19:1). To the eluate 200 μℓ of water was added and washed with 500 μℓ of hexane. Aliquots of 100 μℓ of the aqueous layer were injected.
I-14. The sample was mixed with 50 μℓ of a methanolic solution of the internal standard (10 mg/ℓ). Acetonitrile containing 2.5% (2 mℓ) water was added gradually to the sample which was then vigorously vortexed. After centrifugation, the supernatant was treated with 5 mℓ of charcoal slurry. After mixing and centrifugation, nearly dry charcoal sediment was extracted with 3 mℓ of ethyl acetate. The ethyl acetate extract was evaporated under nitrogen at 40°C. The residue was dissolved in 100 μℓ of the mobile phase and 50 μℓ were injected.
I-15. The sample was treated with an equal volume of acetonitrile and the mixture was saturated with ammonium sulfate. After centrifugation, an aliquot of 0.7 mℓ of the acetonitrile phase was washed with hexane and then treated with 50 mg of a mixture of washed Dowex-50W cation exchange resin and Dowex-1 anion exchange resin. Following centrifugation, the acetonitrile supernatant was used for injection.
I-16. The sample was spiked with 75 μℓ of the internal standard solution (10 mg/ℓ in methanol) followed by the addition of 1 mℓ of 180 m*M* HCl. The mixture was vortexed and extracted with 10 mℓ of ether. The ether layer was washed with 1 mℓ of 95 m*M* sodium hydroxide and evaporated at 45°C. The residue was dissolved in 250 μℓ of ammonium sulfate solution (76 m*M* ammonium sulfate added to 1 ℓ of mobile phase) and the solution washed with 1 mℓ of heptane. A 50-μℓ aliquot of the aqueous phase was injected.

Elution — E-1. Methanol-water (95:5).
E-2. Acetonitrile-water (68.5:31.5).
E-3. (A) 1 mℓ/ℓ Trifluoroacetic acid; (B) acetonitrile. Initial: A/B = 35/55; Final (linear at 15 min): 5/95.
E-4. Acetonitrile-water (72:28).
E-5. Acetonitrile-methanol-water (47:22:33).
E-6. (A) Acetonitrile-water (55:45); (B) acetonitrile-water (75:25).
E-7. Acetonitrile-water (75:25).
E-8. Acetonitrile-trifluoroacetic acid 1 mℓ/ℓ water (70:30).
E-9. Acetonitrile-water (71:29).
E-10. Acetonitrile-water (72:28).
E-11. Acetonitrile-water (49:51). Recycled.

E-12. Acetonitrile-water 945:55). Recycled 5 to 10 times.
E-13. Acetonitrile-water (43:57).
E-14. Acetonitrile-10 m*M* phosphate buffer, pH 7 (43:57).
E-15. Acetonitrile-methanol-water (45:30:25).
E-16. Acetonitrile-water (70:3)).
E-17. Acetonitrile-water-methanol-0.757 *M* ammonium sulfate (470:350:180:1.3).

REFERENCES

1. **Kahan, B. D.,** Individualization of cyclosporine therapy using pharmacokinetic and pharmacodynamic parameters, *Transplantation,* 40, 457, 1985.
2. **Burkle, W. S.,** Cyclosporine pharmacokinetics and blood levels monitoring, *Drug Intel. Clin. Pharm.,* 19, 101, 1985.
3. **Allwood, M. C. and Lawrance, R.,** High pressure liquid chromatographic determination of cyclosporin A in plasma, *J. Clin. Hosp. Pharm.,* 6, 195, 1981.
4. **Sawchuk, R. J. and Cartier, L. L.,** Liquid-chromatographic determination of cyclosporin A in blood and plasma, *Clin. Chem.,* 27, 1368, 1981.
5. **Yee, G. C., Gmur, D. J., and Kennedy, M. W.,** Liquid-chromatographic determination of cyclosporine in serum with use of a rapid extraction procedure, *Clin. Chem.,* 28, 2269, 1982.
6. **Nussbaumer, K., Niederberger, W., and Keller, H. P.,** Determination of cyclosporin A in blood and plasma by column switching-HPLC after rapid sample preparation, *J. High Resol. Chromatogr. Chromatogr. Commun.,* 5, 424, 1982.
7. **Carruthers, S. G., Freeman, D. J., Koegler, J. C., Howson, W., Keown, P. A., Laupacis, A., and Stiller, C. R.,** Simplified liquid-chromotographic analysis for cyclosporin A, and comparison with radioimmunoassay, *Clin. Chem.,* 29, 180, 1983.
8. **Smith, H. T. and Robinson, W. T.,** Semi-automated high-performance liquid chromatographic method for the determination of cyclosporine in plasma and blood using column switching, *J. Chromatogr.,* 305, 353, 1984.
9. **Gmur, D. J., Yee, G. C., and Kennedy, M. S.,** Modified column-switching high-performance liquid chromatographic method for the measurement of cyclosporine in serum, *J. Chromatogr.,* 344, 422, 1985.
10. **Joosthuizen, M. M., Jacobs, D. R., and Myburgh, J. A.,** Preparation, extraction and high performance liquid chromatography of cyclosporin A from plasma that contains interfering compounds, *Ann. Clin. Biochem.,* 22, 402, 1985.
11. **Garraffo, R. and Lapalus, P.,** Simplified liquid chromatographic analysis for cyclosporin A in blood and plasma with use of rapid extraction, *J. Chromatogr.,* 337, 416, 1985.
12. **Hamilton, G., Roth, E., Wallisch, E., and Tichy, F.,** Semi-automated high-performance liquid chromatographic determination fo cyclosporine A in whole blood using one-step sample purification and column-switching, *J. Chromatogr.,* 341, 411, 1985.
13. **Kates, R. E. and Latini, R.,** Simple and rapid high-performance liqud chromatographic analysis of cyclosporine in human blood and serum, *J. Chromatogr.,* 309, 441, 1984.
14. **Lensmeyer, G. L. and Fields, B. L.,** Improved liquid-chromatographic determination of cyclosporine, with concomitant detection of a cell-bound metabolite, *Clin. Chem.,* 31, 196, 1985.
15. **Shihabi, Z. K., Scaro, J., and David, R. M.,** A rapid method for cyclosporine A determination by HPLC, *J. Liq. Chromatogr.,* 8, 2641, 1985.
16. **Takada, K., Shibata, N., Yoshimura, H., Yoshikawa, H., and Muranishi, S.,** High performance liquid chromatographic determination of cyclosporin A in body fluids, *Res. Commun. Chem. Pathol. Pharmacol.,* 48, 369, 1985.

CYCLOSPORINE[a] (continued)

17. **Kabra, P. M., Wall, J. H., and Blanckaert, N.,** Solid-phase extraction and liquid chromatography for improved assay of cyclosporine in whole blood or plasma, *Clin. Chem.*, 31, 1717, 1985.
18. **Aravind, M. K., Miceli, J. N., and Kauffman, R. E.,** Measurement of cyclosporine by high-performance liquid chromatography following charcoal adsorption from whole bood, *J. Chromatogr.*, 344, 328, 1985.
19. **Hoffman, N. E., Rustum, A. M., Quebbeman, E. J., Hamid, A. A. R., and Ausman, R. K.,** HPLC determination of cyclosporine in whole blood, *J. Liq. Chromatogr.*, 8, 2511, 1985.
20. **Kahn, G. C., Shaw, L. M., and Kane, M. D.,** Routine monitoring of cyclosporine in whole blood and in kidney tissue using high performance liquid chromatography, *J. Anal. Toxicol.*, 10, 28, 1986.

CYPROHEPTADINE

Gas Chromatography

Specimen (mℓ)	Extraction	Column (m × mm)	Packing (mesh)	Oven temp (°C)	Gas (mℓ/min)	Det.	RT (min)	Internal standard (RT)	Deriv.	Other compounds (RT)	Ref.
Plasma, urine (1)	I-1	0.91 × 2	3% SP-2250 Supelcoport (80/100)	230	He (30)	NPD	4.2	Ethyldesmethyl cyproheptadine (5.3)	—	—	1

Liquid Chromatography

Specimen (mℓ)	Extraction	Column (cm × mm)	Packing (μm)	Elution	Flow (mℓ/min)	Det. (nm)	RT (min)	Internal standard (RT)	Other compounds (RT)	Ref.
Plasma (0.1)	I-2	30 × 3.9	μ-Bondapak-C_{18} (10)	E-1	1.5	ABS (228)	9.4	Desmethyldoxepin (5.2)	Desmethyl-cyproheptadine (7.8)	2
Plasma, urine (1)	I-3	15 × 4.6	Ultrasphere-C_8 (5)	E-2	1.8	ABS (254)	6.5	Hydroxyzine (5.2)	—	3

Extraction — I-1. To the sample were added 50 μℓ of a methanolic solution of the internal standard (1 μg/mℓ) and 1 mℓ of 0.1 *N* sodium hydroxide. The mixture was extracted with 8 mℓ of petroleum ether-isoamyl alcohol (99:1). The organic layer was extracted with 1 mℓ of 0.1 *N* HCl. The aqueous phase was made alkaline with 0.2 mℓ of 1 *N* NaOH and extracted with 8 mℓ of petroleum ether-isoamyl alcohol. The organic extract was evaporated at 40°C under nitrogen. The residue was dissolved in 50 μℓ of heptane and a 3-μℓ aliquot was injected.
I-2. To the sample were added 50 μℓ of the internal standard solution (1 μg/mℓ in 0.1 *M* HCl) and 100 μℓ of 1.5 *M* sodium hydroxide. The mixture was extracted twice with 3-mℓ portions of hexane-isoamyl alcohol (99:1). The combine dorganic extracts were back extracted into 100 μℓ of 0.05 *M* sulfuric acid. An aliquot of 90 μℓ of the aqueous layer was injected.
I-3. To the sample were added 50 μℓ of the internal standard solution (1 mg/mℓ in methanol) and 0.1 mℓ of 3 *N* sodium hydroxide. The mixture was extracted twice with 3-mℓ portions of ethyl acetate. The combined organic extracts were evaporated at room temperature with a stream of nitrogen. The residue was dissolved in 200 μℓ of the mobile phase and an aliquot of 20 μℓ of this solution was injected.

Elution — E-1. Acetonitrile-methanol-0.1 *M* phosphate buffer + 5 m*M* pentanesulfonic acid pH 4.7 (15:41:44).
E-2. Acetonitrile-0.05 *M* acetate buffer, pH 3.5 (20:80).

REFERENCES

1. **Hucker, H. B. and Hutt, J. E.,** Determination of cyproheptadine in plasma and urine by GLC with a nitrogen-sensitive detector, *J. Pharm Sci.*, 72, 1069, 1983.
2. **Novak, E. A., Stanley, M., McIntyre, I. M., and Hryhorczuk, L. M.,** High-performance liquid chromatographic method for quantification of cyproheptadine in serum or plasma, *J. Chromatogr.*, 339, 457, 1985.
3. **Foda, N. H., June, H. W., and McCall, J. W.,** Quantitation of cyproheptadine in plasma and urine by HPLC, *J. Liq. Chromatogr.*, 9, 817, 1986.

CYPROTERONE ACETATE

Liquid Chromatography

Specimen (mℓ)	Extraction	Column (cm × mm)	Packing (μm)	Elution	Flow (mℓ/min)	Det. (nm)	RT (min)	Internal standard (RT)	Other compounds (RT)	Ref.
Plasma (0.5)	I-1	30 × 3.9	μ-Bondapak-C_{18} (10)	E-1	2.0	ABS (282)	5.3	17-α-Hydroxypregen-4,6-diene-3,20-dione 17-butanoate (7.5)	—	1
Plasma, Urine (0.5)	I-2	30 × 3.9	μ-Bondapak-C_{18} (10)	E-2	NA	ABS (282)	0.72[a]	Cyproterone propionate (1)	15-Hydroxyeproterone acetate (0.41) Cyproproterone (0.56)	2

CYPROTERONE ACETATE (continued)

[a] Relative retention times (internal standard = 1).

Extraction — I-1. To the sample were added 0.5 mℓ of 0.25 *M* sodium hydroxide and 100 μℓ of the internal standard solution (12.5 mg/ℓ in methanol). The mixture was extracted with 10 μℓ of ethyl acetate. The organic layer was evaporated at 40°C under vacuum. The residue was chromatographed on 0.5 g of silica gel with 4 mℓ of 5% ethyl acetate-hexane, followed by 5 mℓ of ethyl acetate. The ethyl acetate fraction was evaporated. The residue dissolved in 100 μℓ of methanol for injection.
I-2. The sample was spiked with 100 μg of the internal standard and extracted with 5 mℓ of ether. The ether layer was washed with 1 mℓ of 0.25 *M* sodium hydroxide and 1 mℓ of water, dried over sodium sulfate, and evaporated. The residue was reconstituted with 1 mℓ of ether and applied to a Florisil (300 mg) column. The column was washed with *n*-heptane (5 mℓ), ether (5 mℓ), and then eluted with methanol-diethyl ether (1:10, 5 mℓ). The eluate was evaporated and reconstituted in 0.3 mℓ of the mobile phase for injection.

Elution — E-1. Methanol-water (70:30).
E-2. Acetonitrile-water (65:35).

REFERENCES

1. **Cannell, G. R., Mortimer, R. H., and Thomas, M. J.,** High-performance liquid chromatographic estimation of cyproterone acetate in human plasma, *J. Chromatogr.*, 226, 492, 1981.
2. **Frith, R. G. and Phillipou, G.,** 15-Hydroxycyproterone acetate and cyproterone acetate levels in plasma and urine, *J. Chromatogr.*, 338, 179, 1985.

CYTOSINE ARABINOSIDE

Liquid Chromatography

Specimen (mℓ)	Extraction	Column (cm × mm)	Packing (μm)	Elution	Flow (mℓ/min)	Det. (nm)	RT (min)	Internal standard (RT)	Other compounds (RT)	Ref.
Plasma, urine	I-1	50 × 4	Aminex-A 27 (13)[a,b]	E-1	0.7	ABS (270)	15	—	Cytosine (10) Uridine (30) Ara-U[c] (68)	1
Plasma	I-2	25 × 4.6[d]	Nucleone-C_{18} (10)	E-2	NA	ABS (280)[e]	15		Cytidine (13) Deoxycytidine (16.5)	2

									5-Methylcytidine (18) Uridine (21.5) Uric Acid (23) Ara-U (29) Deoxyuridine (34)	
Leukemia cells	I-3	25 × 4.6	Partisil 10 SAX (10)	E-3; grad	3.0	ABS (280)	—	—	Ara CTP (16.1)[f] Cytosine triphosphate (13.5) 3-Deazauridine triphosphate (25.4) Uridine triphosphate (18.4)	3
Plasma, CSF	I-4	30 × 5	Spherisorb-ODS (5)	E-4	1.6	ABS (270)	6	—	Ara U (9)	4
Cells	I-3	25 × 4.6	Partisil SAX (10)	E-5[g]	0.5	ABS (254)	—	—	Ara CTP (28.6) 2′-Deoxyeytidine 5′-triphosphate (26.8)	5
Plasma, urine, CSF (1)	I-5	1.15 × 4.6 2.25 × 4.6[b]	Ultrasphere-ODS (5) Partisil SCX (10)	E-6	0.8	ABS (280)	25.7	Adenine arabinoside (35.7)	Ara U (13.4) Cytidine (22) Deoxycytidine (28.5)	6
Serum (0.1)	I-6	10 × 4.6	Ranin C_{18} (5)[i]	E-7	2.0	ABS (254)	5	—	Ara U (7)	7
CSF (0.05)	I-4	25 × 4.6	Ultrasphere-Octyl (5)	E-8	1.6	ABS (281)	4.5	—	Ara U (6.7)[j]	8

[a] Column temp = 65°C.
[b] Separation is also achieved with a C_{18} column after sample purification with boronate affinity gel.
[c] 1-β-D Uracil arabinoside.
[d] Two columns are combined.
[e] Ara-U is detected at 264 nm.
[f] 1-β-D-Arabinofuranosylcytosine 5′-triphosphate.
[g] An alternative separation system with gradient elution is also described.

Extraction — I-1. Plasma samples were filtered by ultracentrifugation through Amicon ultrafilters (2500 GE).
I-2. Plasma was filtered by centrifugation through Amicon ultrafilters (Centriflo, F25). The filtrate (1 mℓ) was treated with ice cold 8 *M* perchloric acid and allowed to stand in an ice bath for 10 min, and centrifuged. Aliquots of 500 μℓ were injected.
I-3. Perchloric acid extract of cells.

CYTOSINE ARABINOSIDE (continued)

I-4. Direct injection of sample without any prior workup.
I-5. The plasma sample was spiked with an appropriate volume of the internal standard solution (1 μg/mℓ) and filtered with MPS-1 Micropartition system for 15 to 20 min. Aliquots of 100 μℓ of the ultra filtrate were injected.
I-6. The sample was deproteinized with 200 μℓ of acetonitrile. The supernatant was evaporated under a nitrogen stream. The residue was reconstituted with 100 μℓ of water for injection.

Elution — E-1. 0.25 *M* Sodium citrate + 0.08 *M* sodium tetraborate buffer, pH 9.3.
E-2. O.2 *M* Potassium dihydrogen phosphate, pH 2.
E-3 (A) 0.005 *M* Ammonium dihydrogen phosphate, pH 2.8; (B) 0.75 *M* ammonium dihydrogen phosphate, pH 3.7. From 35%(B) to 100%(B) in 30 min by a concave gradient (curve 9).
E-4. 0.05 *M* Phosphate buffer, pH 7.
E-5. Methanol-0.5 *M* ammonium dihydrogen phosphate, pH 3.5 (2:98).
E-6. Methanol-2.5 m*M* potassium dihydrogen phosphate, pH 3.2 (2.5:97.5).
E-7. 0.017 *M* Ammonium formate buffer.
E-8. Methanol-0.01 *M* potassium phosphate buffer, pH 7 (1.5:98.5).

REFERENCES

1. **Pallavicini, M. G. and Mazrimas, J. A.,** High-performance liquid chromatographic analysis of cytosine arabinoside and metabolites in biological samples, *J. Chromatogr.*, 183, 449, 1980.
2. **Linssen, P., Drenthe-Schonk, A., Wessels, H., and Haanen, C.,** Determination of 1-β-D-arabinofuranosylcytosine and 1-β-D-arabinofuranosyluracil in human plasma by high-performance liquid chromatography, *J. Chromatogr.*, 223, 371, 1981.
3. **Plunkett, W., Chubb, S., and Barlogie, B.,** Simultaneous determination of 1-β-D-arabinofuranosylcytosine 5′-triphosphate and 3-deazauridine 5′-triphosphate in human leukemia cells by high-performance liquid chromatography, *J. Chromatogr.*, 221, 425, 1980.
4. **Breithaupt, H. and Schick, J.,** Determination of cytarabine and uracil arabinoside in human plasma and cerebrospinal fluid by high-performance liquid chromatography, *J. Chromatogr.*, 225, 99, 1981.
5. **Danks, M. K.,** Two simple high-performance liquid chromatographic methods for simultaneous determination of 2′-deoxycytidine 5%-triphosphate and cytosine arabinoside 5′-triphosphate concentration in biological samples, *J. Chromatogr.*, 233, 141, 1982.
6. **Sinkule, J. A. and Evans, W. E.,** High-performance liquid chromatographic assay for cytosine arabinoside, uracil arabinoside and some related nucleosides, *J. Chromatogr.*, 274, 87, 1983.
7. **Liversidge, G. G., Nishihata, T., and Higuchi, T.,** Simultaneous analysis of 1-β-D-arabinofuranosylcytosine, 1-β-D-arabinofuranosyluracil and sodium salicylate in biological samples by high-performance liquid chromatography, *J. Chromatogr.*, 276, 375, 1983.
8. **Zimm, S., Collins, J. M., Miser, J., Chatterji, D., and Poplack, D. G.,** Cytosine arabinoside cerebrospinal fluid kinetics, *Clin. Pharmacol. Ther.*, 35, 826, 1984.

DACARBAZINE

Liquid Chromatography

Specimen (mℓ)	Extraction	Column (cm × mm)	Packing (μm)	Elution	Flow (mℓ/min)	Det. (nm)	RT (min)	Internal standard (RT)	Other compounds (RT)	Ref.
Plasma, urine (0.5—1)	I-1	30 × 4	μ-Bondapak-C_{18} (10)	E-1; grad	NA	ABS (280)	12.3	3-Methylxanthine (11.6)	2-Azahypoxanthine (6.8) 5-Aminoimidazole-4-carboxamide (4.9)	1
Plasma (1)	I-2	30 × 3.9	μ-Bondapak-phenyl (10)	E-2; grad	2.0	ABS (254)	6.8	—	5-Aminoimidazole-4-carboxamide (2.6) 2-Azahypoxanthine (3.1)	2

Extraction — I-1. The sample was mixed with 100 μℓ of an aqueous internal standard solution (100 μg/mℓ and filtered through a CF 25 membrane cone filter (Amicon).
I-2. The sample was mixed with 5 mℓ of methanol-chloroform (3:1), allowed to stand at 4°C for 15 min, and centrifuged. Aliquots of the supernatant were injected.

Elution — E-1. (A) 0.5 *M* sodium acetate, pH 7 (with 10% phosphoric acid); (B) 25% Acetonitrile in 0.05 *M* sodium acetate, pH 5.5. Isocratic A for 5 min; linear gradient from 0%(B) to 95%(B) over 3 min. This composition was maintained for 6 min.
E-2. (A) 0.1% Ammonium formate buffer, pH 5.5; (B) methanol-water (1:1). Initial 90%(A) at 0.5 min, a 1-min linear gradient from 90%(A) to 40%(A).

REFERENCES

1. **Fiore, D., Jackson, A. J., Didolkar, M. S., and Dandu, V. R.,** Simultaneous determination of dacarbazine, its photolytic degradation product, 2-azahypoxanthine, and the metabolite 5-aminoimidazole-4-carboxamide in plasma and urine by high-pressure liquid chromatography, *Antimicrob. Agents Chemother.*, 27, 977, 1985.
2. **Tate, P. S. and Briele, H. A.,** Reversed-phase high-performance liquid chromatography of 5-(3,3-dimethyl-1-triazeno)imidazole-4-carboxamide and metabolites, *J. Chromatogr.*, 374, 421, 1986.

DANTROLENE

Liquid Chromatography

Specimen (mℓ)	Extraction	Column (cm × mm)	Packing (μm)	Elution	Flow (mℓ/min)	Det. (nm)	RT (min)	Internal standard (RT)	Other compounds (RT)	Ref.
Plasma (0.1)	I-1	30 × 4	μ-Bondapak-C_{18} (10)	E-1	1.0	ABS (400)	4.4	—	—	1
Plasma (0.5)	I-2	30 × 4	LiChrosorb RP-18 (10)	E-2	1.0	ABS (310)	10.5	Benzanilide	5-Hydroxydantrolene	2
Plasma (1)	I-3	25 × 4.6	CP Spher-C_8 (8)	E-3	1.5	ABS (375)	10.9	—	5-Hydroxydantrolene (7.1) F 490 (4.1)[a]	3

[a] Nitroreduced acetylated dantrolene.

Extraction — I-1. The sample was treated with 1 mℓ of acetonitrile. After mixing and centrifugation, 1 mℓ of the supernatant was evaporated at 50°C in a stream of air. The residue was dissolved in 100 μℓ of the mobile phase and aliquots of 10 to 20 μℓ were injected.
I-2. The sample was treated with 2 mℓ of 0.2 *M* acetate buffer (pH 4) and 50 μℓ of a methanolic solution of the internal standard (25 μg/mℓ). The mixture was extracted wtih 5 mℓ of ethyl acetate. The organic layer was evaporated under a stream of nitrogen. The residue was dissolved in 50 μℓ of the mobile phase and 10 to 20-μℓ aliquot of the clear solution was injected.
I-3. The sample was mixed with 0.5 g of ammonium sulfate and extracted wtih 4 mℓ of chloroform-1-butanol (95:5). An aliquot of the organic layer was evaporated at 50°C under nitrogen. The residue was reconstituted with 1 mℓ of the mobile phase.

Elution — E-1. Acetonitrile-20 m*M* glycine buffer, pH 3.5 (50:50).
E.2. Methanol-0.1 *M* acetate buffer, pH 7.4 (50:50).
E-3. Acetonitrile-0.05 *M* phosphate buffer, pH 6.8 (1:2).

REFERENCES

1. **Hackett, L. P. and Dusci, L. J.,** Determination of dantrolene sodium in human plasma using high-performance liquid chromatography, *J. Chromatogr.*, 179, 222, 1979.
2. **Katogi, Y., Tamaki, N., Adachi, M., Terao, J., and Mitomi, M.,** Simultaneous determination of dantrolene and its metabolite, 5-hydroxydantrolene, in human plasma by high-performance liquid chromatography, *J. Chromatogr.*, 228, 404, 1982.
3. **Wuis, E. W., Grutters, A. C. L. M., Vree, T. B., and Van Der Kleyn, E.,** Simultaneous determination of dantrolene and its metabolites, 5-hydroxydantrolene and nitro-reduced acetylated dantrolene (F 490), in plasma and urine of man and dog by high-performance liquid chromatography, *J. Chromatogr.*, 231, 401, 1982.

DAPSONE

Liquid Chromatography

Specimen (mℓ)	Extraction	Column (cm × mm)	Packing (μm)	Elution	Flow (mℓ/min)	Det. (nm)	RT (min)	Internal standard (RT)	Other compounds (RT)	Ref.
Plasma (0.2)	I-1	15 × 4.6	Magnusphere-C_{18} (5)	E-1	1.2	ABS (295)	3.3	—	Monoacetyl-dapsone (5.4)	1
Plasma (1)	I-2	30 × 3.9	μ-Bondapak-C_{18} (10)	E-2	1.5	ABS (254)	3.4	Quinine (7.1)	Monoacetyl-dapsone (4.2) Pyrimethamine (9.3)	2
Plasma, urine (0.5)	I-3	25 × 4	LiChrosorb RP-18 (5)	E-3[a]	1.3	ABS (250)	4.5	*m*-Aminophenyl sulphone (6.2)	Monoacetyl-dapsone (5.3)	3
Serum (2)	I-4	25 × 4	LiChrosorb RP-18 (10)	E-4 grad	1.5	ABS (290)	4.5	Metoprine (31.4)	Monoacetyl-dapsone (8.4) Pyrimethamine (28.3)	4

[a] The mobile phase was maintained at 40°C.

Extraction — I-1. The sample was treated with 20 μℓ of a 1:1 mixture of 60% perchloric acid and methanol. After mixing and centrifugation an aliquot of 30 μℓ of the clear supernatant was injected.
I-2. The sample was spiked with 25 μℓ of an aqueous solution of the internal standard (5 μg/mℓ) and 150 μℓ of 2 *M* sodium hydroxide. The mixture was extracted with 6 mℓ of dichloroethane. The organic extract was evaporated at 60°C under a stream of air. The residue was dissolved in 100 μℓ of the mobile phase and an aliquot of 40 μℓ of this solution was injected.
I-3. To the sample were added 50 μℓ of the internal standard solution (20 μg/mℓ in methanol), 100 μℓ of 1 *M* sodium hydroxide, and 350 μℓ of water. The mixture was extracted with 3 mℓ of dichloromethane. The organic extract was evaporated under nitrogen at 35°C. The residue was reconstituted with 30 to 50 μℓ of the mobile phase and 10 to 15 μℓ of this solution was injected.
I-4. The sample was mixed with the internal standard (700 ng) and 0.2 mℓ of 8 *N* sodium hydroxide. The mixture was extracted with 12 mℓ of 1:2-dichloroethane. The organic layer was evaporated under nitrogen at 40°C. The residue was dissolved in 50 μℓ of the mobile phase just prior to injection.

Elution — E-1. Methanol-0.067 *M* phosphate buffer, pH 5.9 (47:23).
E-2. Methanol-acetonitrile-water (25:15:60).
E-3. Acetonitrile-water-acetic acid (250:730:20). E-4. (A) Methanol; (B) 0.02 *M* phosphate buffer, pH 7.5. Isocratic at 32%(A) for 3 min, linear gradient from 32%(A) to 53%(A) in 8.5 min, isocratic at 53%(A) for 25 min.

DAPSONE (continued)

REFERENCES

1. **Philip, P. A., Roberts, M. S., and Rogers, H. J.,** A rapid method for determination of acetylation phenotype using dapsone, *Br. J. Clin. Pharmacol.*, 17, 465, 1984.
2. **Edstein, M.,** Quantification of antimalarial drugs. II. Simultaneous measurement of dapsone, monoacetyldapsone and pyrimethamine in human plasma, *J. Chromatogr.*, 307, 426, 1984.
3. **Horai, Y. and Ishizaki, T.,** Rapid and sensitive liquid chromatographic method for the determination of dapsone and monoacetyldapsone in plasma and urine, *J. Chromatogr.*, 345, 447, 1985.
4. **Lee, H. S., Ti, T. Y., Lee, P. S., and Yap, C. L.,** Simultaneous estimation of serum concentrations of dapsone, monoacetyldapsone, and pyrimethamine in chinese men on maloprim for malaria prophylaxis using reversed-phase high performance liquid chromatography, *Ther. Drug Monit.*, 7, 415, 1985.

DAUNORUBICIN

Liquid Chromatography

Specimen (mℓ)	Extraction	Column (cm × mm)	Packing (μm)	Elution	Flow (mℓ/min)	Det. (nm)	RT (min)	Internal standard (RT)	Other compounds (RT)	Ref.
Plasma, urine (0.1)	I-1	25 × 3	LiChrosorb Si-60 (10)	E-1	0.6	Fl (480, 560)	4.4	Doxorubicin (6.2)	Daunorubicinol (5.6)	1
Cells	I-2	30 × 3.9	μ-Bondapak-C_{18} (10)	E-2	1.5	Fl (482, 580)	5.6	—	Doxorubicinol (3.4) Doxorubicin (3.9) Doxorubicinone (4.7)	2
Dosage	—	25 × 4.6	μ-Bondapak-C_{18} (10)	E-3	2.0	ABS (254)	6.2	—	Daunorubicinone (7.7) Doxorubicin (4) Doxorubicinone (5.1)	3
Plasma (1)	I-3	25 × 4.6	LiChrosorb RP-2 (5)	E-4	1.0	Fl (475, 557)	9.5	—	Daunomycinol (7.5) Daunomycinone (14) 7-con-O-methylnogarol (8.3)[a]	4

Plasma (0.25)	I-4	30 × 3.9	μ-Bondapak-C_{18} (10)	E-5	2.0	Electro-chem[b,c]	8	Adriamycin (3.8)	Daunorubicinol (4.8) 7-Deoxydaunorubicin-aglycone (22.3) Daunorubicina-glycone (12.6)	5

[a] Detected at ex = 471 nm, em = 550 nm.
[b] Potential = +0.65 V.
[c] A fluorescence detector was also used upstream.

Extraction — I-1. The sample was spiked with 0.1 mℓ of 0.1 *M* borate buffer, pH 9.8, containing 10 μg/mℓ of the internal standard and extracted with 1.8 mℓ of chloroform-methanol (4:1). An aliquot of the organic phase was injected.
I-2. The washed cell pellet was suspended in phosphate buffered saline (pH 7.4), made alkaline with 200 μℓ of Tris buffer, pH 8.4 and extracted twice with 4 volumes of chloroform-methanol (9:1). The combined organic extracts were evaporated at 40°C under vacuum. The residue was dissolved in the mobile phase for injection.
I-3. The sample was treated with an equal volume of acetonitrile-0.1 *M* phosphoric acid (4:1). After vortexing and centrifugation 20 or 100 μℓ of the clear supernatant were injected.
I-4. The sample was spiked with 25 ng of the internal standard and injected into a loop column (3.9 × 2.3 mm) connected to the injector. While still in the load position, the loop column was washed with 1 mℓ of water and then switched to the inject position. The loop column was washed with water (2 mℓ) prior to the injection of next sample.

Elution — E-1. Chloroform-methanol-acetic acid-0.3 m*M* $MgCl_2$ (72:21:4:3).
E-2. Methanol-0.05 *M* NaH_2PO_4 (65:35).
E-3. Methanol-water (650:350) containing 1.15 g of monobasic ammonium phosphate dissolved in 5 mℓ of acetic acid, pH 4.
E-4. Acetonitrile-0.01 *M* phosphoric acid (35:65).
E-5. Acetonitrile-water-acetic acid (28:71:1), pH 4 with 20% sodium acetate.

REFERENCES

1. **Baurain, R., Deprez-De Campeneere, D., and Trouet, A.,** Determination of daunorubicin, doxorubicin and their fluorescent metabolites by high-pressure liquid chromatography: plasma levels in DBA_2 mice, *Cancer Chemother. Pharmacol.*, 2, 11, 1979.
2. **Strauss, J. F., Kitchens, R. L., Patrizi, V. W., and Frenkel, E. P.,** Extraction and quantification of daunomycin and doxorubicin in tissues, *J. Chromatogr.*, 221, 139, 1980.
3. **Haneke, A. C., Crawford, J., and Aszalos, A.,** Quantitation of daunorubicin, doxorubicin, and their aglycones by ion-pair reversed-phase chromatography, *J. Pharm. Sci.*, 70, 1112, 1981.

DAUNORUBICIN (continued)

4. **Brown, J. E., Wilkinson, P. A., and Brown, J. R.,** Rapid high-performance liquid chromatographic assay for the anthracyclines daunorubicin and 7-con-O-methylnogarol in plasma, *J. Chromatogr.*, 226, 521, 1981.
5. **Akpofure, C., Riley, C. A., Sinkule, J. A., and Evans, W. E.,** Quantitation of daunorubicin and its metabolites by high-performance liquid chromatography with electrochemical detection, *J. Chromatogr.*, 232, 377, 1982.

10-DEAZAAMINOPTERIN

Liquid Chromatography

Specimen (mℓ)	Extraction	Column (cm × mm)	Packing (μm)	Elution	Flow (mℓ/min)	Det. (nm)	RT (min)	Internal standard (RT)	Other compounds (RT)	Ref.
Plasma, urine (1)	I-1	10 × 8	Radial-Pak-C_{18} (10)[a]	E-1; grad	2.0	Fl (375, 460)	4.1	10-Ethyl-10-deazaaminopterin (7)	b	1

[a] Protected by Bio-Rad ODS-10 guard cartridge.
[b] Conditions for the determination of metabolites are described.

Extraction — I-1. The sample was diluted 1:1 with 50 m*M* KH_2PO_4, pH 7, spiked with the internal standard, and applied to a prewashed (methanol, water) Sep-Pak-C_{18} cartridge. The cartridge was then washed with 2 mℓ of phosphate buffer and 2 mℓ of water and finally eluted with 2 mℓ of methanol. The eluate was evaporated and the residue dissolved in the intial mobile phase for injection.

Elution — E-1. (A) 50 m*M* KH_2PO_4, pH 7; (B) acetonitrile-50 m*M* KH_2PO_4, pH 7 (60:40). A 4-min linear gradient from 20%(B) to 40%(B).

REFERENCE

1. **Kinahan, J. J., Samuels, L. L., Farag, F., Fanucchi, M. P., Vidal, P. M., Sirotnak, F. M., and Young, C. W.,** Fluorometric high-performance liquid chromatographic analysis of 10-deazaaminopterin, 10-ethyl-10-deazaaminopterin, and known metabolites, *Anal. Biochem.*, 150, 203, 1985.

3-DEAZAGUANINE

Liquid Chromatography

Specimen (mℓ)	Extraction	Column (cm × mm)	Packing (μm)	Elution	Flow (mℓ/min)	Det. (nm)	RT (min)	Internal standard (RT)	Other compounds (RT)	Ref.
Plasma (0.25)	I-1	30 × 4	μ-Bondapak-C_{18} (10)	E-1	1.0	ABS (254)	11	Fluorouridine (16.5)	—	1

Extraction — I-1. To the sample were added 75 μℓ of 6% trichloroacetic acid and 100 μℓ of water. After mixing and centrifugation an aliquot of 20 μℓ of the supernatant was injected.

Elution — E-1. 5 m*M* Ammonium formate.

REFERENCE

1. **Chandrasekaran, S. and Ardalan, B.,** Determination of 3-deazaguanine in mice plasma by high-performance liquid chromatography, *J. Chromatogr.*, 309, 403, 1984.

3-DEAZAURIDINE

Liquid chromatography

Specimen (mℓ)	Extraction	Column (cm × mm)	Packing (μm)	Elution	Flow (mℓ/min)	Det. (nm)	RT (min)	Internal standard (RT)	Other compounds (RT)	Ref.
Plasma (0.4)	I-1	15 × 4.6	Ultrasphere-ODS (5)	E-1	2.0	ABS (280, 254)	4	2′-O-Methyl-3-deazauridine (7.5)	—	1

Extraction — I-1. To the sample were added 80 μℓ of the internal standard solution (50 μg/mℓ) and 800 μℓ of methanolic silver acetate (1.2 g/100 mℓ). The mixture was vortexed for 2 min and incubated at 37°C for 15 min, cooled, and centrifuged. An aliquot of 800 μℓ was applied to an anion exchange resin column (Aminex A-25, acetate form). The column was washed with 3 mℓ of 5 m*M* Tris-acetate (pH 8) containing 10% methanol and eluted with 0.5 mℓ of 2 *M* acetic acid. Aliquots of 20 μℓ of this eluate were injected.

3-DEAZAURIDINE (continued)

Elution — E-1. Methanol-0.1 *M* ammonium acetate, pH 4 (5:95).

REFERENCE

1. **Lin, K. T., Momparler, R. L., and Rivard, G. E.,** Sample preparation and estimation of plasma concentration of 3-deazauridine by high-performance liquid chromatography, *Ther. Drug Monit.*, 5, 491, 1983.

DEBRISOQUINE

Gas Chromatography

Specimen (mℓ)	Extraction	Column (m × mm)	Packing (mesh)	Oven temp (°C)	Gas (mℓ/min)	Det.	RT (min)	Internal standard (RT)	Deriv.	Other compounds (RT)	Ref.
Microsomal incubation	I-1	1.8 × 2	3% OV-1 GasChrom Q (100/120)	200	He (15)	MS-CI[a]	—	[$^2H_9$]-4-Hydroxy-debrisoquine	b	4-Hydroxy-debrisoquine (3)	1, 2

Liquid Chromatography

Specimen (mℓ)	Extraction	Column (cm × mm)	Packing (μm)	Elution	Flow (mℓ/min)	Det. (nm)	RT (min)	Internal standard (RT)	Other compounds (RT)	Ref.
Urine (1)	I-2	25 × 4.5	Brownlee RP-8 (5)	E-1	2.0	ABS (208)	7	—	4-Hydroxydebri-soquine (5)	3
Urine (1)	I-3	25 × 4	μ-Bondapak-C_{18} (10)	E-2	2.0	ABS (220)	11.5	—	4-Hydroxydebri-soquine (3.6)	4

[a] Methane as the reagent gas.
[b] Condensation with hexafluoroacetylacetone to produce a substituted pyrimidinyltetrahydroisoquinoline.

Extraction — I-1. Incubations were terminated by the addition of 1 *M* NaOH, the mixture was spiked with deuterated analogue of 4-hydroxy debrisoquine and washed three times with chloroform, the pH of the mixture was then adjusted to 8.5 and treated with 100 μℓ of hexafluoroacetylacetone and 1 mℓ of toluene. The mixture was incubated for 2 hr in a boiling water bath. The toluene layer was evaporated in a stream of nitrogen.
I-2. Urine is filtered through a Millipore (0.22 μm) filter and an aliquot of 50 μℓ is injected.
I-3. A 3-mℓ CBA Bond-Elut column was washed with 1 mℓ of acetonitrile-0.1M HCl (40:60) followed by 1 mℓ of water. The sample was applied to this column, followed by 1 mℓ of water, 1 mℓ of acetonitrile-water (50:50), 1 mℓ of water, and 0.5 mℓ of 0.1 *M* HCl. Finally, the column was eluted with 1 mℓ of acetonitrile-0.1 *M* HCl (40:60), and an aliquot of 20 μℓ of the eluate was injected.

Elution — E-1. Acetonitrile-0.008 *M* KH_2PO_4 buffer, pH 5 (45:55).
E-2. Acetonitrile-0.1 *M* NaH_2PO_4 (10:90).

REFERENCES

1. **Murray, S., Kahn, G. C., Boobis, A. R., and Davies, D. S.,** Molecular aspects of debrisoquine metabolism studied by gas chromatography mass spectrometry with electron capture negative ion chemical ionisation, *Int. J. Mass Spectrom. Ion Phys.*, 48, 89, 1983.
2. **Kahn, G. C., Boobis, A. R., Murray, S., Brodie, M. J., and Davies, D. S.,** Assay and characterisation of debrisoquine 4-hydroxylase activity of microsomal fractions of human liver, *Br. J. Clin. Pharmacol.*, 13, 637, 1982.
3. **Westwood, B. E., Harman, P. J., and Mashford, M. L.,** Liquid chromatographic assay for debrisoquine and 4-hydroxydebrisoquine in urine, *J. Chromatogr.*, 374, 200, 1986.
4. **Harrison, P. M., Tonkin, A. M., Dixon, S. T., and McLean, A. J.,** Determination of debrisoquine and its 4-hydroxy metabolite in urine by high-performance liquid chromatography, *J. Chromatogr.*, 374, 204, 1986.

4-DEMETHOXYDAUNORUBICIN

Liquid Chromatography

Specimen (mℓ)	Extraction	Column (cm × mm)	Packing (μm)	Elution	Flow (mℓ/min)	Det. (nm)	RT (min)	Internal standard (RT)	Other compounds (RT)	Ref.
Plasma (1)	I-1	30 × 2	μ-Bondapak Phenyl (10)[a]	E-1	0.4	Fl[b] (254, 550)	24	Doxorubicin (13)	13-Dihydro-4-demethoxy-daunorubicin (17)	1

4-DEMETHOXYDAUNORUBICIN (continued)

Liquid Chromatography

Specimen (mℓ)	Extraction	Column (cm × mm)	Packing (μm)	Elution	Flow (mℓ/min)	Det. (nm)	RT (min)	Internal standard (RT)	Other compounds (RT)	Ref.
Blood, plasma, urine (1)	I-2	30 × 4	μ-Bondapak-C_{18} (10)	E-2	2.0	Fl (254, 530)	4.2	Doxorubicin (2.2)	13-Dihydro-4-demethoxy-daunorubicin (3.2) 4-De-methoxy-rubicinone (9.2)	2

[a] Protected by a Whatman pellicular-ODS precolumn.
[b] A UV absorbance detector (254 nm) was also used.

Extraction — I-1. To the sample were added 10 to 40 ng of the internal standard in 0.1 mℓ of water and 2 mℓ of 0.05 *M* borate buffer, pH 8.4. The mixture was extracted with 10 mℓ of a chloroform-1-heptanol (9:1) mixture. The organic layer was back extracted into 0.3 mℓ of 0.3 *M* phosphoric acid containing 10 μg/mℓ of desipramine. The aqueous phase was washed with 2 mℓ of hexane and a 170-μℓ aliquot of the aqueous phase was injected.
I-2. The sample was mixed with 1.3 μg of the internal standard and 1 mℓ of borate buffer, pH 8.5 and extracted twice with 5-mℓ portions of ethyl acetate-butanol (3:2). The combined organic extracts were evaporated under vacuum. The residue was reconstituted with 200 μℓ of the mobile phase for injection.

Elution — E-1. Acetonitrile-0.05 *M* KH_2PO_4 (35:65).
E-2. Acetonitrile-methanol-phosphate buffer, pH 4.7 (40:10:50).

REFERENCES

1. **Moro, E., Bellotti, V., Jannuzzo, M. G., Stegnjaich, S., and Valzelli, G.,** High-performance liquid chromatographic method for pharmacokinetic studies on the new anthracycline 4-demethoxydaunorubicin and its 13-dihydro derivative, *J. Chromatogr.*, 274, 281, 1983.
2. **Pizzorno, G., Trave, F., Mazzoni, A., Russello, O., and Nicolin, A.,** Contemporary detection of 4-demethoxydaunorubicin and its metabolites 13-dihydro-4-demethoxydaunorubicin and 4-demethoxy-daunorubicinone by reverse phase high-performance liquid chromatography, *J. Liq. Chromatogr.*, 8, 2557, 1985.

DEMOXEPAM

Liquid Chromatography

Specimen (mℓ)	Extraction	Column (cm × mm)	Packing (μm)	Elution	Flow (mℓ/min)	Det. (nm)	RT (min)	Internal standard (RT)	Other compounds (RT)	Ref.
Serum (1)	I-1	15 × 4.6	Suplecosil LC-18 (5)	E-1	1.0	Fl[a] (380, 460)	9	—	b	1

[a] The column eluent was passed through a Teflon coil (3.8m × 1.1mm) exposed to a mercury-xenon irradiation lamp prior to detection.
[b] Conditions for the determination of phenothiazines are described.

Extraction — I-1. The sample was mixed with an equal volume of methanol and centrifuged. An aliquot of 20 μℓ of the supernatant was injected.

Elution — E-1. Methanol-0.1 *M* Phosphate buffer, pH 8 (3:2).

REFERENCE

1. **Brinkman, U. A. T., Welling, P. L. M., De Vries, G., Scholten, A. H. M. T., and Frei, R. W.,** Liquid chromatography of demoxepam and phenothiazines using a post-column photochemical reactor and fluorescence detection, *J. Crhomatogr.*, 217, 463, 1981.

DENZIMOL

Gas Chromatography

Specimen (mℓ)	Extraction	Column (m × mm)	Packing (mesh)	Oven temp (°C)	Gas (mℓ/min)	Det.	RT (min)	Internal standard (RT)	Deriv.	Other compounds (RT)	Ref.
Plasma (1)	I-1	10 × 0.53	50% Phenylmethyl silicone	270	He (20)	NPD	1.4	Rec 15-1624[a] (2)	—	β-Hydroxy metabolite[b]	1

DENZIMOL (continued)

Liquid Chromatography

Specimen (mℓ)	Extraction	Column (cm × mm)	Packing (μm)	Elution	Flow (mℓ/min)	Det. (nm)	RT (min)	Internal standard (RT)	Other compounds (RT)	Ref.
Urine (0.5)	I-2	10 × 4.6	Hypersil-C_{18} (5)[c]	E-1; grad	2.2	ABS (214)	4.8	Rec 15-1624[a] (5.5)	β-Hydroxy-metabolite (1.9)	1

[a] Higher homologue of denzimol.
[b] N-[β-[4-(β-phenyl-β-hydroxyethyl)phenyll]-β-hydroxyethyl]imidazole.
[c] Column temp = 35°C.

Extraction — I-1. The sample was mixed with 100 μℓ of the internal standard solution (1 μg/mℓ in water), 1 mℓ of water and 3 mℓ of 0.01 *M* KH_2PO_4 (pH 4.6) and applied to a prewashed (5 mℓ methanol, 20 mℓ water) Sep-Pak-C_{18} cartridge. The cartridge was then washed with 30 mℓ of water and eluted with 5 mℓ of methanol. The eluate was evaporated at 60°C under a gentle stream of nitrogen. The residue was dissolved in 0.1 mℓ of a mixture of methanol-diethyl ether (1:1). Aliquots of 1 to 2 μℓ of this solution were injected.
I-2. The sample was incubated with 100 μℓ of β-glucuronidase-arylsulfatase and 100 μℓ or 2 *M* acetate buffer (pH 5.5) at 45°C for 16 hr. Finally, 0.5 mℓ of methanol and 100 μℓ of the aqueous working internal standard solutions were added. After mixing and centrifugation, an aliquot of 20 μℓ was injected.

Elution — E-1. (A) Acetonitrile; (B) 0.01 *M* potassium dihydrogen phosphate (pH 4.6). A linear gradient from 35% to 70% (A) in 5 min, then isocratically for 10 min.

REFERENCE

1. **Bertin, D., Cova, A., Reschiotto, C., and Tajana, A.,** Determination of denzimol, a new anticonvulsant agent, and its main metabolite in biological material by gas chromatography and high-performance liquid chromatography, *J. Chromatogr.*, 378, 147, 1986.

5′-DEOXY-5-FLUOROURIDINE

Liquid Chromatography

Specimen (mℓ)	Extraction	Column (cm × mm)	Packing (μm)	Elution	Flow (mℓ/min)	Det. (nm)	RT (min)	Internal standard (RT)	Other compounds (RT)	Ref.
Plasma (0.2 — 1)	I-1	12.5 × 4	LiChrosorb RP-18 (5)	E-1[a]	1.0	ABS (269)	12.3	3-Methylxanthine (6.7)	—	1
Plasma, urine (0.5)	I-2	12.5 × 4.9	Spherisorb Phenyl (5)	E-2	1.5	ABS (280)	3	5-Bromouracil (2.5)	5-Fluorouracil (1.8)	2

[a] Mobile phase is maintained at 80°C.

Extraction — I-1. The sample was mixed with 50 to 100 μℓ of an aqueous solution of the internal standard (0.2-20 μg/mℓ) and 1 mℓ of 0.3 *M* solution of acetic acid in methanol. The mixture was placed in a boiling water bath for 1 min. After cooling and centrifugation, the supernatant was extracted with 20 mℓ of diethyl ether. The organic layer was evaporated at 45°C under a stream of nitrogen. The residue was dissolved in 50 to 100 μℓ of water and 10 to 25 μℓ of this solution were injected.

I-2. To the sample were added 50 μℓ of an aqueous solution of the internal standard (4 μg/mℓ) and 50 μℓ of 3% phosphoric acid. The mixture was extracted with 5 mℓ of ethylacetate-isopropyl alcohol (85:15). The organic layer was evaporated at 45 to 50°C under a stream of nitrogen. The residue was dissolved in 200 μℓ of methanol-water (5:95) and the solution applied to a 10 × 0.6 cm column packed with 2.5 g of 50 to 100 mesh silica gel, prewashed with 10 mℓ of ethyl acetate-methanol (9:1). The column was eluted with 4 mℓ of ethyl acetate-methanol (9:1). The eluate was evaporated under a stream of nitrogen. The residue was dissolved in 100 μℓ of water and aliquots of 5 to 20 μℓ of this solution were injected.

Elution — E-1. Water-methanol-acetonitrile (97:1.5:1.5).

E-2. Water.

REFERENCES

1. **Sommadossi, J. P. and Cano, J. P.,** Determination of a novel fluoropyrimidine, 5′-deoxy-5-fluorouridine, in plasma by high-performance liquid chromatography, *J. Chromatogr.*, 225, 516, 1981.
2. **Schaaf, L. J., Ferry, D. G., Hung, C. T., Perrier, D. G., and Edwards, I. R.,** Analysis of 5′-deoxy-5-fluorouridine and 5-fluorouracil in human plasma and urine by high-performance liquid chromatography, *J. Chromatogr.*, 342, 303, 1985.

DESFEROXAMINE

Liquid Chromatography

Specimen (mℓ)	Extraction	Column (cm × mm)	Packing (μm)	Elution	Flow (mℓ/min)	Det. (nm)	RT (min)	Internal standard (RT)	Other compounds (RT)	Ref.
Serum, urine (1)	I-1	20 × 4	Zorbax silica (10)	E-1	2.0	ABS (229, 440)	2.7	—	Ferroxamine (11.1) Aluminoxamine (18.3)	1

Extraction — I-1. The sample after the addition of 0.3 g of sodium chloride was extracted wtih 1 mℓ of benzyl alcohol. Aliquots of the organic layer were injected.

Elution — E-1. Acetonitrile-methanol-*n*-butanol-water-acetic acid-10 *M* sodium hydroxide (40:40:10:10:0.2:0.1).

REFERENCE

1. **Kruck, T. P. A., Kalow, W., and McLachlan, D. R. C.,** Determination of desferoxamine and a major metabolite by high-performance liquid chromatography. Application to the treatment of aluminum-related disorders, *J. Chromatogr.*, 341, 123, 1985.

DESIPRAMINE

Liquid Chromatography

Specimen (mℓ)	Extraction	Column (cm × mm)	Packing (μm)	Elution	Flow (mℓ/min)	Det. (nm)	RT (min)	Internal standard (RT)	Other compounds (RT)	Ref.
Plasma (2)	I-1	25 × 4.6	Partisil-ODS-2 (10)[a,b]	E-1	2.7	ABS (254)	—	2-Hydroxyimipramine (11)	2-Hydroxydesipramine (8.5)	1

[a] Protected by a 2.3 × 0.39 cm precolumn packed wtih Porasil.
[b] Column temp = 43°C.

Extraction — I-1. To the sample were added 150 μℓ of a methanolic solution of the internal standard (1 μg/mℓ) and 2 mℓ of 1 *M* carbonate buffer, pH 11. The mixture was extracted wtih 10 mℓ of dichloromethane-isoamyl alcohol (98:2). The organic layer was back extracted into 400 μℓ of 0.05 *M* phsophate buffer, pH 2.5. An aliquot of 350 μℓ of the aqueous phase was injected.

Elution — E-1. Acetonitrile-0.05 *M* phosphate buffer, pH 4.7 (25:75).

REFERENCE

1. **Wong, S. H. Y., McCauley, T., and Kramer, P. A.,** Determination of 2-hydroxydesipramine by high-performance liquid chromatography, *J. Chromatogr.*, 226, 147, 1981.

DEXAMETHASONE

Gas Chromatography

Specimen (mℓ)	Extraction	Column (m × mm)	Packing (mesh)	Oven temp (°C)	Gas (mℓ/min)	Det.	RT (min)	Internal standard (RT)	Deriv.	Other compounds (RT)	Ref.
Plasma (1)	I-1	0.58 × 3[a]	Supelcoport 1.5% SP-2100 (80-100)	268	He (40)	MS-EI	4.5	[$^{13}C_6$ - $^{2}H_3$]-Dexamethasone	Trimethylsilyl	—	1
Urine (20)	I-2	25 × NA[b]	SE-54	T.P.[c]	He (NA)	MS-NCI[d]	6.4; 6.7[e]	6-α-Methylprednisolone (7.3)	Oxidation	—	2

Liquid Chromatography

Specimen (mℓ)	Extraction	Column (cm × mm)	Packing (μm)	Elution	Flow (mℓ/min)	Det. (nm)	RT (min)	Internal standard (RT)	Other compounds (RT)	Ref.
Plasma, tissue (3)	I-3	30 × 3.9	μ-Bondapak-C_{18} (10)	E-1	3.0	ABS (254)	13.3	Cyheptamide (15.6)	—	3
Plasma (2)	I-4	15 × 4.6	Ultrasphere-ODS (5)	E-2	1.3	ABS (254)	12	Methylprednisolone (10)	—	4
Plasma, urine, saliva (0.5 — 2)	I-5	NA	Macherey & Nagel-C_{18} (10)[f]	E-3	2.0	ABS (232)	24	6-α-Methylprednisolone (21)	Hydrocortisone (13) Dexamethasone phosphate (13)	5

DEXAMETHASONE (continued)

[a] Only 20 cm length of the column is packed.
[b] Wide bore.
[c] Initial temp = 200; rate = 25°C/min from 200 — 250°C; 5°C/min from 250 — 300°C.
[d] Negative chemical ionization. Methane (0.4 torr) as the reagent gas.
[e] A secondary oxidation product.
[f] Column temp = 40°C.

Extraction — I-1. The sample was spiked with 10 μℓ of methanol containing 20 ng of the internal standard and was applied to a Sep-Pak-C_{18} cartridge. The cartridge was washed with 8 mℓ of water and eluted with 2 mℓ of methanol. The eluate was evaporated under a stream of nitrogen and the residue was dissolved in 100 μℓ of 10% methanol in dichloromethane. The solution was applied to a 25 × 0.4 cm column packed with LiChrosorb Si 100 (5 μm). The column was eluted with 30% dichloromethane-2.5% methanol, 0.5% ethanol-0.2% acetic acid in *n*-hexane at the rate of 2.5 mℓ/min. The column effluent was monitored at 240 nm. About 5 mℓ of the eluate corresponding to the dexamethasone elution time was collected, evaporated under a stream of nitrogen. The residue was treated with 10 μg of sodium acetate in 10 μℓ of methanol and the solvent was evaporated. The residue was then dissolved in 50 μℓ of acetone and the solution again evaporated. The residue was treated with 2 μℓ of N,O-*bis*(trimethylsily)acetamide and 5 μℓ of pyridine. The reaction mixture was incubated at 90°C for 90 min. Excess reagents were removed with a stream of nitrogen, the residue dissolved in 20 μℓ of hexane and 2 to 4-μℓ aliquot of the solution was injected.
I-2. The sample was spiked with 5 μg of the internal standard in methanol and extracted with 100 mℓ of dichloromethane. The organic extract was dried over sodium sulfate and evaporated. The residue was dissolved in 200 μℓ of pyridine and treated with 15 mg of chromium trioxide at room temperature for 3 hr. Excess pyridine was removed with nitrogen and the residue extracted with 200 μℓ of ethyl acetate. The extract was passed through a Pasteur pipette packed with 4 cm Sephadex LH-2O. The column was eluted with 3 mℓ of ethyl acetate. The eluate was evaporated under nitrogen, the residue dissolved in 100 μℓ of n-dodecane, and 1-2 μℓ of this solution was injected.
I-3. The sample was mixed with 50 μℓ of 1 *M* sodium hydroxide and 7 mℓ of *n*-heptane. After centrifugation the heptane alyer was discarded. The aqueous phase was saturated with sodium chloride (300 mg) and extracted with 10 mℓ of dichloromethane containing 1 mg/ℓ of the internal standard. The organic extract was evaporated at 70°C under a stream of nitrogen. The residue was dissolved in 20 μℓ of tetrahydrofuran and the entire extract was injected.
I-4. A 1-mℓ Bond-Elut silica extraction column was connected in series with a 1-mℓ BondElut C_{18} column. The columns were washed with 5 mℓ of methanol and 4 mℓ of water. The sample was spiked with 38 μℓ of water containing 76 ng of the internal standard and passed through the column. After the sample had passed through, the silica column was removed. The C_{18} column was washed with 3 mℓ of 5% aqueous methanol, then with 4 mℓ of chloroform. Finally the column was eluted with 2 mℓ of methanol-chloroform (3:1). The eluate was evaporated at 36°C under nitrogen. The residue was dissolved in 50 to 100 μℓ of the mobile phase and a 50-μℓ aliquot was injected.
I-5. The sample was mixed with 100 μℓ of the internal standard solution and 1 g of ammonium sulphate. The mixture was extracted twice with 3-mℓ portions of ethyl acetate.

Elution — E-1. Methanol-acetic acid-1 butanol-water (38:22:60:880).
E-2. Acetonitrile-water (280:720) containing 200 μℓ of triethylamine.
E-3. Acetonitrile-0.05 *M* phosphate buffer, pH 2 (30:70).

REFERENCES

1. **Minagawa, K., Kasuya, Y., Baba, S., Knapp, G., and Skelly, J. P.,** Determination of dexamethasone in human plasma and urine by electron-impact mass spectrometry, *J. Chromatogr.*, 343, 231, 1985.
2. **Her, G. R. and Watson, J. T.,** Quantitative methodology for corticosteroids based on chemical oxidation to electrophilic products for electron capture-negative chemical ionization using capillary gas chromatography-mass spectrometry, *Anal. Biochem.*, 151, 292, 1985.
3. **Cham, B. E., Sadowski, B., O'Hagan, J. M., de Wytt, C. N., Bochner, F., and Eadie, M. J.,** High performance liquid chromatographic assay of dexamethasone in plasma and tissue, *Ther. Drug Monit.*, 2, 373, 1980.
4. **Plezia, P. M. and Berens, P. L.,** Liquid-chromatographic assay of dexamethasone in plasma, *Clin. Chem.*, 31, 1870, 1985.
5. **Derendorf, H., Rohdewald, P., Hochhaus, G., and Mollmann, H.,** HPLC determination of glucocorticoid alcohols, their phosphates and hydrocortisone in aqueous solutions and biological fluids, *J. Pharm. Biomed. Anal.*, 4, 197, 1986.

DEXTROMETHORPHAN

Liquid Chromatography

Specimen (mℓ)	Extraction	Column (cm × mm)	Packing (μm)	Elution	Flow (mℓ/min)	Det. (nm)	RT (min)	Internal standard (RT)	Other compounds (RT)	Ref.
Dosage	—	30 × 4	μ-Bondapak-C_{18} (10)	E-1	2.0	ABS (254)	6	Testosterone propionate (12)	—	1
Plasma (1)	I-1	25 × 4.6	Ultrasphere-ODS (5)	E-2	1.3	Fl (220)[a]	10.3	Valoxazine (8.1)	—	2
Urine (2)	I-2	25 × 4.6	Spherisorb Phenyl (5)	E-3	1.2	ABS (280)	12.5	3-Methoxy-17-methyl-10-oxo-9α, 13α, 14α-morphinan (10.8)	Dextrophan (8.5) 3-Hydroxy-9α, 13α, 14α-morphinan (6.5) 3-Methoxy-9α, 13α, 14α-morphinan (10)	3
Dosage	—	25 × 4.6	IBM-CN (5)	E-4	1.5	ABS (280)	10	—	—	4[b]
Plasma, urine (3)	I-3	25 × 2.1	Partisil-ODS-3 (5)	E-5	0.3	Fl (200)[a]	10	—	—	5
Plasma, urine (2)	I-4	25 × 4.6	Ultrasphere-ODS[c] (5)	E-6	0.7	Fl (200)[a]	11.8	Levallorphan (7.9)	d	6

DEXTROMETHORPHAN (continued)

[a] No emission filter was used.
[b] The authors find liquid chromatographic procedure better than their described gas chromatographic procedure.
[c] Use of a number of alternative columns is described.
[d] Different extraction and chromatographic conditions are described for the determination of metabolites.

Extraction — I-1. The sample was incubated at 37°C for 2 hr after the addition of 0.1 mℓ of β-glucronidase and 18% acetic acid to adjust the pH 5 to 5.3. After cooling the pH of the sample was adjusted to 10.4 with 100 μℓ of a saturated solution of sodium carbonate and 50 μℓ of the internal standard solution (10 μg/mℓ) were added. The mixture was extracted twice with 10% 1-butanol in butyl chloride. The combined organic extracts were back extracted into 300 μℓ of 1% acetic acid and aliquots of 50 to 200 μℓ of the aqueous phase were injected with an autosampler.
I-2. The sample was adjusted to pH 11 to 11.5 with concentrated ammonium hydroxide and 50 μℓ of the internal standard solution in 0.01 *M* HCl (20 μg/mℓ) was added. The mixture was applied to a ClinElut disposable column. After 5 min, the column was eluted with 10% n-butyl alcohol-hexane (10 mℓ). The eluate was extracted with 0.4 mℓ of 0.1 *M* HCl. Aliquots (40 μℓ) of the aqueous layer were injected.
I-3. The sample was mixed with 0.5 mℓ of 6 *M* HCl and then with 0.4 mℓ of 10 *M* sodium hydroxide. The mixture was extracted with 10 mℓ of carbon tetrachloride. An aliquot of the organic layer (8 mℓ) was back extracted with 0.5 mℓ of 3% acetic acid. An aliquot of the aqueous layer was injected.
I-4. The sample was spiked with 30 μℓ of an aqueous solution of the internal standard (1 μg/mℓ), made alkaline with 0.5 mℓ of saturated sodium carbonate and extracted with 20 mℓ of hexane containing 0.1% triethylamine. The organic layer was evaporated under nitrogen at 50°C. The residue was dissolved in 300 μℓ of the mobile phase and an aliquot of 150 μℓ was injected.

Elution — E-1. Acetonitrile-water (55:45) containing 0.01 *M* ammonium nitrate and 0.005 *M* dioctylsulfosuccinate, pH 3.3.
E-2. Acetonitrile-0.04 *M* acetate buffer, pH 4.3 (35:65).
E-3.Acetonitrile-10 m*M* phosphate buffer, pH 4 (55:45).
E-4. Acetonitrile-water (25:75) containing 0.05 *M* potassium nitrate, pH 3 with 60% perchloric acid.
E-5. Acetonitrile-water (45:55) containing 0.01 *M* monobasic ammonium phosphate and 0.005 *M* sodium lauryl sulfonate, pH 3 with phosphoric acid.
E-6. Acetonitrile-0.022 *M* sodium acetate-0.09 *M* perchloric acid-0.0196 *M* n-nonylamine, pH 4.3 (40:60).

REFERENCES

1. **Kubiak, E. J. and Munson, J. W.,** Determination of dextromethorphan hydrobromide by high-performance liquid chromatography using ion-pair formation, *J. Pharm. Sci.*, 69, 1380, 1980.
2. **Gillilan, R., Lanman, R. C., and Mason, W. D.,** High pressure liquid chromatographic determination of dextrorphan in human plasma, *Anal. Lett.*, 13, 381, 1980.
3. **Park, Y. H., Kullberg, M. P., and Hinsvark, O. N.,** Quantitative determination of dextromethorphan and three metabolites in urine by reverse-phase high-performance liquid chromatography, *J. Pharm. Sci.*, 73, 24, 1984.
4. **Gibbs, V. and Zaidi, Z.,** Evaluation of high-performance liquid chromatography and gas chromatography for quantitation of dextromethorphan hydrobromide in cough-cold syrup preparations, *J. Pharm. Sci.*, 73, 1248, 1984.

5. **Achari, R. G., Ederma, H. M., Chin, D., and Oles, S. R.,** Determination of dextromethorphan in biological fluids by liquid chromatography by using semi-microbore columns, *J. Pharm. Sci.*, 73, 1821, 1984.
6. **East, T. and Dye, D.,** Determination of dextromethorphan and metabolites in human plasma and urine by high-performance liquid chromatography with fluorescence detection, *J. Chromatogr.*, 338, 99, 1985.

DIAMIDINES

Thin-Layer Chromatography

Specimen (mℓ)	Extraction	Plate (Manufacturer)	Layer (mm)	Solvent	Post-separation treatment	Det. (nm)	Rf[a]	Internal standard (Rf)	Other compounds (Rf)	Ref.
Plasma, urine (1)	I-1	20 × 20 cm (Laboratory)	Silica 60 HR (0.2)	S-1	—	Fl (365,450)[b]	1 = 0.75 2 = 0.58 3 = 0.52	c	—	1

Note: (1) 2-(4-Amidinophenyl) indole-6-carboxamidine;
(2) 2-(4-amidinophenyl) benzo [b] thiophene-6-carboxamidine;
(3) 2-(4-amidinophenyl)-1-benzofurane-5-carboxamidine.

[a] Different compounds are referred to by numbers. See *Note*.
[b] Optimal settings for compound 1, for compound 2 and 3 ex = 313 nm, em = 405 nm.
[c] One diamidine is used as an internal standard for the assay of another diamidine.

Extraction — I-1. The sample was mixed with an equal volume of 0.9% sodium chloride and 20 to 50 μℓ of the internal standard solution. The mixture was allowed to stand at room temperature for 3 hr. Then 0.1 g of guanidine hydrochloride, 1 mℓ of *n*-octanol and 0.5 mℓ of 2 M sodium hydroxide were added. After mixing and centrifugation the organic layer was back extracted into 50 μℓ of 0.1 *M* HCl. The aqueous phase was washed with 0.1 mℓ of *n*-heptane. An aliquot of the aqueous layer was spotted on the TLC plate.

Solvent — S-1. *n*-Butanol saturated with 2 *M* HCl.

REFERENCE

1. **Glutch, W. P., Kaliwoda, G., and Dann, O.,** Determination of fluorescent trypanocidal diamidines by quantitative thin-layer chromatography, *J. Chromatogr.*, 378, 183, 1986.

2,4-DIAMINO-6-(2,5-DIMETHOXYBENZYL)-5-METHYLPYRIDO-[2,3-d]PYRIMIDINE

Liquid Chromatography

Specimen (mℓ)	Extraction	Column (cm × mm)	Packing (μm)	Elution	Flow (mℓ/min)	Det. (nm)	RT (min)	Internal standard (RT)	Other compounds (RT)	Ref.
Plasma (1)	I-1	25 × 4.6	Zorbax TMS (6)	E-1	1.5	ABS (254)	11	—	—	1

Thin-Layer Chromatography

Specimen (mℓ)	Extraction	Plate (Manufacturer)	Layer (mm)	Solvent	Post-separation treatment	Det. (nm)	Rf	Internal standard (Rf)	Other compounds (Rf)	Ref.
Plasma (1)	I-1	20 × 20 cm (Merck)	Silica gel 60 (0.25)	S-1	—	Fl (Reflectance) (340, 400)	0.35	—	—	1

Extraction — I-1. The sample was mixed with 1 mℓ of pH 10 bicarbonate buffer and extracted twice with 4-mℓ portions of dichloromethane. The combined extract was passed through a prewashed (25 mℓ methanol, 15 mℓ dichloromethane) Sep-Pak silica cartridge. The cartridge was washed with 10 mℓ of dichloromethane and eluted with 6.5 mℓ of methanol. The eluate was evaporated at 55°C under nitrogen. The residue was reconstituted with methanol for liquid chromatography or with chloroform-methanol (9:1) for TLC analysis.

Elution — E-1. Acetonitrile-0.005 *M* 1-octane sulfonic acid (35:65).

Solvent — S-1. Chloroform-2-propanol-ammonium hydroxide (25:20:0.1).

REFERENCE

1. **Foss, R. G. and Sigel, C. W.,** Lipid-soluble inhibitors of dihydrofolate reductase III: quantitative thin-layer and high-performance liquid chromatographic methods for measuring plasma concentrations of the antifolate, 2,4-diamino-6-(2,5-dimethoxybenzyl)-5-methylpyrido-[2,3-d]pyrimidine, *J. Pharm. Sci.*, 71, 1176, 1982.

3,4-DIAMINOPYRIDINE

Liquid Chromatography

Specimen (mℓ)	Extraction	Column (cm × mm)	Packing (μm)	Elution	Flow (mℓ/min)	Det. (nm)	RT (min)	Internal standard (RT)	Other compounds (RT)	Ref.
Plasma (0.5)	I-1	30 × 4	Micropak-C_{18} (10)	E-1	1.3	ABS (228)	3.8	4-Aminopyridine (4.8)	—	1

Extraction — I-1. To the sample were added 300 mg of potassium carbonate and 50 μℓ of the internal standard solution (10 mg/ℓ in water). The mixture was extracted with 5 mℓ of dichloromethane. The organic phase was re-extracted with 50 μℓ of 0.1 *M* HCl and an aliquot of 20 μℓ of the aqueous phase was injected.

Elution — E-1. Acetonitrile-0.05 *M* phosphate buffer, pH 7.4 + 0.02 *M* tetramethylammonium chloride (23:77).

REFERENCE

1. **Lamiable, D., Millart, H., Lemeignan, M., and Vistelle, R.,** High-performance liquid chromatographic determination of 3,4-diaminopyridine in human plasma, *J. Chromatogr.*, 309, 222, 1984.

4,4′-DIAMINODIPHENYL SULFONE

Liquid Chromatography

Specimen (mℓ)	Extraction	Column (cm × mm)	Packing (μm)	Elution	Flow (mℓ/min)	Det. (nm)	RT (min)	Internal standard (RT)	Other compounds (RT)	Ref.
Plasma (1)	I-1	30 × 3.9	μ-Bondapak-C_{18} (10)	E-1	2.0	ABS (296)	7.5	4,4′-Diaminodiphenyl sulfone 2′-sulfonamide (6.7)	4′-4′-Diaminodiphenyl sulfone 4-mono-N-acetate[a] (9.9)	1

[a] Conditions for the determination of clofazimine and rifampicin are also described.

4,4′-DIAMINODIPHENYL SULFONE (continued)

Extraction — I-1. The sample was treated with 1 g of ammonium sulfate and 7 mℓ of water-chloroform-dimethylformamide (1:1:5). After centrifugation the supernatant was filtered.

Elution — E-1. Acetonitrile-water (20:80).

REFERENCE

1. **Gidoh, M., Tsutsumi, S., and Takitani, S.,** Determination of three main antileprosy drugs and their main metabolites in serum by high-performance liquid chromatography, *J. Chromatogr.*, 223, 379, 1981.

DIAZEPAM

Gas Chromatography

Specimen (mℓ)	Extraction	Column (m × mm)	Packing (mesh)	Oven temp (°C)	Gas (mℓ/min)	Det.	RT (min)	Internal standard (RT)	Deriv.	Other compounds (RT)	Ref.
Plasma (1)	I-1	1.4 × 3.3 (Steel)	10% OV-101 Gas-Chrom Q (80/100)	220	N_2 (60)	ECD	3.7	Medazepam (2.1)	—	N-Desmethyl-diazepam (4.4) Oxazepam (2.7) 3-Hydroxy-diazepam (6.8)	1
Plasma (0.2)	I-2	1.8 × 2	3% SP-2250 Supelcoport (100/120)	310	N_2 (25)	ECD	2.1	Medazepam (1.5)	—	Oxazepam (1.8) N-Desmethyl-diazepam (2.8) 3-Hydroxy-diazepam (3.5)	2

Liquid Chromatography

Specimen (mℓ)	Extraction	Column (cm × mm)	Packing (μm)	Elution	Flow (mℓ/min)	Det. (nm)	RT (min)	Internal standard (RT)	Other compounds (RT)	Ref.
Serum (1)	I-3	30 × 3.9	μ-Bondapak-C_{18} (10)	E-1	2.0	ABS (240)	14	5-(*p*-Methyl-phenyl)-5-phenyl-hydantoin (7)	Demoxepam (5.2) N-Desmethyl-chlordiazepoxide (6.5) N-Desmethyl-diazepam (10.6)	3
Plasma (0.5—1)	I-4	30 × 4	μ-Bondapak-C_{18} (10)	E-2	2.0	ABS (254)	10.8[a]		Oxazepam (6)[a] Temazepam (7.2)[a] N-Desmethyldiazepam (9.2)[a]	4
Tissue homogenate	I-5	25 × 4.6	LiChrosorb RP-8 (10)[b]	E-3	4.0	ABS (232)	2.7	—	—	5
Serum (1)	I-6	25 × 4.6	Spherisorb-ODS (5)	E-4	1.0	ABS (254)	4	Prazepam (4.5)	Oxazepam (1) N-Desmethyldiazepam (3.2)	6
Blood, plasma, serum (0.1)	I-7	15 × 4.6	Technicon-C_8 (5)	E-5	1.3	ABS (240)	9.6	Methylnitrazepam (4)	Oxazepam (4.8) Temazepam (6) N-Desmethyldiazepam (7.6)	7
Animal feed (0.3 g)	I-8	25 × 4.5	Zorbax-ODS (NA)[c]	E-6	1.0	ABS (242)	7.8	—	—	8
Brain tissue	I-9	20 × 4.6	LiChrosorb RP-8 (5)[d]	E-7	1.1	ABS (229)	28	Clonazepam (12)	Oxazepam (14) N-Desmethyldiazepam (23)	9
Serum (1)	I-10	5 × 4.6	Shim-pack FLC-C_8 (NA)	E-8	0.6	ABS (254)	9	Estazolam (5)	Oxazepam (6) Temazepam (7) N-Desmethyldiazepam (8)	10

[a] Capacity factors.
[b] Column temp = 40°C.
[c] Protected by a 5-cm Brownlee RP-8 guard column.
[d] Protected by a 3-cm Brownlee RP-8 (10 μm) guard column.

DIAZEPAM (continued)

Extraction — I-1. The sample was spiked with 0.5 mℓ of the internal standard solution (2.5 μg/mℓ in methanol-water, 1:4) and extracted with 4 mℓ of benzene. The organic layer was evaporated at 50°C under nitrogen. The residue was dissolved in 25 μℓ of benzene. An aliquot of 2 μℓ of this solution was injected.
I-2. The sample was spiked with 20 μℓ of methanol containing 250 ng of the internal standard and extracted with 1 mℓ of benzene. Portions of 4 μℓ of the benzene extract were injected.
I-3. The sample was mixed with 1 mℓ of phosphate buffer (0.4 *M*, pH 7.4) and extracted with 10 mℓ of chloroform containing 2 mg/ℓ of the internal standard. The organic layer was evaporated at 40°C under nitrogen. The residue was dissolved in 0.2 mℓ of the mobile phase and aliquots of 7 μℓ of this solution were injected.
I-4. The sample was mixed with 0.2 to 1 mℓ of 1 *M* H_3BO_3-Na_2CO_3-KCl buffer (pH 9) and extracted with 5 mℓ of benzene-dichloromethane (9:1). An aliquot of 4.8 mℓ of the organic layer was evaporated at 60°C under nitrogen. The residue was dissolved in 50 μℓ of the mobile phase and a 25-μℓ aliquot of this solution was injected.
I-5. Homogenate (20 mℓ) was applied to an Extrelut column and allowed to stand for 15 min. The column was eluted with 40 mℓ of diethyl ether. An aliquot of 25 μℓ of the eluate was evaporated under nitrogen. The residue was dissolved in 5 mℓ of methanol.
I-6. The sample, sodium phosphate (pH 13, 1 mℓ) was added to an extraction tube containing 100 μg of the internal standard. The mixture was extracted with 10 mℓ of chloroform. The organic layer was evaporated at 35°C under nitrogen. The residue was dissolved in 100 μℓ of methanol and 10-μℓ aliquots were injected.
I-7. Bond-Elut 1-mℓ C_{18} column was washed repeatedly with methanol, water, and with 100 μℓ of 0.1 *M* sodium borate buffer (pH 9.5) and 50 μℓ of 1000 U/mℓ heparin. The sample mixed with 10 μℓ of an aqueous solution of the internal standard (10 μg/mℓ) was applied to the column. The column was washed with 2 column volumes of water followed by 50 μℓ of methanol. The column was eluted with 200 μℓ of methanol followed by another aliquot of 100 μℓ of methanol. The eluate was evaporated under nitrogen. The residue was reconstituted with 25 μℓ of the mobile phase. A 10-μℓ volume of this solution was injected.
I-8. The sample was mixed with 6 mℓ of methanol. After centrifugation 1 mℓ of the supernatant was applied to a prewashed (methanol, water) Baker 1-mℓ C_{18} column and the eluate collected. The column was rinsed with 100 μℓ of methanol. The combined eluates were mixed.
I-9. The weighed brain sample was homogenized twice with 5-mℓ portions of ether. The combined ether supernatants were evaporated. the residue dissoved in 5 mℓ of water and the solution applied to a prewashed (twice with methanol, twice with water) 3-mℓ Bond-Elut C_{18} column. The column was washed twice with 3 mℓ volumes of water and eluted with two 500-μℓ aliquots of methanol. An aliquot of the internal standard solution was mixed with the combined eluates and an aliquot of 50 μℓ of the resulting solution was injected.
I-10. The sample was mixed with 50 μℓ of the internal standard solution (3.2 μg/ℓ) and 2 mℓ of 0.1 *M* sodium hydroxide. The mixture was extracted with 8 mℓ of diethyl ether. An aliquot of 4 mℓ of the ether phase was evaporated at 40°C. The residue was dissolved in 100 μℓ of the mobile phase and 50 μℓ of this solution were injected.

Elution — E-1. Acetonitrile-methanol-0.2 *M* acetate buffer, pH 5 (20:22.5:50).
E-2. Methanol-water (550:450).
E-3. Methanol-water (60:40).
E-4. Methanol-water-acetic acid (56:42:2).
E-5. Methanol-2 m*M* KH_2PO_4-acetonitrile (53:46:1).
E-6. Acetonitrile-water (70:30).
E-7. Methanol-0.01 *M* acetic acid (52.5:47.5).
E-8. Methanol-5 m*M* phosphate buffer, pH 6 (53:47).

REFERENCES

1. **Valetine, J. L. and Psaltis, P.**, Simultaneous gas chromatographic determination of diazepam and its major metabolites in human plasma, urine, and saliva, *Anal. Lett.*, 15, 1665, 1982.
2. **Loscher, W.**, Rapid gas chromatographic measurement of diazepam and its metabolites desmethyldiazepam, oxazepam, and 3-hydroxydiazepam (temazepam) in small samples of plasma, *Ther. Drug Monit.*, 4, 315, 1982.
3. **Foreman, J. M., Griffiths, W. C., Dextraze, P. G., and Diamond, I.**, Simultaneous assay of diazepam, chlordiazepoxide, N-desmethyldiazepam, N-desmethylchlordiazepoxide, and demoxepam in serum by high performance liquid chromatography, *Clin. Biochem.*, 13, 122, 1980.
4. **Cotler, S., Puglisi, C. V., and Gustafson, J. H.**, Determination of diazepam and its major metabolites in man and in the cat by high-performance liquid chromatography, *J. Chromatogr.*, 222, 95, 1981.
5. **Haberstumpf, H., Mayer, U., and GoBler, K.**, Mikromethode fur die quantitative Bestimmung von Diazepam in verschiedenen Schichten der Kalberaugenlinse mit der Hochdruck-Flussigkeits-Chromatographie, *Fresenius Z. Anal. Chem.*, 307, 400, 1981.
6. **Ratnaraj, N., Goldberg, V. D., Elyas, A., and Lascelles, P. T.**, Determination of diazepam and its major metabolites using high-performance liquid chromatography, *Analyst*, 106, 1001, 1981.
7. **Rao, S. N., Dhar, A. K., Kutt, H., and Okamoto, M.**, Determination of diazepam and its pharmacologically active metabolites in blood by Bond Elut™ column extraction and reversed-phase high-performance liquid chromatography, *J. Chromatogr.*, 231, 341, 1982.
8. **Pakuts, A. P., Downie, R. H., and Matula, T. I.**, A rapid HPLC analysis of diazepam in animal feed, *J. Liq. Chromatogr.*, 6, 2557, 1983.
9. **Komiskey, H. L., Rahman, A., Weisenburger, W. P., Hayton, W. L., Zobrist, R. H., and Silvius, W.**, Extraction, separation, and detections of ^{14}C-diazepam and ^{14}C-metabolites from brain tissue of mature and old rats, *J. Anal. Toxicol.*, 9, 131, 1985.
10. **Tada, K., Moroji, T., Sekiguchi, R., Motomura, H., and Noguchi, T.**, Liquid-chromatographic assay of diazepam and its major metabolites in serum, and application to pharmacokinetic study of high doses of diazepam in schizophrenics, *Clin. Chem.*, 31, 1712, 1985.

DIAZOXIDE

Liquid Chromatography

Specimen (mℓ)	Extraction	Column (cm × mm)	Packing (μm)	Elution	Flow (mℓ/min)	Det. (nm)	RT (min)	Internal standard (RT)	Other compounds (RT)	Ref.
Plasma, urine (0.1)	I-1	10 × 4.6	LiChrosorb RP-8 (5)	E-1	1.0	ABS (270)	5.8[a]	—	Chlorthalidon (3.2)[a]	1

[a] Capacity factor.

DIAZOXIDE (continued)

Extraction — I-1. The sample was mixed with 400 μℓ of 0.33 *N* perchloric acid and allowed to stand for 5 min. After centrifugation, 100 μℓ of the supernatant was injected.

Elution — E-1. Methanol-0.005 *M* sodium acetate (175:325).

REFERENCE

1. **Vree, T. B., Lenselink, B., Huysmans, F. T. M., Fleuren, H. L. J.,and Thien, Th. A.,** Rapid determination of diazoxide in plasma and urine of man by means of high-performance liquid chromatography, *J. Chromatogr.*, 164, 228, 1979.

DIBEKACIN

Liquid Chromatography

Specimen (mℓ)	Extraction	Column (cm × mm)	Packing (μm)	Elution	Flow (mℓ/min)	Det. (nm)	RT (min)	Internal standard (RT)	Other compounds (RT)	Ref.
Serum	I-1	25 × 4.6	LiChrosorb RP-18 (10)	E-1	1.0	Fl[a] (340, 440)	5	Netilmicin (18)	—	1

[a] The effluent of the column is treated with *o*-phthalaldehyde/mercaptoethanol reagent prior to detection.

Extraction — I-1. The sample was spiked with the internal standard and treated with trichloroacetic acid solution (5% final concentration) and the supernatant was applied to a Dowex 50 W-X8 (100 to 200 mesh) microcolumn. The column was washed with 10 vol of 0.1 *M* acetate-0.01 *M* Na_2SO_4 (pH 7.4) buffer, and eluted with 0.1 *N* NaOH. The eluate was quickly neutralized with *p*-toluene sulfonic acid (40 μ*M*) and acetic acid (40 μ*M*).

Elution — E-1. Acetonitrile-0.02 *M p*-toluene sulfonic acid + 0.02 *M* acetate + 0.2 *M* sodium sulfate + 0.04 *M* acetic acid (5:95).

REFERENCE

1. **Rollman, B., Van Der Auwera, P., and Tulkens, P. M.,** Dibekacin assay in serum by automated fluorescence polarization immunoassay (Abbott Tdx): comparison with high-performance liquid chromatography, substrate-labelled fluorescent immunoassay and radioimmunoassay, *J. Pharm. Biomed. Anal.*, 4, 53, 1986.

DIBUCAINE

Gas Chromatography

Specimen (mℓ)	Extraction	Column (m × mm)	Packing (mesh)	Oven temp (°C)	Gas (mℓ/min)	Det.	RT (min)	Internal standard (RT)	Deriv.	Other compounds (RT)	Ref.
Serum (0.5—2)	I-1	0.8 × 2	1.5% OV-17 Chromosorb W (80/100)	250	NA	MS-EI[a]	2.2		[$^2H_9$]-Dibucaine	—	1

[a] The chemical ionization mode was also used.

Extraction — I-1. The sample was mixed with an aqueous solution of the internal standard (200 ng/mℓ of sample) and 2 mℓ of 2 *M* sodium carbonate. The mixture was extracted with 9 mℓ of *n*-heptane containing 1% isopentyl alcohol. After mixing, the organic layer was separated and re-extracted with 0.4 mℓ of 0.1 *N* H_2SO_4. The aqueous layer was made alkaline and re-extracted with 1.4 mℓ of *n*-heptane. The organic phase was evaporated at 55°C under nitrogen. The residue was dissolved in 15 μℓ of methanol. An aliquot of this solution was injected.

REFERENCE

1. **Alkalay, D., Carlsen, S., and Wagner, W. E., Jr.,** Quantitation of the local anesthetic dibucaine with gas chromatography/mass spectrometry, *Anal. Lett.*, 14, 1745, 1981.

(±)-[[6,7-DICHLORO-2-(4-FLUOROPHENYL)-2-METHYL-1-OXO-5-INDANYL]OXY]ACETIC ACID

Gas Chromatography

Specimen (mℓ)	Extraction	Column (m × mm)	Packing (mesh)	Oven temp (°C)	Gas (mℓ/min)	Det.	RT (min)	Internal standard (RT)	Deriv.	Other compounds (RT)	Ref.
Plasma, urine (1)	I-1	1.2 × 4	3% OV-17	280	Ar: 95- Methane: 5	ECD	1.5	4-Chloroanalog (2.6)	Methyl	—	1

(±)-[[6,7-DICHLORO-2-(4-FLUOROPHENYL)-2-METHYL-1-OXO-5-INDANYL]OXY]ACETIC ACID (continued)

Extraction — I-1. The sample was spiked with 1 μg of the internal standard, made acidic with 1 mℓ of 2 *M* HCl, and extracted with 15 mℓ of toluene. The organic layer was back extracted into 2 mℓ of 0.1 *M* NaOH. The aqueous layer was made acidic with 0.25 mℓ of 2 *M* HCl and extracted with 5 mℓ of dichloromethane. The organic layer was treated with 100 μℓ of ethereal diazomethane and the solvents were evaporated. The residue was dissolved in 1.5 mℓ of toluene and a 4-μℓ aliquot was injected.

REFERENCE

1. **Weidner, L. L. and Zacchei, A. G.,** Gas chromatographic determination of a diuretic-antihypertensive agent, (±-[[6,7-Dichloro-2-(4-fluorophenyl)-2-methyl-1-oxo-5-indanyl]oxy]acetic Acid, in biological fluids, *J. Pharm. Sci.*, 73, 268, 1984.

DICHLOROMETHYLENE DIPHOSPHONIC ACID

Liquid Chromatography

Specimen (mℓ)	Extraction	Column (cm × mm)	Packing (μm)	Elution	Flow (mℓ/min)	Det. (nm)	RT (min)	Internal standard (RT)	Other compounds (RT)	Ref.
Serum, urine (5)	I-1	10 × 4.6	AG1-X8 Anion exchange resin (100/200 mesh)	E-1; grad	NA	a	15	b	—	1

[a] A flame phosphorus detector (525 nm) was used.
[b] Extraction efficiency was monitored with C_{14} labeled drug.

Extraction — I-1. The sample was spiked with [^{14}C] labeled drug and treated with 4 mℓ of 25% trichloroacetic acid. The supernatant was treated with 100 μℓ of 0.5 *M* sodium dihydrogen phosphate and 50 μℓ of 2.5 *M* calcium chloride. The pH of the mixture was adjusted to 12 to 12.5 with 25% NaOH. The resulting precipitate was isolated by centrifugation, washed with water, and dissolved in 5 mℓ of 2 *M* HCl. The solution was heated on a boiling water bath for 30 min and the pH adjusted to 7.0 with 25% and finally with 2.5% NaOH. The precipitate was isolated by centrifugation and dissolved in 2 mℓ of EDTA solution. An aliquot was used to determine radioactivity and another aliquot was injected.

Elution — E-1. (A) 0.025 *M* HCl; (B) 1 *M* HCl.

REFERENCE

1. **Chester, T. L., Lewis, E. C., Benedict, J. J., Sunberg, R. J., and Tettenhorst, W. C.,** Determination of (dichloromethylene)diphosphonate in physiological fluids by ion-exchange chromatography with phosphorus-selective detection, *J. Chromatogr.*, 225, 17, 1981.

DICLOFENAC

Gas Chromatography

Specimen (mℓ)	Extraction	Column (cm × mm)	Packing (mesh)	Oven temp (°C)	Gas (mℓ/min)	Det.	RT (min)	Internal standard (RT)	Deriv.	Other compounds (RT)	Ref.
Urine (1)	I-1	2 × 3	3% OV-17 Gas-Chrom Q (80/100)	245	Ar: 90- Methane: 10 60)	ECD	—	CGP 7406 (9.5)	Methyl	4-Hydroxy-diclofenac (7) 5-Hydroxy-diclofenac (7)	1
Plasma (0.5)	I-2	2 × 3	1.5% OV-17 Shimalite W (80/100)	250	N_2 (50)	ECD	NA[a]	Aldrin	Methyl	—	2–4
Plasma (1)	I-3	0.5 × 3	3% OV-1 Chromosorb W (80/100)	200	He (30)	MS-EI	1	4′-Methoxydiclofenac (2.5)	Cyclization	—	5
Urine (0.01—0.3)	I-1	30 × 0.3	Carbowax 40M	T.P.[b]	He (2)	ECD	4	4′-Hydroxy-5-chloro-diclofenac (8.1)	Methyl	4′-Hydroxy-diclofenac (6.8) 5-Hydroxy-diclofenac (7.2) 3′-Hydroxy-diclofenac (8.1) 4,5-Dihydroxy-diclofenac (14.3)	6

DICLOFENAC (continued)

Liquid Chromatography

Specimen (mℓ)	Extraction	Column (cm × mm)	Packing (μm)	Elution	Flow (mℓ/min)	Det. (nm)	RT (min)	Internal standard (RT)	Other compounds (RT)	Ref.
Serum, urine (0.1)	I-4	30 × 4	μ-Bondapak-C_{18} (10)	E-1	0.8	ABS (254)	7.5	Acetaminophen (5)	—	7
Plasma (1)	I-5	15 × 4.6	Supelcosil LC-18 (5)	E-2	2.0	ABS (215)	3.9	CGP 4287 (4.5)	4′-Hydroxy-diclofenac (1.8) 5-Hydroxy-diclofenac (2.1) 4′-5-Dihydroxy-diclofenac (1.2) 3′-Hydroxy-diclofenac (1.2)	8
Plasma (1)	I-6	25 × 4	LiChrosorb RP-8 (10)[c]	E-3	1.3	ABS (282)	5.8	2-(*p*-cyclohexen-1′-yl-phenyl) propionic acid	3′-Hydroxy-diclofenac (3.1)[d] 4′-Hydroxy-diclofenac (3.2)[c] 4′,5-Dihydroxy-diclofenac (3.3) 5-Hydroxy-diclofenac (3.8)	9

Thin-Layer Chromatography

Specimen (mℓ)	Extraction	Plate (Manufacturer)	Layer (mm)	Solvent	Post-separation treatment	Det. (nm)	Rf	Internal standard (Rf)	Other compounds (Rf)	Ref.
Plasma (1)	I-7	10 × 20 cm (Camag)	Silica gel-60 F_{254}[e]	S-1	—	Reflectance (290)	0.64	—	—	10

[a] Methyl ester of diclofenac elutes at 16 min when column oven temperature = 200°C and the carrier gas flow rate = 40 mℓ/min.
[b] Initial temp = 200°C; initial time = 1 min; rate = 30/min; final temp = 230°.

[c] Protected by a 5 × 0.47 cm precolumn filled with Co:Pell ODS (30 to 38 μm).
[d] The 3′- and 4′-hydroxy metabolites in urine are determined together by a separate procedure.
[e] High performance TLC plates.

Extraction — I-1. The sample was spiked with 0.5 mℓ of an aqueous solution of the internal standard pH adjusted with 3 mℓ of 1 *M* acetate buffer, pH 5 and 100 mg of ascorbic acid was added. The mixture was extracted with 3.5 mℓ of etherdichloromethane (3:1). The organic layer was evaporated at 45°C under nitrogen. The residue dissolved in 2 mℓ of sodium hydroxide to which 0.2 mℓ of tetrabutylammonium hydrogen sulfate and 50 μℓ of iodomethane in 3.5 mℓ of dichloromethane were added. The mixture was mixed at room temperature for 20 min. After centrifugation, the dichloromethane layer was collected and evaporated to dryness. The residue was dissolved in 0.2 mℓ of hexane and 3 μℓ injected.
I-2. The sample was acidified with 1 mℓ of 2.7 *M* phosphoric acid and extracted with 5 mℓ of benzene. A 4-mℓ aliquot of benzene extract was back extracted into 2 mℓ of 0.1 *N* NaOH. The aqueous phase was acidified with 0.2 mℓ of 43% phosphoric acid and re-extracted with 3 mℓ of benzene. A 2.5-mℓ aliquot of the benzene extract was evaporated at 40°C under reduced pressure. The residue was dissolved in 0.15 mℓ of methanol containing 0.1% sulfuric acid. The solution was incubated in a sealed ampoule at 60°C for 1 hr. After cooling, the reaction mixture was mixed with 0.4 mℓ of 25% potassium hydrogen carbonate and 2 mℓ of *n*-hexane. After centrifugation 1.5 mℓ of the hexane layer was evaporated at a temperature below 40°C under a stream of nitrogen. The residue was dissolved in an aliquot of benzene containing aldrin (25 ng/mℓ), of which a 1-μℓ aliquot was injected.
I-3. The sample was mixed with 50 μℓ of a methanolic solution of the internal standard (22 μg/mℓ) and 1 mℓ of 1 *M* phosphoric acid. The mixture was extracted with 7 mℓ of benzene. The benzene layer was back extracted into 1 mℓ of 0.08 *M* sodium carbonate buffer. The aqueous layer was made acidic 1 mℓ of 1 *M* phosphoric acid and re-extracted with 7 mℓ of benzene. The benzene layer was evaporated at 50°C under nitrogen. The residue was dissolved in 1 mℓ of *n*-hexane and 100 μℓ of pentafluoropropionic anhydride were added. The mixture was allowed to stand at room temperature for 30 min and then evaporated at 40°C under nitrogen. The residue was dissolved in 25 μℓ of chloroform for injection.
I-4. The sample was mixed with 40 μg of the internal standard, 0.5 mℓ of 0.1 *N* HCl and 0.5 mℓ of water. The mixture was extracted with 5 mℓ of benzene. The organic layer was evaporated at 60°C and the residue dissolved in 1 mℓ of the mobile phase. Aliquots of 10 to 25 μℓ were injected.
I-5. The sample was mixed with 150 μℓ of an aqueous solution of the internal standard (1 μg/mℓ) and 4 mℓ of 2.5 *N* phosphoric acid. The mixture was extracted with 5 mℓ of hexane-isopropyl alcohol (9:1). The organic phase was evaporated at 37°C under a stream of nitrogen. The residue was dissolved in 150 μℓ of the mobile phase and an aliquot of 50 μℓ was injected.
I-6. The sample was mixed with 100 μℓ of an aqueous solution of the internal standard (12.5 μg/mℓ) and 2 mℓ of 0.83 *M* phosphoric acid. The mixture was extracted with 4 mℓ of hexane-isopropyl alcohol (9:1). The organic layer was evaporated under nitrogen at room temperature. Just prior to analysis, the residue was reconstituted in 300 μℓ of the mobile phase and an aliquot of 70 μℓ was injected.
I-7. The sample was acidified with 0.1 mℓ of 3 *N* HCl and applied to a Sep-Pak C_{18} cartridge. The cartridge was washed with water and ethanol-water (35:65) and finally eluted with 2 mℓ of methanol. The eluate was evaporated at 80°C under nitrogen. The residue was dissolved in 100 μℓ of ethyl acetate and an aliquot of 40 μℓ was spotted with an autospotter.

Elution — E-1. Methanol-water (1:2).
E-2. Methanol-acetonitrile-0.02 *M* acetate buffer, pH 7 (25:20:55).
E-3. Methanol-phosphate buffer, pH 7 (60:40).

Solvent — S-1. Dichloromethane-methanol-tetrahydrofuran (85:15:0.5).

DICLOFENAC (continued)

REFERENCES

1. **Schweizer, A., Willis, J. V., Jack, B., and Kendall, M. J.,** Determination of total monohydroxylated metabolites of diclofenac in urine by electron-capture gas-liquid chromatography, *J. Chromatogr.*, 195, 421, 1980.
2. **Ikeda, M., Kawase, M., Hiramatsu, M., Hirota, K., and Ohmori, S.,** Improved gas chromatographic method of determining diclofenac in plasma, *J. Chromatogr.*, 183, 41, 1980.
3. **Jack, D. B. and Willis, J. V.,** Comments to the article "improved gas chromatographic method of determining diclofenac in plasma", *J. Chromatogr.*, 223, 484, 1981.
4. **Ikeda, M., Kawase, M., Kishie, T., and Ohmori, S.,** Supplementary data for improved gas chromatographic method of determining diclofenac in plasma. Behavior of the methyl ester and the indolone derivative of diclofenac in gas-liquid chromatography with electron-capture detection, *J. Chromatogr.*, 223, 486, 1981.
5. **Kadowaki, H., Shiino, M., Uemura, I., and Kobayashi, K.,** Sensitive method for the determination of diclofenac in human plasma by gas chromatography-mass spectrometry, *J. Chromatogr.*, 308, 329, 1984.
6. **Schneider, W. and Degen, P. H.,** Simultaneous determination of diclofenac sodium and its hydroxy metabolites by capillary column gas chromatography with electron-capture detection, *J. Chromatogr.*, 217, 263, 1981.
7. **Said, S. A. and Sharaf, A. A.,** Pharmacokinetics of diclofenac sodium using a developed HPLC method, *Arzneim. Forsch.*, 31, 2089, 1981.
8. **Chan, K. K. H., Vyas, K. H., and Wnuck, K.,** A rapid and sensitive method for the determination of diclofenac sodium in plasma by high-performance liquid chromatography, *Anal. Lett.*, 15, 1649, 1982.
9. **Godbillon, J., Gauron, S., and Metayer, J. P.,** High-performance liquid chromatographic determination of diclofenac and its monohydroxylated metabolites in biological fluids, *J. Chromatogr.*, 338, 151, 1985.
10. **Schumacher, A., Geissler, H. E., and Mutschler, E.,** Quantitative Bestimmung von Diclofenac-Natrium aus Plasma durch Absorptionsmessung mit Hilfe der direkten Auswertung von Dunnschichtchromatogrammen, *J. Chromatogr.*, 181, 512, 1980.

DICLOFENSINE

Gas Chromatography

Specimen (mℓ)	Extraction	Column (m × mm)	Packing (mesh)	Oven temp (°C)	Gas (mℓ/min)	Det.	RT (min)	Internal standard (RT)	Deriv.	Other compounds (RT)	Ref.
Plasma (1)	I-1	2 × 2	3% OV-17 Chromosorb W	265	N_2 (45)	ECD	2.8	Imipramine (2)	Demethylation; heptafluoro-butyryl	—	1

Liquid Chromatography

Specimen (mℓ)	Extraction	Column (cm × mm)	Packing (μm)	Elution	Flow (mℓ/min)	Det. (nm)	RT (min)	Internal standard (RT)	Other compounds (RT)	Ref.
Plasma (1)	I-2	10 × 8	Radial-Pack-C_{18} (5)	E-1	1.8	Fl (254, 389)	3.5	N-Ethylnordiclofensine (4.2)	Nordiclofensine[a] (5) O-Demethyldiclofensine (6.4) O-Demethylnordiclofensine (10.4)	2

[a] An additional step (N- and O-alkylation) is required for the determination of metabolites.

Extraction — I-1. The sample was mixed with 100 μℓ of an aqueous solution of the internal standard (10 μg/mℓ) and 0.2 mℓ of 1 *M* sodium hydroxide. The mixture was extracted with 1.5 mℓ of diethyl ether. The organic layer was evaporated. The residue was treated with 0.5 mℓ *n*-heptane, 10 mg of sodium carbonate, and 200 μℓ of methylchloroformate. The mixture was refluxed at 100°C for 30 min, then evaporated under nitrogen, and the residue treated with 0.5 mℓ of hydrogen bromide in glacial acetic acid. The mixture was again heated at 100°C for 10 min and cooled. The mixture was then treated with 1 mℓ of concentrated ammonia and extracted with 300 μℓ of *n*-heptane. The organic layer was treated with 50 μℓ of triethylamine and 5 μℓ of heptafluorobutyric anhydride and allowed to stand at room temperature for 1 hr. The reaction mixture was then washed with 2 mℓ of 0.1 *M* sodium hydroxide and the organic layer was used for chromatography.
I-2. The sample was treated with the residue of a 20-μℓ aliquot of an ethanolic solution of the internal standard (0.2 μg/mℓ) and 1 mℓ of 0.1 *M* sodium hydroxide. The mixture was extracted twice with 4-mℓ aliquots of diethyl ether. The combined ether layers were evaporated at 20 to 30°C under nitrogen. The residue was dissolved in 0.3 mℓ of 0.05 *M* sulfuric acid, treated with 0.3 mℓ of mercuric acetate reagent (1.5 g of mercuric acetate in 100 mℓ of acetate buffer, ph 6), and the mixture incubated at 100°C for 30 min. After cooling, the reaction mixture was exposed to high intensity light from a Pyro-Lux R-57 lamp for 20 min. After cooling, the reaction mixture was diluted with 1.4 mℓ of the mobile phase and an aliquot of 100 μℓ of the final solution was injected.

Elution — E-1. Methanol-acetonitrile-tetrahydrofuran-0.25 *M* triethylammonium phosphate (pH 2.5)-0.25 *M* acetic acid (125:375:25:150:350).

REFERENCES

1. **Hojabri, H., Dadgar, D., and Glennon, J. D.,** Gas-liquid chromatographic analysis with electron-capture detection of diclofensine in human plasma following derivatization, *J. Chromatogr.*, 311, 189, 1984.
2. **Strojny, N. and de Silvan, J. A. F.,** Determination of diclofensine, an anti-depressant agent, and its major metabolites in human plasma by high-performance liquid chromatography with fluorometric detection, *J. Chromatogr.*, 341, 313, 1985.

DICYCLOMINE

Gas Chromatography

Specimen (mℓ)	Extraction	Column (m × mm)	Packing (mesh)	Oven temp (°C)	Gas (mℓ/min)	Det.	RT (min)	Internal standard (RT)	Deriv.	Other compounds (RT)	Ref.
Plasma (1)	I-1	1.8 × 2	5% OV-225 ChromosorbHP (80/100)	245	N_2 (30)	NPD	5.2	Chlorcyclizine (8.9)	—	—	1

Extraction — I-1. The sample was mixed with 6 μℓ of a methanolic solution of the internal standard (5 μg/mℓ) and 0.5 mℓ of 1 *M* sodium hydroxide. The mixture was extracted with 5 mℓ of diethyl ether. The ether layer was back extracted into 1 mℓ of 1 *M* HCl. The aqueous layer was made alkaline with 1.2 mℓ of 1 *M* sodium hydroxide and re-extracted with 5 mℓ of ether. The organic layer was evaporated under nitrogen. The residue was dissolved in 10 μℓ of ethyl acetate containing 5% methanol for injection.

REFERENCE

1. **Beretta, E. and Vanazzi, G.,** Determination of nanogram amounts of dicyclomine with gas chromatography and nitrogen-selective detection, *J. Chromatogr.*, 308, 341, 1984.

1,3-DIDECANOYL-2-[6-(5-FLUOROURACIL-1-YL)CARBONYLAMINO]GLYCERIDE

Liquid Chromatography

Specimen (mℓ)	Extraction	Column (cm × mm)	Packing (μm)	Elution	Flow (mℓ/min)	Det. (nm)	RT (min)	Internal standard (RT)	Other compounds (RT)	Ref.
Plasma (0.1)	I-1	25 × 4	LiChrosorb ODS (10)	E-1	1.0	ABS (260)	5.8	—	—	1

Extraction — I-1. The sample was mixed with 1 mℓ of 0.1 *N* HCl and extracted twice with 4 mℓ aliquots of ethyl acetate. The combined organic layers were evaporated at room temperature under vacuum. The residue was dissolved in 200 μℓ of methanol of which 50 μℓ were injected.

Elution — E-1. Methanol-water-tetrahydrofuran (400:10:4).

REFERENCE

1. **Takada, K., Yoshikawa, H., and Muranishi, S.,** Determination of a novel 5-fluorouracil derivative in rat and human plasma by high-performance liquid chromatography, *J. Chromatogr.*, 232, 192, 1982.

DIETHYLCARBAMAZINE

Gas Chromatography

Specimen (mℓ)	Extraction	Column (m × mm)	Packing (mesh)	Oven temp (°C)	Gas (mℓ/min)	Det.	RT (min)	Internal standard (RT)	Deriv.	Other compounds (RT)	Ref.
Plasma, urine (0.5—1)	I-1	NA	2% Carbowax 20M, 5% KOH Chromosorb G (100/120)	160	N_2 (40)	NPD	6	Phenmetrazine (9)	—	—	1
Blood (1)	I-2	2.7 × 4	5% SP-2401DB Supelcoport (80/100)	180	N_2 (40)	NPD	3.8	1-Diethylcarbamyl-4-ethylpiperazine (4.5)	—	Desethyl-carbamazine (7)	2

Extraction — I-1. To the sample were added 100 μℓ of an aqueous solution of the internal standard (20 μg/mℓ) and 500 μℓ of 2 *M* sodium hydroxide. The mixture was extracted twice with 5-mℓ portions of ethyl acette. The combined organic extracts were evaporated at room temperature under a stream of nitrogen. The residue was dissolved in 200 μℓ of hexane and aliquots of 5 μℓ of this solution were injected.

I-2. The sample was mixed with 100 μℓ of a methanolic solution of the internal standard (2 μg/mℓ) and 3 mℓ of 0.1 *M* sodium hydroxide. The mixture was extracted twice with 3-mℓ portions of ethyl acetate-methyl ethyl ketone (4:1). The combined organic layer was back extracted into 1 mℓ of 0.1 *M* citric acid. The aqueous layer was washed with 2 mℓ of ethyl acetate followed by 2 × 3 mℓ of hexane, was made alkaline with 1 mℓ of 2 *M* sodium hydroxide and re-extracted three times with a mixture of toluene-dichloromethane (3:2). The pooled organic layer was evaporated under nitrogen. The residue was reconstituted with 100 μℓ of acetone and two 4-μℓ aliquots of this solution were injected.

REFERENCES

1. **Allen, G. D., Goodchild, T. M., and Weatherley, B. C.,** Determination of 1-diethyl-carbamoyl-4-methylpiperazine (diethylcarbamazine) in human plasma and urine, *J. Chromatogr.*, 164, 521, 1979.
2. **Nene, S., Anjaneyulu, B., and Rajagopalan, T. G.,** Determination of diethyl-carbamazine in blood using gas chromatography with alkali flame ionization detection, *J. Chromatogr.*, 308, 334, 1984.

DIETHYLSTILBESTROL

Gas Chromatography

Specimen (mℓ)	Extraction	Column (m × mm)	Packing (mesh)	Oven temp (°C)	Gas (mℓ/min)	Det.	RT (min)	Internal standard (RT)	Deriv.	Other compounds (RT)	Ref.
Plasma, tissue (1)	I-1	25 × Na[a]	OV-101	T.P.[b]	He (2)	MS-EI	3.6, 4, 4[c]	Dimethylstilbestrol (3.2, 4)	Trifluoro-acetyl	Dienestrol (3,6, 4.2) Indensterol (4.6) Hydroxy-diethylstilbestrol (3,6, 4.6) 4′Methoxydiethyl-stilbestrol (3.8, 4.6)	1
Urine (10)	I-2	15 × 0.32	OV-73 (0.25 μm)[d]	T.P.[e]	H_2[f]	ECD	g	—	Pentafluoro-benzyl	—	2

Liquid Chromatography

Specimen (mℓ)	Extraction	Column (cm × mm)	Packing (μm)	Elution	Flow (mℓ/min)	Det. (nm)	RT (min)	Internal standard (RT)	Other compounds (RT)	Ref.
Dosage	—	30 × 4	μ-Bondapak-C_{18} (10)	E-1	h	ABS (254)	*cis* = 8 *trans* = 5	Chloro-*m*-cresol (3)	—	3
Pure compounds	—	25 × 4.6	Jasco Chiralpak OT (+) NA	E-2	0.5	ABS[i] (254)	12		(+) IndenestrolA (10) (−) IndenestrolA (12) E,E-Dienestrol (12) Z-2-, Dienestrol (10)	4
Urine (2)	I-3	15 × 4.6	Hypersil-ODS (5)[j,k]	E-3[l]	2.0	ABS (240)	*trans* = 5	—	—	5—7
Plasma (1)	I-4	12 × 4.6	Nucleosil-C_{18} (3)	E-4	0.6	Elec-trochem[m]	10	—	n	8, 9

[a] Narrow bore capillary column.
[b] Initial temp = 150°C; rate = 15°C/min; final temp = 225°C.
[c] Multiple retention times refer to isomers.
[d] Film thickness.
[e] Initial temp = 180°C; rate = 4°C/min; final temp = 270°C.
[f] Inlet pressure = 1.6 atm.
[g] The derivatized diethylstilbestrol elutes at 270°C.
[h] Constant pressure = 80 kg/cm ^{2}.
[i] The effluent of the UV detector was also analyzed by thermospray mass spectrometry.
[j] An alternative column LiChrosorb Diol was used for purification.
[k] Column temp = 30°C.
[l] The column was washed with methanol for 2 min after the elution of diethyl-stilbestrol (8 min).
[m] +1.0 V.
[n] Determination of monoconjugates (glucuronide, sulfate and glucuronide-sulfate) is described with a number of alternative systems.

Extraction — I-1. To the sample were added 50 μℓ of a methanolic solution of the internal standard (1 μg/mℓ) and 1 mℓ of 0.1 *M* citrate-0.062 *M* phosphate buffer (92:8, pH 5.8). The mixture was extracted with 3 mℓ of dichloromethane. The organic layer was evaporated at 60°C under nitrogen. The residue was treated with 250 μℓ of dichloromethane-trifluoroacetic anhydride (3:1) for 20 min at room temperature. The excess reagent was evaporated at room temperature under nitrogen. The residue was reconstituted with 30 to 40 μℓ of dichloromethane for injection.
I-2. The sample was hydrolyzed enzymatically with glucuronidase-sulfatase overnight at 40°C. The hydrolyzed sample was applied to a Sep-Pak column, rinsed with 10 mℓ of 40% methanol and eluted with 3 mℓ of tetra hydrofuran. The eluate was evaporated, the residue dissolved in ether, and dipentafluorobenzyl ether was prepared using pentafluorobenzyl bromide. The sample after derivatization was evaporated and the residue dissolved in 100 μℓ of cyclohexane-1% tetrahydrofuran. This solution was applied to a 100 × 3 mm glass column packed with Spherisorb S-5-W. The column was eluted with cyclohexane-1% tetrahydrofuran at a flow rate of 260 μℓ/min and the column effluent was monitored at 230 nm. The detector outlet was connected to the injector of the gas-chromatograph. The fraction corresponding to the derivatized diethylstilbestrol was injected by a switching valve.
I-3. The sample was hydrolyzed with glucuronidase/sulfatase for 2 hr at 37°C and extracted with 10 mℓ of *n*-hexane. The organic layer was evaporated under nitrogen. The residue was dissolved in 0.3 mℓ of isooctane-ethanol (97:3) and an aliquot of 250 μℓ of this solution was injected on a 15 × 0.46 cm diol column. The column was eluted with isoctane-ethanol (97:3) (2 mℓ/min) and the fraction corresponding to the retention time of diethylstillbestrol was collected, evaporated, and the residue dissolved in 0.3 mℓ of methanol-water (6:4). An aliquot of 250 μℓ of this solution was injected on the reversed phase column.
I-4. The sample was treated with 0.5 mℓ of an aqueous solution of 0.03 *M* tetrabutylammonium phosphate solution followed by slow addition of 2 mℓ of methanol. After centrifugation, the supernatant was collected and the residue was further extracted with 2 mℓ of methanol. The combined supernatant was stored at 4°C. An aliquot of 1 mℓ of this extract was diluted with 3 mℓ of water, pH adjusted to pH 2 with 20% phosphoric acid, and applied to a preconditioned LiChroCart-C_{18} column. The column was washed with 2 mℓ of 0.01 *M* phosphoric acid and 1 mℓ of 15% acetonitrile. The column was eluted with 1 mℓ of the mobile phase containing 65% acetonitrile.

DIETHYLSTILBESTROL (continued)

Elution — E-1. Methanol-water (75:25).
E-2. Methanol-water (9:1).
E-3. Methanol-water (60:40).
E-4. Acetonitrile-0.005 *M* pentane sulfonic acid, pH 3.3 (50:50).

REFERENCES

1. **Abramson, F. P. and Lutz, M. P.,** Application of a capillary gas chromatographic-selected-ion recording mass spectrometric technique to the analysis of diethylstilbestrol and its phosphorylated precursors in plasma and tissues, *J. Chromatogr.*, 339, 87, 1985.
2. **Grob, K., Jr., Neukom, H. P., and Etter, R.,** Coupled high-performance liquid chromatography-capillary gas chromatography as a replacement for gas chromatography-mass spectrometry in the determination of diethylstilbestrol in bovine urine, *J. Chromatogr.*, 357, 416, 1986.
3. **Lea, A. R., Kayaba, W. J., and Hailey, D. M.,** Analysis of diethylstilbestrol and its impurities in tablets using reversed-phase high-performance liquid chromatography, *J. Chromatogr.*, 177, 61, 1979.
4. **Parker, C. E., Levy, L. A., Smith, R. W., Yamaguchi, K., and Gaskell, S. J.,** Separation and detection of enantiomers of stilbestrol analogues by combined high-performance liquid chromatography-thermospray mass spectrometry, *J. Chromatogr.*, 344, 378, 1985.
5. **Jansen, E. H. J. M., Zoontjes, P. W., Both-Miedema, R., Van Blitterswijk, H., and Stephany, R. W.,** Screening of parts-per-billion levels of diethylstilbestrol in bovine urine by high-performance liquid chromatography with ultraviolet detection, *J. Chromatogr.*, 347, 379, 1985.
6. **Jansen, E. H. J. M., Van Den Berg, R. H., Van Blitterswijk, H., Both-Miedema, R., and Stephany, R. W.,** A highly specific detection method for diethylstilbestrol in bovine urine by radioimmunoassay following high-performance liquid chromatography, *Food Add. Contam.*, 2, 271, 1985.
7. **Jansen, E. H. J. M., Van Den Berg, R. H., Zomer, G., Both-Miedema, R., Enkelaar-Willemsen, C., and Stephany, R. W.,** Combination of high-performance liquid chromatography and chemiluminescent immunochemical detection of hormonal anabolics and their metabolites, *Anal. Chim. Acta*, 170, 21, 1985.
8. **Oelschlager, V. H., Rothley, D., and Dunzendorfer, U.,** Direkte Bestimmung von Diethylstilbestrol und seinem Monokonjugaten in Plasma, *Arzneim. Forsch.*, 36, 759, 1986.
9. **Oelschlager, V. H., Rothley, D., and Dunzendorfer, U.,** Plasmaspiegel von Fosfestrol und seinem Monophosphat, von Diethylstilbestrol und seinem Monoglucuronid nach intravenoser Gabe bei Patienten mit metastasierendem Prostatakarzinom, *Arzneim. Forsch.*, 34, 1333, 1984.

DIFLORASONE DIACETATE

Liquid Chromatography

Specimen (mℓ)	Extraction	Column (cm × mm)	Packing (μm)	Elution	Flow (mℓ/min)	Det. (nm)	RT (min)	Internal standard (RT)	Other compounds (RT)	Ref.
Dosage	I-1	10 × 4.6	Perkin-Elmer Silica gel (3)	E-1	2.5	ABS (254)	11	Isoflupredone acetate (24)	a	1

[a] The chromatogram shows the separation of the related steroids.

Extraction — I-1. A weighed amount of the sample was mixed with 30 mℓ of the internal standard solution (40 μg/mℓ in chloroform). After centrifugation, aliquots (10 μℓ) of the chloroform layer were injected.

Elution — E-1. Butyl chloride-dichloromethane-tetrahydrofuran-acetic acid (350:125:10:15).

REFERENCE

1. **Shaw, M. C. and Vanderwielen, A. J.,** Liquid chromatographic assay for diflorasone diacetate in cream and ointment formulations, *J. Pharm. Sci.*, 73, 1606, 1984.

DIFLUNISAL

Liquid Chromatography

Specimen (mℓ)	Extraction	Column (cm × mm)	Packing (μm)	Elution	Flow (mℓ/min)	Det. (nm)	RT (min)	Internal standard (RT)	Other compounds (RT)	Ref.
Plasma (0.5)	I-1	15 × 4.6	LiChrosorb RP-8 (5)[a]	E-1	1.4	ABS (254)	4.1	Naproxen (2.6)	—	1
Plasma, urine	I-1	25 × 4.6	Ultrasphere-ODS (5)[b,c]	E-2	1.0	Fl (315, 389)	7	Naproxen (10)	—	2
Plasma, urine (0.5)	I-2	15 × 4.5	Hypersil-ODS (5)	E-3	1.3	ABS (251)	6	Flufenamic acid (8.5)	—	3
Plasma, serum, urine (0.1)	I-3	25 × 4	Spherisorb-ODS (5)[d]	E-4	1.0	ABS (254)	7	1-Hydroxy-2-naphthoic acid (5)	—	4
Plasma, urine (0.05)	I-4	30 × 4	μ-Bondapak-C_{18} (10)	E-5	2.0	ABS[e] (254)	10.1	Desmethyl-naproxen (4.5)	Difunisal ether glucuronide (3.2) Diflunisal ester glucuronide (6)	5
Urine (1)	I-5	30 × 5	Hypersil-ODS (5)	E-6; gra-dient	2.0	ABS[f] (254)	27	5-(4′-Fluoro-phenyl) sali-cylic acid (25)	Diflunisal ether glucuronide (10) Diflunisal ester glucuronide (16)	6

[a] Column temp = 32°C.
[b] Protected by a Brownlee RP-8 guard column.
[c] Column temp = 50°C.
[d] Column temp = 35°C.
[e] A fluorescence detector (ex = 235 nm; em = 370 nm) was also used to monitor glucuronides.
[f] A fluorescence detector (ex = 215 nm; em = 440 nm) was also used.

Extraction — I-1. The sample was mixed with 0.2 mℓ of a methanolic solution of the internal standard (0.25 mg/mℓ) and 0.7 mℓ of 1.5 *N* HCl. The mixture was extracted with diethyl ether-*n*-hexane (1:1). The organic layer was evaporated at 30°C under nitrogen. The residue was dissolved in 1 mℓ of methanol and aliquots of 10 μℓ were injected.
I-2. The sample was mixed with 0.5 mℓ of the internal standard solution (0.5 mg/mℓ in 0.001 *M* sodium bicarbonate) and 0.5 mℓ of acetone. After centrifugation 25-μℓ aliquots of the supernatant were injected.
I-3. The sample was mixed with 20 μℓ of the internal standard solution (150 μg/mℓ in methanol) and 1 mℓ of 1% sulfuric acid containing 1.42% sodium sulfate. The mixture was extracted with 3 mℓ of diethyl ether. The organic layer was evaporated under nitrogen at room temperature. The residue was dissolved in 200 μℓ of the mobile phase and 50-μℓ aliquots were injected.
I-4. The sample was mixed with 150 μℓ of the internal standard solution (50 mg/ℓ in acetonitrile-20% trichloroacetic acid, 1:3). After centrifugation, an aliquot of 10 μℓ of the supernatant was injected.
I-5. The sample was mixed with 1 mℓ of water, 3 drops of glacial acetic acid, and 30 μg of the internal standard. After centrifugation, aliquots of the clear solution were analyzed. (Alternative procedure involving hydrolysis of the ester glucuronide is described.)

Elution — E-1. Methanol-water (50:50) containing 0.01 *M* tetramethylammonium hydrogen sulfate and tris(hydroxymethyl)aminomethane.
E-2. Methanol-0.05 *M* phosphate buffer, pH 3 (64:36).
E-3. Isopropanol-ethyl acetate-0.08 *M* potassium nitrate in 2% acetic acid (25:20:55).
E-4. Methanol-0.1 *M* phosphoric acid-tetrahydrofuran (50:33.3:16.7).
E-5. Acetonitrile-0.01 *M* citrate buffer, pH 3 (30:70).
E-6. (A) Methanol-2% acetic acid (30:70); (B) Methanol-2% acetic acid (65:35). Linear gradient from 60% (A) to 0% (A) in 25 min.

REFERENCES

1. **Van Loenhout, J. W. A., Ketelaars, H. C. J., Gribnau, F. W. J., Van Ginneken, C. A. M., and Tan, Y.,** Rapid high-performance liquid chromatographic method for the quantitative determination of diflunisal in plasma, *J. Chromatogr.*, 182, 487, 1980.
2. **Ray, J. E. and Day, R. O.,** High-performance liquid chromatographic analysis of diflunisal in plasma and urine: application to pharmacokinetic studies in two normal volunteers, *J. Pharm. Sci.*, 72, 1403, 1983.
3. **Balali-Mood, M., King, I. S., and Prescott, L. F.,** Rapid estimation of diflunisal in plasma and urine by high-performance liquid chromatography and a comparison with a fluorometric method, *J. Chromatogr.*, 229, 234, 1982.
4. **Midskov, C.,** Rapid reversed-phase high-performance liquid chromatographic assay of diflunisal in biological fluids, *J. Chromatogr.*, 278, 439, 1983.
5. **Veenendaal, J. R. and Meffin, P. J.,** Direct analysis of diflunisal ester and ether glucuronides by high-performance liquid chromatography, *J. Chromatogr.*, 307, 432, 1984.
6. **Musson, D. G., Lin, J. H., Lyon, K. A., Tocco, D. J., and Yeh, K. C.,** Assay methodology for quantification of the ester and ether glucuronide conjugates of diflunisal in human urine, *J. Chromatogr.*, 337, 363, 1985.

α-DIFLUOROMETHYLORNITHINE

Liquid Chromatography

Specimen (mℓ)	Extraction	Column (cm × mm)	Packing (μm)	Elution	Flow (mℓ/min)	Det. (nm)	RT (min)	Internal standard (RT)	Other compounds (RT)	Ref.
Plasma, urine cells	I-1	30 × 4	DC-6a resin[a] (NA)	E-1	0.5	Fl[b] (340, 440)	30	—	—	1

[a] Two columns are used. While one column is being used for analysis, the other is being regenerated after previous injection.
[b] The column eluent is treated with *o*-phthalaldehyde-mercaptoethanol reagent prior to fluorescence detection.

Extraction — I-1. The sample was treated with one half volume of 20% trichloroacetic acid, allowed to stand for 30 min at 0°C and then centrifuged. The supernatant was further diluted with 0.2 *M* (pH 2.2) lithium citrate buffer.

Elution — E-1. 0.668 *M* Lithium citrate buffer, pH 4.6, with HCl.

REFERENCE

1. **Grove, J., Fozard, J. R., and Mamont, P. S.,** Assay of α-difluoromethylornithine in body fluids and tissues by automatic amino-acid analysis, *J. Chromatogr.*, 223, 409, 1981.

DIGITOXIN

Thin-Layer Chromatography

Specimen (mℓ)	Extraction	Plate (Manufacturer)	Layer (mm)	Solvent	Post-separation treatment	Det. (nm)	Rf	Internal standard (Rf)	Other compounds (Rf)	Ref.
Serum (2)	I-1	20 × 20 cm (Baker)	Silica 250 (0.25)	S-1	Sp: 3% Chloramine-T-25% ethanolic solution of trichloroacetic acid (1:4)[a]	Visual (360)[a] Radioimmuno-assay	0.24	[^{3}H]Digitoxin	Digitoxigenin bisdigitoxoside (0.41) Digitoxigenin monodigitoxoside(0.58) Digitoxigenin (0.79)	1

[a] Only the standards are sprayed and visualized. The zones of unknowns of corresponding Rf values are marked.

Extraction — I-1. The sample was spiked with 0.1 mℓ of a solution of tritiated digitoxin in phosphate buffered saline (0.15 *M* sodium chloride, 0.01 *M* dipotassium hydrogen phosphate, 5% bovine serum albumin) and extracted three times with 5-mℓ aliquots of dichloromethane. The combined organic extracts were evaporated under nitrogen. The residue was dissolved in 0.7 mℓ of chloroform-methanol (3:1) and spotted on a reversed phase (KC_{18} 200 μm) TLC plate. The plate was developed with dioxane-methanol-water (2:5:3). After drying only the channels of standards were sprayed and the plate was heated at 100°C for 8 min. The plate was visualized under long wave UV light and the zones corresponding to different compounds were marked, scraped and eluted three times with 4-mℓ portions of ethanol. The pooled eluates were evaporated. The residue dissolved in chloroform-methanol and the solution applied on silica plates.

Solvent — S-1. Isopropyl ether-methanol (9:1) three times developments.

REFERENCE

1. **Graves, P. E., Perrier, D., and Marcus, F. I.,** Quantitation of digitoxin and the *bis*- and monodigitoxosides of digitoxigenin in serum, *J. Chromatogr.*, 278, 397, 1983.

DIGOXIN

Liquid Chromatography

Specimen (mℓ)	Extraction	Column (cm × mm)	Packing (μm)	Elution	Flow (mℓ/min)	Det. (nm)	RT (min)	Internal standard (RT)	Other compounds (RT)	Ref.
Dosage	—	60 × 4	Partisil silica (10)	E-1[a]	1.0	ABS (254)	36.4	—	Digitoxigenin (10.2) Gitoxigenin (15.6) Digoxigenin (18.4) Digitoxin (20.4)	1
Serum, tissue (1)	I-1	30 × 3.9	μ-Bondapak-C_{18} (10)	E-2	3.0	ABS[b] (254)	NA	Nalorphine, Ethoxzolamide	—	2
Serum (1)	I-2	25 × 3.2	Spherisorb-ODS (5)	E-3	1.0	ABS[b] (254)	8.6		Digoxigenin (4.9) Digoxigenin mono-digoxitoxoside (5.4) Digoxigenin *bis*-digoxitoxoside (6.8)	3
Urine (20)	I-3	15 × 4.5	LiChrosorb Si 60[c] (5)	E-4	1.5	ABS (220)	14.5	—	Digoxigenen (6) Digoxigenin mono-digitoxoside (6) Digoxigenin didigitoxoside (7.5)	4
Urine (10)	I-4	25 × 4.6	Partisil silica (10)	E-5	1.8	ABS (254)	53.7	Digitoxigenin (16.8)	Digoxigenin (24.6) Digoxigenin mono-digitoxoside (30.6) Digoxigenin di-digitoxoside (40.8)	5

Liquid Chromatography

Specimen (mℓ)	Extraction	Column (cm × mm)	Packing (μm)	Elution	Flow (mℓ/min)	Det. (nm)	RT (min)	Internal standard (RT)	Other compounds (RT)	Ref.
Pure compounds	I-5	15 × 0.5	SC-01-ODS (5)	E-6	0.008	ABS (230)	38	Gitoxin (50)	Digoxigenin (15)[d] β-Methyldigoxin	6
Serum (0.5)	I-6	12.5 × 4	LiChrosorb RP-18 (5)[e]	E-7	1.0	Fl[f] (360, 480)	7	—	Digoxigenin (3) Digoxigenin monoglucoside (3.2) Digoxigenin *bis*-glucoside (4.5) Lantosid (6)	7
Serum (1)	I-7	25 × 4.6	Brownlee-C_8 (5)	E-8	1.0	ABS[b] (220)	71	—	Digoxigenin (21) 3-Ketodigoxigenin (28) 3-Epidigoxigenin (35) Digoxigenin monodigitoxoside (45) Digoxigenin *bis*-digitoxoside (59)	8, 9
Pure compounds	—	25 × 4	Nucleosil-C_{18} (10)	E-9	1.0	ABS (220)	12	—	Digoxigenin (4.1) Digoxigenin monodigitoxoside (4.9) Digoxigenin *bis*-digitoxoside (7.7)[g]	10
Pure compounds	—	15 × 4.6	Spherisorb-ODSII (3)[h]	E-10	0.3	Fl; (360, 425)	38	—	Dihydrodigoxin (36.5) Dihydrodigoxigenin (16) Digoxigenin (16) Digoigenin monodigitoxoside (17) Digoxigenin *bis*-digitoxoside (21.5) Spironolactone (58)	11

DIGITOXIN (continued)

[a] Retention data in a number of alternative solvents given.
[b] The fraction corresponding to digoxin was collected evaporated and subjected to radioimmunoassay.
[c] A reversed phase packing LiChrosorb RP-8 was also used.
[d] Separation of a number of digoxin and digitoxin metabolites is shown.
[e] Protected by a 4 × 4 mm precolumn packed with LiChrosorb RP-18 (10 μm).
[f] The column effluent is treated with 37% HCl at 70°C for 1.2 min prior to detection.
[g] Condition for separate and simultaneous separation of digitoxin and its metabolites are also described.
[h] Protected by a direct connect guard column packed with 37 μm RP-18 packing.
[i] The column effluent was treated with 1.1 m*M* hydrogen peroxide in a 0.1% ascorbic acid solution and concentrated hydrochloric acid prior to detection.

Extraction — I-1. The sample was mixed with 100 μℓ of an aqueous solution of 20 μg/mℓ of ethoxzolamide and 50 μℓ of 1.5 *M* HCl. The mixture was extracted with 12 mℓ of dichloromethane. The organic extract was evaporated with a stream of nitrogen. The residue was reconstituted in 150 μℓ of methanol and 3 μℓ of a 2-mg/mℓ solution of nalorphine was added as a marker. Of this, 100 μℓ was injected onto the column.
I-2. The sample was treated with 1.5 mℓ of acetonitrile and allowed to stand at 4°C for 5 min. After centrifugation the supernatant was diluted with 1 mℓ of water and again centrifuged. A 1-mℓ aliquot of the supernatant was added to 1.8 mℓ of water and injected into a 2.6-mℓ loop injector. The eluent corresponding to the retention volume of digoxin was collected, evaporated and quantitated by radioimmunoassay.
I-3. The sample was extracted twice with 20 and 15 mℓ volumes of dichloromethane. Aliquots (10 mℓ) from each extract were combined and evaporated under nitrogen in the presence of 20 μℓ of 1-pentanol. The residue was dissolved in 250 μℓ of mobile phase and aliquots of 100 μℓ were injected. Reversed phase chromatography was used to separate dihydrodigoxin.
I-4. The sample was mixed with 0.5 mℓ of the internal standard solution (20 μg/mℓ in dichloromethane) and extracted with 20 mℓ of dichloromethane. The organic phase was washed with 15 mℓ of a 5% sodium bicarbonate solution and then evaporated at 50°C under a stream of nitrogen. The residue was treated with 200 μℓ of a solution of dinitrobenzoyl chloride in pyridine (85 mg/mℓ). The mixture was allowed to stand at room temperature for 10 min with gentle shaking. The mixture was then evaporated with a stream of nitrogen at 50°C, the residue treated with 2 mℓ of 5% sodium bicarbonate solution containing 2 mg/mℓ 4-dimethylaminopyridine and extracted with 1 mℓ of chloroform. The organic layer was washed four times with 3-mℓ portions of 0.05 *M* HCl containing 5% sodium chloride. Aliquots of the chloroform extract were injected.
I-5. A 15-mg amount of 3,5-dinitrobenzoyl chloride was added to a solution of cardiac steroid in 0.2 mℓ of dry pyridine, the mixture shaken for 2 hr and evaporated with a stream of nitrogen. The residue was dissolved in 1.5 mℓ of ethyl acetate and washed four times with 1 mℓ of 5% sodium bicarbonate solution which contained 2.5 mg of 4-dimethylaminopyridine. The organic layer was further washed with 1 mℓ of 1% HCl and four times with 1 mℓ of water. The ethyl acetate layer was evaporated under a stream of nitrogen and the residue dissolved in the mobile phase for injection.
I-6. The sample is applied to a clean up cartridge (10 × 4 mm Lichrosorb, RP-2, 30 μm) replacing the loop of the injector. The loop column is washed with 0.5 mℓ of water, followed by 0.5 mℓ of 25% methanol while still in load position. The loop column is then switched to inject position.
I-7. The sample was applied to a C_{18}-Bond Elut column which was prewashed with 6 mℓ of methanol and 3 mℓ of water. The column was washed with 1 mℓ of water and eluted with 3 mℓ of methanol. The eluate was evaporated under a stream of air and the residue reconstituted with 200 mℓ of 20% isopropanol for injection.

Elution — E-1. (A) Tetrahydrofuran-methanol-hexane (15:7.5:77.5); (B) tetrahydrofuran-methanol-isopropanol-hexane (15:40:20:25). Isocratic 90%(A) + 10%(B).
E-2. Acetonitrile-0.025 *M* KH_2PO_4, pH 6.9 (30:70).
E-3. Acetonitrile-ethyl alcohol-water (25:3.3:71.7).
E-4. *n*-Heptane-1-pentanol-acetonitrile-water (64:26:9:1).
E-5. Acetonitrile-dichloromethane-hexane (3:3:8).
E-6. Acetonitrile-methanol-water (3:1:1).
E-7. (A) Methanol-water (65 + 35); (B) 0.2 mℓ 0.24% H_2O_2 + 200 μℓ phosphoric acid/ℓ of water. Isocratic 90%(A) + 10%(B).
E-8. Isopropanol-water (20:80).
E-9. Acetonitrile-methanol-water (20:20:60).
E-10. Methanol-ethanol-isopropanol-water (52:3:1:45).

REFERENCES

1. **Linley, P. A. and Mohamed, A. G. M.,** Separation of digitalis cardenolides by HPLC, *J. High Resol. Chromatogr. Chromatogr. Commun.*, 4, 239, 1981.
2. **Wagner, J. G., Dick, II, M., Behrendt, D. M., Lockwood, G. F., Sakmar, E., and Hees, P.,** Determination of myocardial and serum digoxin concentrations in children by specific and nonspecific assay methods, *Clin. Pharmacol. Ther.*, 33, 577, 1982.
3. **Loo, J. C. K., McGilveray, I. J., and Jordan, N.,** The estimation of serum digoxin by combined HPLC separation and radioimmunological assay, *J. Liq. Chromatogr.*, 4, 879, 1981.
4. **Eriksson, B. M., Tekenbergs, L., Magnusson, J. O., and Molin, L.,** Determination of tritiated digoxin and metabolites in urine by liquid chromatography, *J. Chromatogr.*, 223, 401, 1981.
5. **Bockbrader, H. N. and Reuning, R. H.,** Digoxin and metabolites in urine: a derivatization-high-performance liquid chromatographic method capable of quantitating epimers of dihydrodigoxin, *J. Chromatogr.*, 310, 85, 1984.
6. **Fujii, Y., Oguri, R., Mitsuhashi, A., and Yamazaki, M.,** Micro HPLC separation of 3,5-dinitrobenzoyl derivatives of cardiac glycosides and their metabolites, *J. Chromatogr. Sci.*, 21, 495, 1983.
7. **Reh, E. and Jork, H.,** Determination of digitalis-glycosides by HPLC and reaction-detection, *Fresenius Z. Anal. Chem.*, 322, 365, 1985.
8. **Gault, M. H., Longerich, L., Dawe, M., and Vasdev, S. C.,** Combined liquid chromatography/radioimmunoassay with improved specificity for serum digoxin, *Clin. Chem.*, 31, 1272, 1985.
9. **Gault, M. H., Longerich, L. L., Loo, J. C. K., Ko, P. T. H., Fine, A., Vasdev, S. C., and Dawe, M. A.,** Digoxin biotransformation, *Clin. Pharmacol. Ther.*, 35, 74, 1984.
10. **Plum, J. and Daldrup, T.,** Detection of digoxin, digitoxin, their cardioactive metabolites and derivatives by high-performance liquid chromatography and high-performance liquid chromatography-radioimmunoassay, *J. Chromatogr.*, 377, 221, 1986.
11. **Kwong, E. and McErlane, K. M.,** Development of a high-performance liquid chromatographic assay for digoxin using post-column fluorogenic derivatization, *J. Chromatogr.*, 377, 233, 1986.

DIHYDRALAZINE

Gas Chromatography

Specimen (mℓ)	Extraction	Column (m × mm)	Packing (mesh)	Oven temp (°C)	Gas (mℓ/min)	Det.	RT (min)	Internal standard (RT)	Deriv.	Other compounds (RT)	Ref.
Plasma (1)	I-1	1.5 × 4	3% OV-225 GasChrom Q (230-270)	250	N_2 (50)	ECD	6	6-Trifluoromethyl-dihydralazine (4)[a]	Conversion to: methoxytetra-zolophthal-azine	—	1

Liquid Chromatography

Specimen (mℓ)	Extraction	Column (cm × mm)	Packing (μm)	Elution	Flow (mℓ/min)	Det. (nm)	RT (min)	Internal standard (RT)	Other compounds (RT)	Ref.
Plasma (2)	I-2	25 × 4.6	Partisil ODS-2 (10)	E-1	2.0	ABS (230)[b]	6	Methylhydra-lazine (3.5)	—	2
Plasma (1)	I-3	20 × 2.1	μ-Bondapak-C_{18} (10)[c,d]	E-2	0.9	Fl (230, 430)[e]	7	Methyl hydra-lazine (4)	Hydralazine (3)	3

[a] Two peaks because of isomers. The major peak at 4 min is used for quantitation of dihydralazine.
[b] Ditetrazolophlhalazine derivatives are detected.
[c] Protected by a 50 × 2.1 mm precolumn packed with Co:Pell ODS.
[d] Column temp = 55°C.
[e] Detected as methoxytetrazolophthalazine.

Extraction — I-1. The sample was treated with 4 mℓ of 1.5 *N* sulfuric acid and 50 μℓ of a solution of the internal standard (1 μg/mℓ in 0.1 *N* HCl). The mixture was incubated at 90°C with agitation for 25 min. After cooling to room temperature, 0.1 mℓ of 50% sodium nitrite solution was added and allowed to stand for 15 min at room temperature. Then, the mixture was adjusted to pH 4.5 with 4 mℓ of potassium phthalate buffer, pH 13.8 and extracted with 5 mℓ of toluene. The organic layer was evaporated under a stream of nitrogen at 40°C. A solution of 8.6 μmol sodium methylate in 1 mℓ of toluene (with 5% methanol) was added to the residue and left at 50°C for 1 hr, 3 mℓ of buffer solution pH 7 were then added and shaken for 10 min. After centrifugation, aliquots of 5 μℓ of the organic phase were injected.

DIHYDRALAZINE (continued)

I-2. The sample was mixed with the internal standard (40 ng/mℓ), 2 mℓ of 2 *M* HCl, 2 mℓ of water and 200 μℓ of 50% sodium nitrite. The mixture was allowed to stand at room temperature for 15 min, then adjusted to pH 10 with 2.5 *M* sodium hydroxide and extracted with 10 mℓ of benzene. The organic phase was evaporated at 37°C. The dry residue was redissolved in 100 μℓ of methanol and a portion (50 to 70 μℓ) injected.
I-3. The sample was treated with 100 μℓ of 50% sodium nitrite, 2 mℓ of 0.02 *M* HCl and 10 μℓ of the internal standard solution (1 μg/mℓ in 0.1 *M* HCl). The mixture was allowed to stand at room temperature for 15 min, then 3 mℓ of phosphate buffer, pH 8 added, and extracted with 4 mℓ of chloroform. The organic phase was evaporated under a stream of nitrogen. To the residue 1 mℓ of 8.6 m*M* sodium methylate in toluene-methanol (95:5) was added and the mixture incubated at 50°C for 45 min. The cooled mixture was treated with 3 mℓ of pH 7 phosphate buffer and 1 mℓ of chloroform. After mixing and centrifugation, the organic layer was evaporated, the residue dissolved in 100 μℓ of the mobile phase and an aliquot of 70 μℓ injected.

Elution — E-1. Methanol-0.01 *M* KH_2PO_4, pH 3 (60:40)
E-2. Acetonitrile-1.8 m*M* phosphoric acid (15:85).

REFERENCES

1. **Degen, P. H., Brechbuhler, S., Schneider, W., and Zbinden, P.,** Determination of apparent dihydralazine in plasma by gas-liquid chromatography and electron-capture detection, *J. Chromatogr.*, 233, 375, 1982.
2. **Waller, A. R., Chasseaud, L. F., and Taylor, T.,** High-performance liquid chromatographic determination of dihydralazine in human plasma, *J. Chromatogr.*, 173, 202, 1979.
3. **Rouan, M. C. and Campestrini, J.,** Liquid chromatographic determination of dihydralazine and hydralazine in human plasma and its application to pharmacokinetic studies of dihydralazine, *J. Pharm. Sci.*, 74, 1270, 1985.

DIHYDROCRISTINE

Liquid Chromatography

Specimen (mℓ)	Extraction	Column (cm × mm)	Packing (μm)	Elution	Flow (mℓ/min)	Det. (nm)	RT (min)	Internal standard (RT)	Other compounds (RT)	Ref.
Plasma (1)	I-1	25 × 4	LiChrosorb RP-8 (10)	E-1	1.0	Fl[a] (295, 350)	7.5	Dihydro-ergotamine (5.8)	—	1
Plasma (5)	I-2	15 × 4	LiChrosorb Si-60 (5)	E-2	0.5	Fl (285, 345)	6[b]	Dihydro-ergosine (7.5)	—	2

DIHYDROCRISTINE (continued)

[a] An absorbance detector (223 nm) was also used.
[b] The drug analyzed is dihydroergotoxine, which is a mixture of dihydroergocornine, dihydroergocryptine and dihydroergocristine. All the three drugs elute together.

Extraction — I-1. The sample was mixed with 50 μℓ of an aqueous solution of the internal standard (1 μg/mℓ), 30 μℓ of 5 *M* sodium hydroxide and extracted with 7 mℓ of chloroform. The organic layer was evaporated at 40°C under a stream of nitrogen. The residue was reconstituted in 100 μℓ of the mobile phase and 10 to 30 μℓ were injected. I-2. The sample was made alkaline with diethylamine and extracted with 7 mℓ of benzene-toluene-ethylacetate (3:1:1). The organic layer was evaporated under vacuum. The residue was dissolved in 70 μℓ of the mobile phase and a 50-μℓ aliquot of this solution was injected.

Elution — E-1. Acetonitrile-phosphate buffer pH 7.2 (60:40). E-2. Acetonitrile-methanol-ammonia (12:1:0.008).

REFERENCES

1. **Zecca, L., Bonini, L., and Bareggi, S. R.,** Determination of dihydroergocristine and dihydroergotamine in plasma by high-performance liquid chromatography with fluorescence detection, *J. Chromatogr.*, 272, 401, 1983.
2. **Zorz, M., Marusic, A., Smerkolj, R., and Prosek, M.,** Quantitative determination of low concentrations of DHETX m.s. in human plasma by high performance liquid chromatography with fluorescence detection, *J. High Resolut. Chromatogr. Chromatogr. Commun.*, 6, 306, 1983.

DIHYDROERGOTAMINE

Thin-Layer Chromatography

Specimen (mℓ)	Extraction	Plate (Manufacturer)	Layer (mm)	Solvent	Post-separation treatment	Det (nm)	Rf	Internal standard (Rf)	Other compounds (Rf)	Ref.
Plasma (2)	I-1	20 × 20 cm (Merck)	Silica gel 60 (0.25)	S-1	—	Fl reflectance (264, 390)[a]	0.41	Dihydroergo-kryptine (0.51)	Dihydrocris-tine(0.54)	1

[a] ZeissFl-39 filter.

Extraction — I-1. The sample was spiked with an ethanolic solution of the internal standard, made alkaline with 1 mℓ of 0.1 *N* of sodium hydroxide, and extracted twice with 3-mℓ portions of dichloromethane. The combined organic extract was evaporated under nitrogen in darkness. The residue was reconstituted in ethanol and applied to a TLC plate.

Solvent — S-1. Ethanol-benzene-chloroform (1:2:4) containing 1 mℓ of ammonia per 200 mℓ of solvent.

REFERENCE

1. **Riedel, E., Kreutz, G., and Hermsdorf, D.,** Quantitative thin-layer chromatographic determination of dihydroergot alkaloids, *J. Chromatogr.*, 229, 417, 1982.

9-(1,3-DIHYDROXY-2-PROPOXY)METHYL GUANINE

Liquid Chromatography

Specimen (mℓ)	Extraction	Column (cm × mm)	Packing (μm)	Elution	Flow (mℓ/min)	Det. (nm)	RT (min)	Internal standard (RT)	Other compounds (RT)	Ref.
Liver homogenate	I-1	25 × 4.6	Partisil SCX (10)[a,b]	E-1	NA	ABS (254)	13.3	—	c	1

[a] Connected in series with a 50 × 4.6 mm 10 μm reversed phase column.
[b] Protected by a 30 × 4.6 mm guard column packed with a mixture of Co:Pell ODS and SCX media.
[c] Retention times of a number of mono and diesters are given.

Extraction — I-1. The tissue homogenate was diluted with acetonitrile or methanol. After centrifugation, aliquots of the supernatant were injected.

Elution — E-1. Methanol-0.001 *M* phosphate buffer, pH 2.5.

REFERENCE

1. **Benjamin, E. J., Firestone, B. A., and Schneider, J. A.,** A dual-column HPLC method for the simultaneous determination of DHPG (9-[(1,3-Dihydroxy-2-Propoxy)Methyl]Guanine) and its mono and diesters in biological samples, *J. Chromatogr. Sci.*, 23, 168, 1985.

2,6-DIISOPROPYLPHENOL

Liquid Chromatography

Specimen (mℓ)	Extraction	Column (cm × mm)	Packing (µm)	Elution	Flow (mℓ/min)	Det. (nm)	RT (min)	Internal standard (RT)	Other compounds (RT)	Ref.
Blood (1)	I-1	20 × 5	Hypersil-ODS (5)	E-1	1.5	ABS (276)	7.5	Thymol (5)	—	1

Extraction — I-1. The sample was mixed with an aliquot of a methanolic solution of the internal standard and 1 mℓ of 0.1 *M* KH_2PO_4. The mixture was extracted with 5 mℓ of cyclohexane. To an aliquot of the organic layer (4.5 mℓ), 60 µℓ of a solution of 2,6-dichloroquinone-4-chloroimide (1 mg/mℓ in isopropanol) and 50 µℓ of 24% tetramethylammonium hydroxide diluted 1:10 with isopropanol were added. The reaction was allowed to proceed at room temp for 20 min,then the reaction mixture was diluted with 1 mℓ of water. After mixing and centrifugation, the organic layer was discarded, aqueous sodium chloride (25%, 1 mℓ) was added and extracted with 5 mℓ of ether. The organic layer was evaporated under a stream of nitrogen. The residue was dissolved in 0.5 mℓ of acetonitrile-water-ammonia (80:20:0.05) for injection.

Elution — E-1. Acetonitrile-water-trifluoroacetic acid (80:20:0.1).

REFERENCE

1. **Adam, H. K., Douglas, E. J., Plummer, G. F., and Cosgrove, M. B.,** Estimation of ICI 35,868 (Diprivan®) in blood by high-performance liquid chromatography, following coupling with Gibbs' reagent, *J. Chromatogr.*, 223, 232, 1981.

DILITAZEM

Gas Chromatography

Specimen (mℓ)	Extraction	Column (m × mm)	Packing (mesh)	Oven temp (°C)	Gas (mℓ/min)	Det.	RT (min)	Internal standard (RT)	Deriv.	Other compounds (RT)	Ref.
Plasma (2)	I-1	2 × 4	1% OV-17 Chromosorb W (100/120)	280	N_2 (40)	NPD	4	N-Butyryl-des-acetyldiltiazem (6)	—	—	1

Plasma (1)	I-2	2 × 2	GasChrom Q (80/100)	A_2-methane (60)	ECD	9.2	Loxapine (3.7)	Trimethyl-silyl	Desacetyl-dilitazem (6.6)	2

Liquid Chromatography

Specimen (mℓ)	Extraction	Column (cm × mm)	Packing (μm)	Elution	Flow (mℓ/min)	Det. (nm)	RT (min)	Internal standard (RT)	Other compounds (RT)	Ref.
Plasma (1)	I-3	25 × 4.6	Zorbax CN (6)	E-1	1.5	ABS (237)	6.7	Verapamil (8)	Desacetyl-diltiazem (5)	3
Plasma (2)	I-4	30 × 3.9	μ-Bondapak-C_{18} (10)[a,b]	E-2	1.5	ABS (240)	4	Prazepam (8)	Desacetyl-diltiazem (3)	4
Urine (1)	I-5	25 × 4.6	Spherisorb-ODS (5)	E-3	2.5	ABS (210)	10.1	Loxapine (13)	Desacetyl-diltiazem (8.9) Nordiltiazem (7.4)[d]	5
Plasma (1—2)	I-6	30 × 3.9	μ-Bondapak-C_{18} (10)	E-4	1.8	ABS (254)	9	Desipramine (10)	Desacetyl-diltiazem (6)	6
Plasma (1)	I-7	12.5 × 4.6	Spherisorb-ODS-II (5)	E-5	1.2	ABS (237)	7.6	Propionyl-de-sacetyl- (10.8)	Desacetyl-diltiazem (6) Nordiltiazem (4.9) Desacetyl-nordiltiazem (3.9) Des-O-methyl-desacetyl-diltiazem (2.9)	7

[a] A silica saturation column was used between the injector and the pump.
[b] A 60 × 2 mm Perisorb RP-18 column (30 to 40 μm) was used before the analytical column.
[c] Flow gradient; 2.5 mℓ/min from 0 to 5.5 min, 3 mℓ from 5.5 to 13 min.
[d] Separation of four additional metabolites is shown.

DILITAZEM (continued)

Extraction — I-1. The lyophalized sample was mixed with 1 mℓ of 0.1 *M* pH 7 phosphate buffer, 50 μℓ of the internal standard solution and extracted with 5 mℓ of *n*-hexane. The organic layer was evaporated at 50°C under a stream of nitrogen, the residue dissolved in 25 μℓ of ethyl acetate and 2 μℓ of the solution injected.
I-2. The sample was mixed with 100 μℓ of an aqueous solution of the internal standard (5 μg/mℓ) and 3 mℓ of phosphate buffer, pH 7.5. The mixture was extracted with 6 mℓ of 50% ether in ethyl acetate. The organic phase was evaporated under nitrogen. The residue was treated with 25 μℓ of N-methyl-N-(trimethylsilyl) trifluoroacetamide and 100 μℓ benzene. This mixture was heated at 70°C for 1 hr and then evaporated under a stream of nitrogen. The residue was dissolved in 50 μℓ of methanol and 5 μℓ of the resulting solution was injected.
I-3. The sample was spiked with 50 μℓ of an aqueous solution of the internal standard (15 μg/mℓ) and extraced with 5 mℓ of *tert*-butyl ether. The organic layer was back extracted into 80 μℓ of 0.05 *M* sulfuric acid. A 50-μℓ aliquot of the aqueous layer was injected.
I-4. The sample was mixed with 1 mℓ of the internal standard solution (0.4 mg/mℓ in 0.1 *M* pH 9 borate buffer), and extracted with 7 mℓ of hexane-2-propanol (98:2). The organic layer was evaporated to dryness under a nitrogen stream. The residue was reconstituted with 200 μℓ of the mobile phase and aliquots of this solution were injected with an autosampler.
I-5. The sample was mixed with 1 mℓ of 0.1 *M* acetate buffer, pH 5 and 100 μℓ of internal standard (100 μg/mℓ). The mixture was extracted with 10 mℓ of chloroform containing 1% isoamyl alcohol. The organic phase was evaporated at 60°C under a nitrogen stream. The aqueous phase was passed through a Sep-Pak-C_{18} cartridge prewashed with 2.5 mℓ of water, 2 mℓ of methanol, and 2.5 mℓ of water) and then washed with 2 mℓ of 0.1 *M* phosphate buffer and eluted with 3 mℓ of acetoritrile. The eluate was mixed with the residue of the chloroform extract and again evaporated. The residue dissolved in 0.3 mℓ of 0.01 *M* HCl and a 150 μℓ aliquot was injected.
I-6. The sample was spiked with 150 μℓ of methanolic solution of the internal standard (1 μg/mℓ) and extracted with 4 mℓ of hexane-isoamyl alcohol (98:2). The organic layer was back extracted with 100 μℓ of 1 *M* HCl. A 20- to 80- μℓ aliquot of this aqueous layer was injected.
I-7. The sample was spiked with 100 μℓ of an aqueous standard (1 μg/mℓ) and extracted with 5 mℓ of methyl-*tert*-butyl ether. The organic phase was back extracted with 1.5 mℓ of 0.01 *M* HCl. The aqueous phase was evaporated at 30°C under vacuum and then under a stream of nitrogen. The residue was dissolved in 100 μℓ of the mobile phase and 20- to 50- μℓ aliquots were injected.

Elution — E-1. Methanol-0.05 *M* ammonium dihydrogen phosphate-triethylamine (45:55:0.25), pH 5 with 1 *M* phosphoric acid.
E-2. Acetonitrile-water (50:50) containing 1.5 g heptane sulfonic acid sodium salt + 8 g of sodium acetate per liter of the solvent. Final pH 6.6 with acetic acid.
E-3. Acetonitrile-0.01 *M* phosphate buffer, pH 3 (72:28).
E-4. Acetonitrile-methanol-0.06 *M* acetate buffer (37:5:58) containing 5 m*M* heptanesulfonic acid, pH 6.45.
E-5. Acetonitrile-water-0.04 *M* ammonium bromide in methanol (36:24:40) containing 0.06 mℓ of triethylamine, pH 8.5.

REFERENCES

1. **Calaf, R., Marie, P., Ghiglione, Cl., Bory, M., and Reynaud, J.,** Dosage du diltiazem plasmatique par chromatographie en phase gazeuse, *J. Chromatogr.*, 272, 385, 1983.
2. **Clozel, J. P., Caille, G., Yaeymans, Y., Theroux, P., Biron, P., and Besner, J. G.,** Improved gas chromatographic determination of diltiazem and deacetyldiltiazem in human plasma, *J. Pharm. Sci.*, 73, 207, 1984.

3. **Verghese, C., Smith, M. S., Aanonsen, L., Pritchett, E. L. C., and Shand, D. G.,** High-performance liquid chromatographic analysis of diltiazem and its metabolite in plasma, *J. Chromatogr.*, 272, 149, 1983.
4. **Wiens, R. E., Runser, D. J., Lacz, J. P., and Dimmitt, D. C.,** Quantitation of diltiazem and desacetyldiltiazem in dog plasma by high-performance liquid chromatography, *J. Pharm. Sci.*, 73, 688, 1984.
5. **Clozel, J. P., Caille, G., Taeymans, Y., Theroux, P., Biron, P., and Trudel, F.,** High-performance liquid chromatographic determination of diltiazem and six of its metabolites in human urine, *J. Pharm. Sci.*, 73, 771, 1984.
6. **Abernethy, D. R., Schwartz, J. B., and Todd, E. L.,** Diltiazem and desacetyldiltiazem analysis in human plasma using high-performance liquid chromatography: improved sensitivity without derivatization, *J. Chromatogr.*, 342, 216, 1985.
7. **Goebel, K. J. and Kolle, E. U.,** High-performance liquid chromatographic determination of diltiazem and four of its metabolites in plasma, *J. Chromatogr.*, 345, 355, 1985.

2,3-DIMERCAPTOPROPANE-1-SULFONIC ACID

Liquid Chromatography

Specimen (mℓ)	Extraction	Column (cm × mm)	Packing (μm)	Elution	Flow (mℓ/min)	Det. (nm)	RT (min)	Internal standard (RT)	Other compounds (RT)	Ref.
Urine (0.1—0.2)	I-1	25 × 4.6	Ultrasphere-C_{18} (5)[a]	E-1; grad	1.0	Fl (356, 350)	12	—	—	1

[a] Protected by 45 × 4.6 guard column packed with 10 μm Ultrasphere-C_{18} packing.

Extraction — I-1. The sample was treated 50 μℓ of a 40 m*M* bromobimane solution in acetonitrile and 0.1 *M* NH_4HCO_3 solution (pH 8) to make the total volume 2 mℓ. The head space of the tube was purged with nitrogen, and the mixture shaken in the dark for 5 min. The excess reagent was extracted by shaking the mixture with 2 mℓ of dichloromethane. The aqueous phase was adjusted to pH 7 and an aliquot of 20 μℓ of this solution was injected.

Elution — E-1. (A) 20 m*M* Tetrabutylammonium bromide in methanol; (B) 20 m*M* tetra-ammonium butyl bromide in water. Isocratic at 45%(B) for 11 min; 45 to 25%(B) in 1 min; isocratic at 25%(B) for 7 min; 25 to 45%(B) in 1 min; equilibration at 45%(B) for 15 min.

REFERENCE

1. **Maiorino, R. M., Weber, G. L., and Aposhian, H. V.,** Fluorometric determination of 2,3-dimercaptopropane-1-sulfonic acid and other dithiols by precolumn derivatization with bromobimane and column liquid chromatography, *J. Chromatogr.*, 374, 297, 1986.

DIMETHINDENE

Gas Chromatography

Specimen (mℓ)	Extraction	Column (m × mm)	Packing (mesh)	Oven temp (°C)	Gas (mℓ/min)	Det.	RT (min)	Internal standard (RT)	Deriv.	Other compounds (RT)	Ref.
Serum, urine (5)	I-1	2 × 2	10% Apiezon L-2% KOH Chromosorb W (80/100)	240	N_2 (30)	FID	16	Docosane (10.5)	—	—	1

Extraction — I-1. To the sample were added 5 μℓ of tetradecane, 0.5 mℓ of 25% ammonia solution, and 5 mℓ of the internal standard solution (0.5 μg/mℓ in *n*-pentane). After mixing and centrifugation the organic layer was collected and the aqueous layer extracted with another 5-mℓ aliquot of pentane. The combined organic layers were evaporated at 45°C under argon. Aliquots of 0.5 μℓ of the residual liquid (due to tetradecane) were injected.

REFERENCE

1. **Wermeille, M. M. and Huber, G. A.,** Gas-liquid chromatographic determination of free dimethindene in human serum and urine at low concentrations, *J. Chromatogr.*, 228, 187, 1982.

N-[*trans*-2-(DIMETHYLAMINO)-CYCLOPENTYL]-N-(3′-4′-DICHLOROPHENYL)PROPANAMIDE

Gas Chromatography

Specimen (mℓ)	Extraction	Column (m × mm)	Packing (mesh)	Oven temp (°C)	Gas (mℓ/min)	Det.	RT (min)	Internal standard (RT)	Deriv.	Other compounds (RT)	Ref.
Serum (1)	I-1	15 × 0.53	OV-17 (1 μm)[a]	240	He (3)	MS-CI[b,c]	10	Homologue	Acetyl	d	1

[a] Film thickness.
[b] A nitrogen selective detector was also used.

[c] Ammonia as the reagent gas.
[d] Separation of a number of possible metabolites and homologues with the use of nitrogen detector is shown.

Extraction — I-1. Extraction cartridges type W (DuPont PREP 1 sample processor) were conditioned with acetone and water. The sample was mixed with 0.5 mℓ of the internal standard (50 ng/mℓ in water) and 1 mℓ of water and applied to the extraction cartridge and extracted automatically with program No. 15. The residue was reconstituted with 100 μℓ of toluene-methanol (8:2), treated with 10 μℓ of acetic anhydride, and the mixture incubated at 100°C for 30 min. Aliquots of the cooled reaction mixture were injected.

REFERENCE

1. **Theis, D. L., Halstead, G. W., Capponi, V. J., Roach, B. L., and Robins, R. H.,** Quantitative determination of N-[*trans*-2-dimethylamino)-cyclopentyl]-N-(3′,4′-dichlorophenyl)propanamide, its $^{2}H_5$-labeled analogue and their N-dealkylated metabolites in dog serum by capillary gas chromatography-mass spectrometry, *J. Chromatogr.*, 375, 299, 1986.

DIMETHYLSULFOXIDE

Gas Chromatography

Specimen (mℓ)	Extraction	Column (m × mm)	Packing (mesh)	Oven temp (°C)	Gas (mℓ/min)	Det.	RT (min)	Internal standard (RT)	Deriv.	Other compounds (RT)	Ref.
Urine (0.5)	I-1	2 × 3	5% PEG 20M Shimalite W (60/80)	T.P.[a]	N_2 (80)	b	7	Methylsulfide (1.5)	—	Dimethyl-sulfone (12)	1

[a] Initial temp = 80°C; rate = 10°C/min; final temp = 200°C.
[b] Flame photometric detector at 394 nm.

Extraction — I-1. The sample was spiked with 5 μℓ of the internal standard and extracted twice with 5-mℓ portions of chloroform. Aliquots 10 μℓ of the combined extract were injected.

REFERENCE

1. **Ogata, M. and Fujii, T.,** Quantitative determination of urinary dimethyl sulfoxide and dimethyl sulfone by the gas chromatograph equipped with a flame photometric detector, *Ind. Health*, 17, 73, 1979.

p-(3,3-DIMETHYL-1-TRIAZENO)BENZOIC ACID

Liquid Chromatography

Specimen (mℓ)	Extraction	Column (cm × mm)	Packing (μm)	Elution	Flow (mℓ/min)	Det. (nm)	RT (min)	Internal standard (RT)	Other compounds (RT)	Ref.
Plasma (0.1)	I-1	25 × 4.6	Erbasil-C_{18} (10)	E-1	1.5	ABS (340)	11	*p*-(3,3-Dimethyl-1-triazeno) carboxamide (8)	*p*-(3-Methyl-1-triazeno)benzoic acid (6)	1

Extraction — I-1. The sample was spiked with the internal standard and treated with 2 volumes of ice-cold methanol. After centrifugation, aliquots of the clear supernatant were injected.

Elution — E-1. Acetonitrile-0.005 *M* tetrabutylammonium hydroxide, pH 7.6, with phosphoric acid (18:82).

REFERENCE

1. **Farina, P., Benfenati, E., Lassiani, L., Nisi, C., and D'Incalci, M.,** High-performance liquid chromatographic assay for the determination of *p*-(3,3-dimethyl-1-triazeno)benzoic acid in mouse plasma, *J. Chromatogr.*, 345, 323, 1985.

DIMINAZENE

Liquid Chromatography

Specimen (mℓ)	Extraction	Column (cm × mm)	Packing (μm)	Elution	Flow (mℓ/min)	Det. (nm)	RT (min)	Internal standard (RT)	Other compounds (RT)	Ref.
Plasma (1)	I-1	10 × 5	Radial-PAK CN (10)[a]	E-1	0.8	ABS (254)	5.1	Imidocarb (6.6)	—	1

[a] Protected by a CN guard column.

Extraction — I-1. The sample was spiked with the internal standard (6 μg) and applied to a prewashed (2 mℓ methanol, 5 mℓ water) Sep-Pak-C_{18} cartridge. After the sample had passed through, the cartridge was washed with 2 mℓ of 20% methanol, and 2 mℓ of methanol. Finally, the cartridge was eluted with 1 mℓ of 0.025 *M* 1-heptanesulphonic acid in 90% methanol. The eluate was mixed and an aliquot of 50 μℓ was injected.

Elution — E-1. Acetonitrile-water (50:50), pH 4.2 with phosphoric acid.

REFERENCE

1. **Aliu, Y. O. and Odegaard, S.,** Paired-ion extraction and high-performance liquid chromatographic determination of diminazene in plasma, *J. Chromatogr.*, 276, 218, 1983.

DIOXYANTHRAQUINONE

Gas Chromatography

Specimen (mℓ)	Extraction	Column (m × mm)	Packing (mesh)	Oven temp (°C)	Gas (mℓ/min)	Det.	RT (min)	Internal standard (RT)	Deriv.	Other compounds (RT)	Ref.
Urine (5)	I-1	1.8 × 3	3.8% SE-30 Chromosorb W (80/100)	280	He (20)	MS-EI	5	—	Trimethylsilyl	Desacetylbisacodyl (6.5) Phenolphthalein (12) Oxyphenisatin (15)	1

Extraction — I-1. The urine sample was hydrolyzed enzymatically by incubation at 37°C after the addition of 1 mℓ of acetate buffer (pH 4.5, 1 *M*) and 0.5 mℓ of Ketodase. The hydrolyzed urine was adjusted to pH 7.5 with 2 mℓ of phosphate buffer (pH 7.5, 1 *M*) and applied to an Extrelut column and allowed to stand for 10 min. The column was eluted with 40- and 20-mℓ portions of diethyl ether. The combined eluents were evaporated at 37°C under a stream of air. The residue was treated with 100 μℓ of trimethylchlorosilane-hexamethyldisilazane and pyridine (1:3:6) at room temp for 15 min and aliquots of this solution were injected.

REFERENCE

1. **Kok, R. M. and Faber, D. B.,** Qualitative and quantitative analysis of some synthetic, chemically acting laxatives in urine by gas chromatography-mass spectrometry, *J. Chromatogr.*, 222, 389, 1981.

DIPHENHYDRAMINE

Gas Chromatography

Specimen (mℓ)	Extraction	Column (m × mm)	Packing (mesh)	Oven temp (°C)	Gas (mℓ/min)	Det.	RT (min)	Internal standard (RT)	Deriv.	Other compounds (RT)	Ref.
Serum (3)	I-1	30 × 0.25	DB-1	T.P.[a]	He[b]	NPD	11	Orphenadrine (12.5)	—	—	1
Plasma (0.25—2)	I-2	1.8 × 2	3% SP-2250 Supelcoport (80/100)	205	He (30)	NPD	3.4	Orphenadrine (4.3)	—	—	2
Serum (1)	I-3	15 × 0.32	007 Methylsilicone (0.5 μm)[c]	180	He[d]	NPD	1.3	Orphenadrine (1.7)	—	—	3

Liquid Chromatography

Specimen (mℓ)	Extraction	Column (cm × mm)	Packing (μm)	Elution	Flow (mℓ/min)	Det. (nm)	RT (min)	Internal standard (RT)	Other compounds RT)	Ref.
Pure compounds	—	30 × 4	μ-Bondapak-C_{18} (10)	E-1	1.2	ABS (254)	9.4	—	2-Methyldiphenyl-methoxy acetic acid (2.4) Benzophenone (4.4) N,N-Didesmethyl di-phenhydramine (6.5) N-Desmethyl-diphenhydramine (7.3)	4

[a] 50°C (1 min) to 180°C for, 5 min, then at a rate of 6°C/min to 210°C.
[b] Column head pressure = 17.4 psi.
[c] Film Thickness.
[d] Column head pressure = 20 psi.

Extraction — I-1. The sample was mixed with 10 μℓ of a methanolic solution of the internal standard (14 μg/mℓ) and 1 mℓ of 1 *M* sodium hydroxide. The mixture was extracted with 7 mℓ of *n*-heptane. The organic layer was evaporated under nitrogen at 55°C. The residue was reconstituted in 40 μℓ of acetone, concentrated to about 10 μℓ, and 1 μℓ was injected.
I-2. The sample was mixed with 100 μℓ of a methanolic solution of the internal standard (1 μg/mℓ) and 1 mℓ of 0.25 *M* sodium hydroxide. The mixture was extracted with 5 mℓ of hexane-isoamyl alcohol (98:2). The organic layer was back extracted with 1.2 mℓ of 0.1 *M* HCl. The aqueous layer was made alkaline with 0.5 mℓ of carbonate-bicarbonate buffer (pH 11.5) and extracted with 200 μℓ of toluene-isoamyl alcohol (85:15). Aliquots of 6 μℓ of the organic layer were injected.
I-3. The sample was mixed with 100 μℓ of an aqueous solution of the internal standard (1 μg/mℓ) and 100 μℓ of 1 *M* sodium hydroxide. The mixture was extracted with 5 mℓ of hexane. The organic layer was evaporated to about 10 μℓ under a stream of dry air; and 1 μℓ of the residual solution was injected.

Elution — E-1. Methanol-water (78:22) containing 1.71 m*M* NaCl, pH = 7.

REFERENCES

1. **Lutz, D., Gielsdorf, W., and Jaeger, H.,** Quantitative determination of diphenhydramine and orphenadrine in human serum by capillary gas chromatography, *J. Clin. Chem. Clin. Biochem.*, 21, 559, 1983.
2. **Abernethy, D. R. and Greenblatt, D. J.,** Diphenhydramine determination in human plasma by gas-liquid chromatography using nitrogen-phosphorus detection: application to single low-dose pharmacokinetic studies, *J. Pharm. Sci.*, 72, 941, 1983.
3. **Meatherall, R. C. and Guay, D. R. P.,** Isothermal gas chromatographic analysis of diphenhydramine after direct injection onto a fused-silica capillary column, *J. Chromatogr.*, 307, 295, 1984.
4. **Bergh, M. L. E. and de Vries, J.,** High-pressure liquid chromatographic separation of diphenhydramine and some of its metabolites: effects of eluent salt concentration on chromatographic characteristics, *J. Liq. Chromatogr.*, 3, 1173, 1980.

DIPIPANONE

Liquid Chromatography

Specimen (mℓ)	Extraction	Column (cm × mm)	Packing (μm)	Elution	Flow (mℓ/min)	Det. (nm)	RT (min)	Internal standard (RT)	Other compounds (RT)	Ref.
Plasma (1)	I-1	25 × 4.5	Spherisorb-ODS (NA)	E-1	2.0	ABS (230)	9	Codeine (4.5)	Cyclizine (6)	1

Extraction — I-1. The sample was mixed with 1 mℓ of internal standard solution (100 ng/mℓ in water) and 1 mℓ of 1 *M* HCl. The mixture was washed with 10 mℓ of ether, made alkaline with 2 mℓ of 1 *M* sodium hydroxide, and extracted twice with 10 mℓ of diethyl ether. The combined extracts were evaporated at 30°C in a stream of nitrogen. The residue was dissolved in 20 μℓ of methanol for injection.

DIPIPANONE (continued)

Elution — E-1. Acetonitrile-1% ammonium acetate (70:30) containing 0.05 *M* triethylamine, pH 7.

REFERENCE

1. **Cathapermal, S. and Caddy, B.,** Determination of dipipanone by high-performance liquid chromatography, *J. Chromatogr.*, 351, 249, 1986.

DIPROBUTINE

Gas Chromatography

Specimen (mℓ)	Extraction	Column (m × mm)	Packing (mesh)	Oven temp (°C)	Gas (mℓ/min)	Det.	RT (min)	Internal standard (RT)	Deriv.	Other compounds (RT)	Ref.
Plasma (2)	I-1	2.1 × 2	10% Carbowax 20M + 2% KOH Chromosorb W (80/100)	100	N_2 (25)	NPD	6	Propyl-1-iso-butyl-1-butyl-amine (7.5)	—	—	1

Extraction — I-1. The sample was treated with 5 μℓ of the internal standard solution (60 m*M* in water) and 1 mℓ of borate buffer, pH 10. The mixture was extracted with 5 mℓ of diethyl ether. The organic phase was back extracted into 2 mℓ of 0.1 *M* sulfuric acid. The aqueous layer was made alkaline with 100 μℓ of 10 *M* sodium hydroxide and again extracted with 5 mℓ of ether. The organic layer was evaporated under a stream of nitrogen. The residue was reconstituted with 50 μℓ of ether and 5 μℓ of this solution were injected.

REFERENCE

1. **Davies, C. L. and Molyneux, S. G.,** Determination of diprobutine in human plasma using gas-liquid chromatography with nitrogen-selective detection, *J. Chromatogr.*, 339, 186, 1985.

n-DIPROPYLACETAMIDE

Gas Chromatography

Specimen (mℓ)	Extraction	Column (m × mm)	Packing (mesh)	Oven temp (°C)	Gas (mℓ/min)	Det.	RT (min)	Internal standard (RT)	Deriv.	Other compounds (RT)	Ref.
Plasma (0.5)	I-1	1.5 × 3	10% DEGS-PS[a] Supelcoport (80/100)	190	N_2 (35)	NPD	2.7	*n*-Tripropyl-acetamide (3.4)	—	—	1

[a] Diethyleneglycol-succinate-phosphate.

Extraction — I-1. The sample was mixed with 25 μℓ of the internal standard solution (10 μg/mℓ in acetone), 0.5 mℓ of 0.5 *M* H_3PO_4. The mixture was extracted with 8 mℓ of diethyl ether. A 6.5-mℓ aliquot of the organic layer was evaporated at room temp under vacuum. The residue was dissolved in 100 μℓ of *n*-hexane and 1 to 2 μℓ of this solution were injected.

REFERENCE

1. **Riva, R., Albani, F., Olivi, F., Pantaloni, M., and Baruzzi, A.,** Quantitative determination of *n*-dipropylacetamide in the plasma of epileptic patients by gas-liquid chromatography with nitrogen-selective detection, *J. Chromatogr.*, 233, 371, 1982.

N,N-DIPROPYL-2-AMINO-5,6-DIHYDROXYTETRALIN

Liquid Chromatography

Specimen (mℓ)	Extraction	Column (cm × mm)	Packing (μm)	Elution	Flow (mℓ/min)	Det. (nm)	RT (min)	Internal standard (RT)	Other compounds (RT)	Ref.
Plasma, brain homogenate (0.05)	I-1	15 × 4.6	Nucleosil-C_{18} (5)[a]	E-1	1.0	Elec-tro-chem	12.5	—	N,N-Dipropyl-2-amino-6,7-di-hydroxytetralin (11)	1

[a] A precolumn packed with Nucleosil-C_{18} is placed between the pump and the injector.

N,N-DIPROPYL-2-AMINO-5,6-DIHYDROXYTETRALIN (continued)

Extraction — I-1. The sample was treated with 50 μℓ of 0.01 *M* sodium *m*-bisulfite, 150 μℓ acetonitrile, and 1 mℓ 0.1 *M* perchloric acid. The supernatant was applied onto a small Sephadex G-10 column which was previously washed with 3 mℓ of 0.02 *M* and 3 mℓ of 0.01 *M* formic acid. After the sample had passed through, the column was washed with 2.5 mℓ of 0.01 *M* formic acid and eluted with another 2 to 5 mℓ formic acid. An aliquot of 2 μℓ of the eluate was injected.

Elution — E-1. Methanol-0.1 *M* citrate-phosphate buffer, pH 4 (23:77).

REFERENCE

1. **Feenstra, M. G. P., Rollema, H., Mulder, T. B. A., Westerink, B. H. C., and Horn, A. S.,** Amperometric detection of low concentrations of dopamine receptor agonists after liquid chromatographic on-column sample enrichment: effect of o-methylation on brain concentrations of dipropyl-5,6-ADTN and dipropyl-6,7-ADTN, *Life Sci.*, 32, 459, 1983.

DIPYRIDAMOLE

Liquid Chromatography

Specimen (mℓ)	Extraction	Column (cm × mm)	Packing (μm)	Elution	Flow (mℓ/min)	Det. (nm)	RT (min)	Internal standard (RT)	Other compounds (RT)	Ref.
Plasma (0.2)	I-1	12.5 × 4.6	LiChrosorb-RP-18 (5)	E-1	1.0	Fl (415, 478)	3.5	Methoxy-dipyridamole (5)	—	1
Plasma, blood (0.1—1)	I-2	30 × 3.9	μ-Bondapak-C_{18} (10)	E-2	2.0	Fl (285, 470)	5.1	RA 433 (3.5)	—	2
Plasma (1)	I-3	30 × 3.9	μ-Bondapak-C_{18} (10)	E-3	1.5	ABS (280)	5.5	Lidocaine (7.5)	—	3
Plasma (1)	I-4	25 × 4.6	Ultrasphere-C_{18} (5)	E-4	2.5	ABS (280)	7.5	Propranolol (6.5)	—	4

Thin-Layer Chromatography

Specimen (mℓ)	Extraction	Plate (Manufacturer)	Layer (mm)	Solvent	Post-separation treatment	Det. (nm)	Rf	Internal standard (Rf)	Other compounds (Rf)	Ref.
Plasma (1)	I-5	20 × 20 (Merck)	Silica gel (0.25)	S-1	D: Parrafin wax (70 g) in petroleum ether (1L)	Fl Reflectance (380, 430)	NA	—	—	5

[a] 2,4,6-Trimorpholinopyrimido-(5,4-d)pyrimidine.

Extraction — I-1. The sample was treated with 1 mℓ of 1 *N* sodium hydroxide and extracted with 10 mℓ of dichloromethane containing 100 ng of the internal standard. The organic phase was evaporated under a stream of nitrogen at 30°C. The residue was reconstituted with 50 μℓ of the mobile phase which was injected.
I-2. The sample was mixed with 25 μℓ of an aqueous solution of the internal standard (1 μg/mℓ) and 1 mℓ of 1 *N* sodium hydroxide. The mixture was extracted with 5 mℓ of diethyl ether. The organic layer was evaporated at room temp under nitrogen. The residue was dissolved in 100 μℓ of the mobile phase and aliquots of this solution were injected.
I-3. The sample was treated with the residue of 0.1 mℓ of a methanolic solution (1 mg/mℓ) of the internal standard and 0.5 mℓ of 0.1 *N* sodium hydroxide. The mixture was extracted with 5 mℓ of ethyl acetate. A 4-mℓ aliquot of the organic layer was evaporated at 40°C under a stream of dry air. The residue was reconstituted with 0.5 mℓ of the mobile phase. A 100-μℓ of this solution was injected.
I-4. The sample was mixed with 50 μℓ of an ethanolic solution of the internal standard (10 μg/mℓ), a spatula full of sodium chloride and 1 mℓ of 1 *M* tris buffer, pH 10. The mixture was extracted with 5 mℓ of diethyl ether. The organic layer was back extracted into 50 μℓ of 0.1 *N* HCl. An aliquot of or the entire aqueous phase was injected.
I-5. The sample was mixed with 1 mℓ of carbonate buffer, pH 10 and the mixture extracted with 5 mℓ of diethyl ether-dichloromethane (8:2). The organic layer was evaporated under nitrogen at 50°C and the residue dissolved in 250 μℓ of dichloromethane. Aliquots of 5 μℓ were applied to the TLC plate.

Elution — E-1. Methanol-0.2 *M* Tris buffer (80:20).
E-2. Methanol-water (65:35) containing 0.005 *M* 1-heptanesulfonic acid with 0.1% acetic acid.
E-3. Acetonitrile-0.01 *M* phosphate buffer, pH 7 (50:50).
E-4. Acetonitrile-0.02 *M* phosphate buffer containing 0.01 *M* N,N,N,N-tetramethylene diamine, pH 2.9. (33:67)

Solvent — S-1. Ethylacetate-methanol-28% ammonia (85:10:5).

DIPYRIDAMOLE (continued)

REFERENCES

1. **Schmid, J., Beschke, K., Roth, W., Bozler, G., and Koss, F. W.,** Rapid, sensitive determination of dipyridamole in human plasma by high-performance liquid chromatography, *J. Chromatogr.*, 163, 239-243, 1979.
2. **Wolfram, K. M. and Bjornsson, T. D.,** High-performance liquid chromatographic analysis of dipyridamole in plasma and whole blood, *J. Chromatogr.*, 183, 57, 1980.
3. **Williams, C., II, Huang, C. S., Erb, R., and Gonzalez, M. A.,** High-performance liquid chromatographic assay for plasma dipyridamole monitoring, *J. Chromatogr.*, 225, 225, 1981.
4. **Rosenfeld, J., Devereaux, D., Buchanan, M. R., and Turpie, A. G. G.,** High-performance liquid chromatographic determination of dipyridamole, *J. Chromatogr.*, 231, 216, 1982.
5. **Steyn, J. M.,** Spectrofluorimetric determination of dipyridamole in serum — a comparison of two methods, *J. Chromatogr.*, 164, 487, 1979.

DIPYRONE

Liquid Chromatography

Specimen (mℓ)	Extraction	Column (cm × mm)	Packing (μm)	Elution	Flow (mℓ/min)	Det. (nm)	RT (min)	Internal standard (RT)	Other compounds (RT)	Ref.
Plasma (1—2)	I-1	61 × 2	Bondapak AX/ Corasil (35—50)	E-1	0.5	ABS (254, 280)[a]	6.3	Sodium salicy- late (11.2)	4-Methylamino- antipyrine (5.2)[h]	1
Plasma (1)	I-2	30 × 3.9	μ-Bondapak-C_{18} (10)	E-2	1.6	ABS (257)	—	4-Propylamino antipyrine (41)	4-Methylamino- antipyrine (14) 4-Aminoantipyrine (17) 4-Formylamino- antipyrine (21) 4-Acetylamino- antipyrine (25)	2
Plasma (0.1)	I-3	25 × 4	Pine SI2 -C_{18} (5)[c]	E-3	1.5	ABS (260)	8.2	Hexobarbital (18)	4-Methylamino- antipyrine (11.4) 4-Aminoantipyrine (9.4)	3

4-Acetylamino-
antipyrine (5.2)
4-Formylamino-
antipyrine (4.4)
Aminopyrine (14.5)

[a] Dipyrone was monitored at 254 nm and the internal standard at 280 nm.
[b] A separate extraction and chromatographic procedure is described for the assay of this active metabolite.
[c] Protected by a 23 × 3.8 precolumn packed with Co:Pell ODS (30 μm).

Extraction — I-1. The sample was mixed with 1 mℓ of methanolic solution of the internal standard and an excess of potassium carbonate powder (0.6 to 1 g). The upper methanol layer was evaporated under nitrogen. The residue was dissolved in 0.5 mℓ of the mobile phase and aliquots of 25 to 50 μℓ were injected.
I-2. The sample was mixed with 200 μℓ of an aqueous solution of the internal standard (50 μg/mℓ) and 100 μℓ of 1 *M* sodium hydroxide. The mixture was extracted twice with 5-mℓ portions of chloroform. The organic phases were evaporated at 40°C under a stream of air. The residue was dissolved in 50 μℓ of methanol and 5- to 10-μℓ aliquots were injected.
I-3. The sample was mixed with 10 μℓ of an aqueous solution of sodium bisulfite (15 mg/mℓ) and 1 mℓ of acetonitrile containing (9 μg/mℓ) of the internal standard. The supernatant was evaporated at room temperature under reduced pressure. The residue was dissolved in 70 μℓ of the mobile phase and an aliquot of 50 μℓ was injected.

Elution — E-1. Methanol-phosphate buffer, pH 5.6 (15:85).
E-2. Methanol-acetate buffer, pH 3 (8:92).
E-3. Acetonitrile-10 m*M* buffer, pH 4.5 containing 1.24 m*M* tetra-*n*-butylammonium bromide (22:78).

REFERENCES

1. **Asmardi, G. and Jamali, F.,** High-performance liquid chromatography of dipyrone and its active metabolite in biological fluids, *J. Chromatogr.*, 277, 183, 1983.
2. **Katz, E. Z., Granit, L., Drayer, D. E., and Levy, M.,** Simultaneous determination of dipyrone metabolites in plasma by high-performance liquid chromatography, *J. Chromatogr.*, 305, 477, 1984.
3. **Itoh, S., Tanabe, K., Furuichi, Y., Suzuka, T., Kubo, K., Yamazaki, M., and Kamada, A.,** Ion-pair high-performance liquid chromatographic analysis of sulpyrine and its metabolites in rabbit plasma, *Chem. Pharm. Bull.*, 32, 3194, 1984.

DISOPYRAMIDE

Gas Chromatography

Specimen (mℓ)	Extraction	Column (m × mm)	Packing (mesh)	Oven temp (°C)	Gas (mℓ/min)	Det.	RT (min)	Internal standard (RT)	Deriv.	Other compounds (RT)	Ref.
Serum (2)	I-1	0.6 × 2	3% OV-17 GasChrom Q (100/120)	245	N_2 (25)	NPD	1.8	*p*-Chlorodisopyramide (3)	Acetyl	Mono-N-dealkyl-disopyramide (5)	1
Plasma (0.5)	I-2	2 × 2	3% OV-1 GasChrom Q (100/120)	200	N_2 (40)	NPD	5.9	Aminopentamide (3.5)	Dehydration	—	2
Plasma (0.1)	I-3	1 × 4	3% SE-30 Supelcoport (80/100)	260	N_2 (40)	NPD	1.6	*p*-Chlorodisopyramide (2.5)	—	—	3
Plasma (1)	I-4	0.6 × 2	3% OV-17 GasChrom Q (100/120)	250	He (30)	NPD	0.9	*p*-Chlorodisopyramide (1.5)	Acetyl	Mono-N-dealkyl-disopyramide	4
Plasma (0.5)	I-5	25 × 0.31	5% Phenylmethyl-silicone (0.17)[a]	T.P.	He (1)	NPD	10.1	*p*-Chlorodisopyramide (12.9)	Dehydration; trifluoro-acetyl	Mono-N-dealkyl-disopyramide (10.6)	5, 6

Liquid Chromatography

Specimen (mℓ)	Extraction	Column (cm × mm)	Packing (μm)	Elution	Flow (mℓ/min)	Det. (nm)	RT (min)	Internal standard (RT)	Other compounds (RT)	Ref.
Serum (0.5)	I-6	30 × 3.9	μ-Bondapak-C_{18} (10)	E-1	2.0	ABS (205)	3.9	*p*-Chlorodisopyramide (7.4)	Lidocaine (2.7) Quinidine (5)	7
Plasma (1)	I-7	10 × 3	α_1-AGP-Silica[c] (13)[d]	E-2	0.5	ABS (261)	R = 7 S = 13	—	R-Mono-N-dealkyl-disopyramide (5) S-Mono-N-dealkyl-disopyramide (6)	8

Plasma (1)	I-8	25 × 4.6	LiChrosorb RP-8 (10)[e]	E-3	1.8	ABS (254)	6	*p*-Chlorodisopyramide (12.8)	Mono-N-dealkylpyramide (3.5) Quinidine (3.6) Lidocaine (3.9)	9
Plasma, urine (1)	I-9	25 × 4.6	Brownlee RP-8 (10)	E-4	2.0	ABS (202)	4.2	*p*-Chlorodisopyramide (5.4)	Mono-N-dealkyl-disopyramide (3.1)	10
Serum (0.1)	I-10		Whatman ODS-3 (5)[f]	E-5	2.0	ABS (254)	3.8	*p*-Chlorodisopyramide (6.9)	—	11
Serum (0.1)	I-11	10 × 8	Radial-Pak CN (10)	E-6	2.0	ABS (210)	7.8	*p*-Chlorodisopyramide (10.5)	Mono-N-dealkyl-disopyramide (4.5)	12
Serum (0.5)	I-12	10 × 8	Radial-Pak C_{18} (10)	E-7	3.0	ABS (254)	4.5	Benzocaine (10)	Quinidine (6) Lidocaine (3.5)	13

Thin-Layer Chromatography

Specimen (mℓ)	Extraction	Plate (Manufacturer)	Layer (mm)	Solvent	Post-separation treatment	Det. (nm)	Rf	Internal standard (Rf)	Other compounds (Rf)	Ref.
Serum (0.1)	I-13	10 × 10 cm (Merck)	Silica gel 60 F_{254} (HPTLC) (0.25)	S-1	—	Reflectance (254)	0.46	*p*-Chloro-disopyramide (0.52)	Mono-N-dialkyl disopyramide (0.26)	14

[a] Film thickness.
[b] Initial temp = 160°C; rate = 5°C/min; final temp = 195°C.
[c] α_1-Acid glycoprotein immobilized on silica particles.
[d] A 50 × 3 mm precolumn packed with 50 μm RP-2 LiChrosorb was used.
[e] Column temp = 40°C.
[f] Protected by a Whatman precolumn.

DISOPYRAMIDE (continued)

Extraction — I-1. The sample was mixed with 50 μℓ of an ethanolic solution of the internal standard (0.1 mg/mℓ) and 50 μℓ of 10 *M* sodium hydroxide. The mixture was extracted with 4 mℓ of dichloromethane. An aliquot of the organic layer (2 mℓ) was treated with 50 μℓ of acetic anhydride and the mixture evaporated at room temperature under nitrogen. The residue was dissolved in 25 μℓ of ethanol and 3 μℓ were injected.
I-2. The sample was mixed with 0.5 mℓ of an aqueous solution of the internal standard (30 μg/mℓ) and 100 μℓ of 2 *N* sodium hydroxide. The mixture was extracted with 7 mℓ of chloroform. The organic layer was dried over anhydrous sodium sulfate, evaporated under nitrogen, the residue treated with 100 μℓ of chloroform and 200 μℓ of trifluoroacetic anhydride, and the mixture incubated at 65°C for 30 min. The excess reagent was evaporated under nitrogen, the residue dissolved in 100 μℓ of toluene, and 2 to 4 μℓ injected.
I-3. The sample was mixed with 50 μℓ of 4 *N* sodium hydroxide and 20 μℓ of an aqueous solution of the internal standard (20 mg/ℓ). The mixture was extracted with 4 mℓ of diethyl ether. The organic layer was evaporated at 60°C, the residue dissolved in 50 μℓ of ether, and 5-μℓ aliquots of the solution were injected.
I-4. The sample was mixed with 50 μℓ of an aqueous solution of the internal standard (100 μg/mℓ), 25 μℓ of 10 *M* sodium hydroxide, and the mixture was extracted with 9 mℓ of diethyl ether. The organic layer was back extracted into 1 mℓ of 0.1 *M* HCl. The aqueous layer was made alkaline with 25 μℓ of 10 *M* sodium hydroxide and re-extracted with 5 mℓ of ether. The organic layer was treated with 100 μℓ of acetic anhydride and evaporated to dryness at 40°C with a stream of nitrogen. The residue was dissolved in 150 μℓ of ethanol and a 1-μℓ aliquot was analyzed.
I-5. The sample was mixed with 0.5 mℓ of the internal standard solution (5.5 μg/mℓ in 0.1 *M* HCl), 0.5 mℓ of 1 *M* sodium hydroxide, and 2 mℓ of water. The mixture was extracted with 6 mℓ of toluene. The organic layer was evaporated at 40°C under a stream of nitrogen. The residue was reconstituted with 0.5 mℓ of toluene and treated with 150 μℓ of trifluoroacetic anhydride. The mixture was incubated at 55°C for 45 min. The excess reagent was removed by evaporation at 40°C under a stream of nitrogen. The residue was dissolved in 100μℓ of toluene and 2-μℓ aliquots were injected.
I-6. The sample was mixed with 0.5 mℓ of the internal standard solution (8 mg/ℓ in water) and 100 μℓ of 1 *M* sodium hydroxide. The mixture was extracted with 3 mℓ of dichloromethane. The organic layer was evaporated at room temperature with a stream of air. The residue was dissolved in 250 μℓ of the mobile phase and aliquots of 15 μℓ were injected.
I-7. The sample was mixed with 100 μℓ of 2 *M* sodium hydroxide and extracted with 6 mℓ of water saturated diethyl ether. The organic layer was evaporated under a stream of nitrogen at 40°C. The residue was dissolved in 125 μℓ of the mobile phase and 50 μℓ were injected.
I-8. The sample was made alkaline with 100 μℓ of 5 *M* sodium hydroxide and extracted with 5 mℓ of chloroform containing 2 mg/ℓ of the internal standard. The organic layer was evaporated with a stream of air. The residue was dissolved in 250 μℓ of the mobile phase and aliquots of 20 μℓ were injected.
I-9. The sample was treated with 0.5 mℓ of an aqueous solution of the internal standard (4 μg/mℓ) and 0.1 mℓ of concentrated ammonium hydroxide. The mixture was extracted with 6 mℓ of diethyl ether. The organic layer was back extracted into 0.2 mℓ of 0.1 *M* acetic acid. Aliquots of 25 to 100 μℓ of the aqueous phase were analyzed by HPLC.
I-10. The sample was mixed with 100 μℓ of an aqueous solution of the internal standard (30 mg/ℓ) and 100 μℓ of 0.1 *M* sodium carbonate. The mixture was extracted with 0.4 mℓ of chloroform. The organic phase was back extracted into 250 μℓ of 0.1 *N* HCl. An aliquot of 50 μℓ of the aqueous phase was injected.
I-11. A Sep-Pak silica cartridge was prepared by rinsing it with 10 mℓ of 1 *N* NaOH followed by 30 mℓ of ethyl acetate. Forty microliters of the internal standard solution (20 μg/mℓ in ethyl acetate) and 100 μℓ of the sample were applied to the column. The cartridge was eluted with 8 mℓ of ethyl acetate. The organic layer was back extracted with 300 μℓ of 0.01 *M* butylamine phosphate, pH 3. An aliquot of 100 μℓ of the aqueous layer was injected.
I-12. The sample was treated with 50 μℓ of 30% trichloroacetic acid. After centrifugation, 100 μℓ of the supernatant was mixed with 25 μℓ of an aqueous solution (25 μg/mℓ) of the internal standard. Aliquots of 50 μℓ of this solution were analyzed.
I-13. The sample was mixed with 100 μℓ of the internal standard solution (6 μg/mℓ in 0.1 *M* HCl) and 200 μℓ of saturated aqueous sodium carbonate. The mixture was extracted twice with 1 mℓ portions of chloroform. The combined organic phases were evaporated under a stream of nitrogen. The residue was dissolved in 50 μℓ of chloroform and 30 μℓ of the solution was applied to the TLC plate.

Elution — E-1. Acetonitrile-30 m*M* phosphate buffer, pH 4.4 (280:720).
E-2. 2-Propanol-phosphate buffer, pH 6.2 (4.3:95:7) containing 1.95 m*M* N, N-dimethyloctylamine.
E-3. Acetonitrile-0.05 *M* phosphate buffer, pH 3 (27:73).
E-4. Acetonitrile-0.05 *M* acetic acid-0.05 *M* ammonium formate water (55:9:13.5:22.5).
E-5. Acetonitrile-1 *M* ammonium dihydrogen phosphate, pH 4.4 acetic acid-water (26:5:4.2:64.8).
E-6. Acetonitrile-0.01 *M* dibutylamine phosphate, pH 3 (25:75).
E-7. Acetonitrile-10 m*M* acetate buffer, pH 4.5 (25:75).

Solvent — S-1. Ethanol + ammonium hydroxide (98:2).

REFERENCES

1. **Bredesen, J. E.,** Gas-chromatographic determination of disopyramide and its mono-N-dealkylated metabolite in serum with use of a nitrogen-selective detector, *Clin. Chem.*, 26, 638, 1980.
2. **Gal, J., Brady, J. T., and Kett, J.,** Gas-chromatographic determination of diospyramide with nitrogen detection, *J. Anal. Toxicol.*, 4, 15, 1980.
3. **Johnston, A. and Hamer, J.,** Gas chromatography and enzyme immunoassay compared for analysis of disopyramide in plasma, *Clin. Chem.*, 27, 353, 1981.
4. **Brien, J. F., Nakatsu, K., and Armstrong, P. W.,** Determination of disopyramide and mono-N-desisopropyl-disopyramide in serum by gas-liquid chromatography with nitrogen-selective detection, *J. Pharmacol. Methods*, 9, 295, 1983.
5. **Kapil, R. P., Abbott, F. S., Kerr, C. R., Edwards, D. J., Lalka, D., and Axelson, J. E.,** Simultaneous quantitation of disopyramide and its mono-dealkylated metabolite in human plasma by fused-silica capillary gas chromatography using nitrogen-phosphorus specific detection, *J. Chromatogr.*, 307, 305, 1984.
6. **Kapil, R. P., Axelson, J. E., Lalka, D., Edwards, D. J., and Kerr, C. R.,** Applicability of capillary gas liquid chromatography to the measurement of free fraction of disopyramide in human plasma, *Res. Commun. Chem. Pathol. Pharmacol.*, 48, 153, 1985.
7. **Flood, J. G., Bowers, G. N., and McComb, R. B.,** Simultaneous liquid-chromatographic determination of three antiarrhythmic drugs: disopyramide, lidocaine, and quinidine, *Clin. Chem.*, 26, 197, 1980.
8. **Hermansson, J., Eriksson, M., and Nyquist, O.,** Determination of (R)- and (S)-disopyramide in human plasma using a chiral α_1-acid glycoprotein column, *J. Chromatogr.*, 336, 321, 1984.
9. **Ahokas, J. T., Davies, C., and Ravenscroft, P. J.,** Simultaneous analysis of disopyramide and quinidine in plasma by high-performance liquid chromatography, *J. Chromatogr.*, 183, 65, 1980.
10. **Charette, C., McGilveray, I. J., and Mainville, C.,** Simultaneous determination of disopyramide and its mono-N-dealkyl metabolite in plasma and urine by high-performance liquid chromatography, *J. Chromatogr.*, 274, 219, 1983.
11. **Swezey, C. B. and Ponzo, J. L.,** Determination of disopyramide phosphate in serum by high-performance liquid chromatography, *Ther. Drug Monit.*, 6, 211, 1984.
12. **Kubo, H., Kinoshita, T., Kobayashi, Y., and Tokunaga, K.,** Rapid method for determination of disopyramide and its mono-N-dealkylated metabolite in serum by high-performance liquid chromatography using Sep-Pak silica extraction, *Anal. Lett.*, 17, 55, 1984.
13. **Taylor, E. H., Nelson, D., Taylor, R. D., and Pappas, A. A.,** Rapid sample preparation and high performance liquid chromatographic determination of total and unbound serum disopyramide, *Ther. Drug Monit.*, 8, 219, 1986.
14. **Simona, M. G. and Grandjean, E. M.,** Simple high-performance thin-layer chromatography method for the determination of disopyramide and its mono-N-dealkylated metabolite in serum, *J. Chromatogr.*, 224, 532, 1981.

DISULFIRAM

Liquid Chromatography

Specimen (mℓ)	Extraction	Column (cm × mm)	Packing (μm)	Elution	Flow (mℓ/min)	Det. (nm)	RT (min)	Internal standard (RT)	Other compounds (RT)	Ref.
Plasma, Urine (1)	I-1	25 × 2.6	Spherisorb silica (5)	E-1	1.2	ABS (254)	7.4	Ethyl-*p*-nitrobenzoate (4.3)	Methyldiethyl-dithio carbamate (3.1)	1
Plasma (2)	I-2	30 × 3.9	μ-Bondapak alkyl-phenyl (10)[a,b]	E-2	1.5	ABS (280)	8.4	*n*-Propyl-diethyl dithiocarbamate (7)	Methyldiethyl-dithio carbamate (5.7)	2
Urine (1)	I-3	30 × 3.9	μ-Bondapak-C_{18} (10)	E-3	2.0	ABS (254)	—	Ethylpropylamine (4.3)[c]	Diethylamine (3.2)[c]	3

[a] Protected by a (50 × 3.9 mm) guard column packed with phenyl/corasil.
[b] Column temp = 37°C.
[c] Retention times of 3,5-dinitrobenzoyl derivatives.

Extraction — I-1. The sample was mixed with 2 mℓ of 0.01 *M* EDTA solution in 1% sodium chloride (pH 8.5) and extracted with 5 mℓ of chloroform containing 1 μg/mℓ of the internal standard. The organic phase was evaporated at room temperature under a stream of nitrogen to about 50 μℓ which was injected. (The remaining aqueous phase was further extracted for the other metabolites.)
I-2. The sample was treated with an equal volume of 0.01 *M* EDTA in 0.05 *M* Tris buffer (pH 8.5) and 20 μℓ of ethyl iodide and an appropriate amount of the internal standard. The mixture was incubated at 40°C for 30 min. After cooling, zinc sulfate (500 mg) was added and the mixture extracted with 4 mℓ of diethyl ether. The ether layer was washed with 9 mℓ of 0.1 *M* carbonate buffer, pH 9, and concentrated to about 100 μℓ under nitrogen, diluted to 1 mℓ with acetonitrile, and reconcentrated to about 200 μℓ. A 15-to 20-μℓ aliquot was injected.
I-3. The sample was mixed with 530 nmol of the internal standard. The amines were converted to 3,5-dinitrobenzoyl derivatives. The derivatives were extracted with ether. (For details, see *J. Chromatogr.*, 117, 187, 1976; *J. Study Alcohol*, 42, 202, 1981.)

Elution — E-1. Heptane-tetrahydrofuran-methanol (97.6:2.2:0.2).
E-2. Acetonitrile-water (52:48).
E-3. Methanol-water (55:45).

REFERENCES

1. **Jensen, J. C. and Faiman, M. D.,** Determination of disulfiram and metabolites from biological fluids by high-performance liquid chromatography, *J. Chromatogr.*, 181, 407, 1980.
2. **Masso, P. D. and Kramer, P. A.,** Simultaneous determination of disulfiram and two of its dithiocarbamate metabolites in human plasma by reversed-phase liquid chromatography, *J. Chromatogr.*, 224, 457, 1981.
3. **Neiderhiser, D. H. and Fuller, R. K.,** High-performance liquid chromatographic method for the determination of diethylamine, a metabolite of disulfiram, in urine, *J. Chromatogr.*, 229, 470, 1982.

DIXYRAZINE

Gas Chromatography

Specimen (mℓ)	Extraction	Column (m × mm)	Packing (mesh)	Oven temp (°C)	Gas (mℓ/min)	Det.	RT (min)	Internal standard (RT)	Deriv.	Other compounds (RT)	Ref.
Plasma, serum (2.5)	I-1	7 × NA[a]	SE-30	280	He (3.5)	MS-EI	2	Perphenazine (2)	Trimethylsilyl	—	1

[a] CKB 2101-104 glass capillary column.

Extraction — I-1. The sample was mixed with a methanolic solution of the internal standard (0.5 μg/mℓ) and 0.25 mℓ of pH 10 borate buffer. The mixture was extracted with 6 mℓ of toluene. The organic phase was back extracted into 1 mℓ of 50 m*M* sulfuric acid. An aliquot of 0.8 mℓ of the aqueous phase was made alkaline with 0.1 mℓ of 4 *M* sodium hydroxide and re-extracted with 3 mℓ of toluene. The organic phase (2.5 mℓ) was evaporated to 1.5 mℓ at 70°C under a stream of nitrogen and treated with 50 μℓ of a solution of 0.01% N,O-*bis*-(trimethylsilyl)-acetamide in toluene. The mixture was allowed to react at 70°C for 10 min. After reaction, the solvents were evaporated at 70°C under a stream of nitrogen, the residue was dissolved in 30 μℓ of toluene and 3 to 6 μℓ of this solution was injected.

REFERENCE

1. **Brante, G., Jonsson, S., and Melander, A.,** Gas chromatographic-mass spectrometric determination of dixyrazine in human blood, *Eur. J. Clin. Pharmacol.*, 20, 307, 1981.

DOBUTAMINE

Liquid Chromatography

Specimen (mℓ)	Extraction	Column (cm × mm)	Packing (μm)	Elution	Flow (mℓ/min)	Det. (nm)	RT (min)	Internal standard (RT)	Other compounds (RT)	Ref.
Plasma (1)	I-1	25 × 4.6	Spherisorb-ODS (5)[a]	E-1	1.2	Electrochem[b]	4.8	LYO 89811 (14.2) LYO 89838 (11.5)	—	1
Plasma (1)	I-2	25 × 4.6	μ-Bondapak-C_{18} (10)[a]	E-2	1.0	Electrochem[c]	9.2	Analogue (5.3)	—	2
Plasma (1)	I-3	15 × 3.9	Resolve RP-18 (5)	E-3	1.5	Fl (195, 330)	2.6	Buphenine (9.9)	—	3

[a] Protected by a 5-cm guard column packed with pericellular RP-18 silica.
[b] 0.55 V.
[c] 0.6 V.

Extraction — I-1. The sample was mixed with 1 mℓ of an aqueous solution of a mixture of the internal standards (100 ng/mℓ each). A 1-mℓ aliquot of this solution was passed through a prewashed 1-mℓ BondElut C_{18} column. The column was washed with 2 volumes of water and eluted with 1-mℓ of 50% methanol-mobile phase. Aliquots of the eluate were injected.
I-2. The sample was treated with 100 μℓ of the internal standard solution (150 ng/mℓ in 0.01 *M* perchloric acid), 50 μℓ of 0.01 *M* sodium bisulfite, and 0.4 mℓ of Tris/EDTA, pH 8.6 buffer. The mixture was treated with 30 mg of alumina for 15 min. The supernate was discarded, alumina washed twice with 1-mℓ portions of water, and eluted with 0.2 mℓ of 0.5 *M* perchloric acid. Aliquots of 20 to 100 μℓ of the eluate were injected with an autosampler. I-3. The sample and 100 μℓ of the internal standard solution (1 μg/mℓ in 50% methanol in mobile phase) were applied to a prewashed (1 mℓ methanol, 2 mℓ water) 1-mℓ BondElut CN column. The column was washed with 1 mℓ of water and eluted with 0.3 mℓ of 50% methanol in mobile phase, pH 2.3.

Elution — E-1. Acetonitrile-0.035 *M* KH_2PO_4 + 0.03 *M* citric acid + 2 m*M* Na_2EDTA, pH 3 (25:75).
E-2. Acetonitrile-0.01 *M* KH_2PO_4, 0.003 citric acid, 0.001 *M* Na_2EDTA, pH 3 (20:80).
E-3. Methanol-0.1 *M* K_2HPO_4 (20:80) containing 9 mℓ/ℓ dibutylamine, pH 2.6.

REFERENCES

1. **Hardee, G. E. and Lai, J. W.,** Determination of dobutamine in plasma by liquid chromatography with electrochemical detection, *Anal. Lett.*, 16, 69, 1983.
2. **Dixon, R., Hsiao, J., and Caldwell, W.,** Cardiotonic agents: a simple HPLC procedure for the quantitation of dobutamine and a new congener in plasma, *Res. Commun. Chem. Pathol. Pharmacol.*, 48, 313, 1985.
3. **Knoll, R. and Brandl, M.,** Rapid and simple method for the routine determination of dobutamine in human plasma by high-performance liquid chromatography, *J. Chromatogr.*, 345, 425, 1985.

L-DOPA

Liquid Chromatography

Specimen (mℓ)	Extraction	Column (cm × mm)	Packing (µm)	Elution	Flow (mℓ/min)	Det. (nm)	RT (min)	Internal standard (RT)	Other compounds (RT)	Ref.
Plasma (1)	I-1	25 × 4.6	Ultrasphere-ODS (5)[a]	E-1	0.8	Electrochem[b]	8.2	Dihydroxy-benzylamine	—	1
Plasma (1)	I-2	15 × 4.6	Ultrasphere-Octyl (5)	E-2	1.2	Electrochem[b]	2.5	α-Methylnor-epinephrine (5.1)	Carbidopa (6.8)	2
Plasma (2)	I-3	15 × 4.6	Cosmosil-C_{18} (5)	E-3	0.46; gra-dient	Coulochem[c]	8.5	—	—	3, 4
Serum (1)	I-4	10 × 8	Radial-Pak-C_{18} (10)[d]	E-4	3.0	Electrochem[e]	2.5	—	Carbidopa (7.5) 3-O-Methyldopa (8.5)	5
Plasma (0.1)	I-5	20 × 4	Nuclosil-C_{18} (5)	E-5	1.0	Coulochem[f]	4	—	3-O-Methyldopa (7)	6

[a] Protected by a 30 × 4.6mm guard column packed with Perisorb RP-18 (30 to 40 µm).
[b] 0.72 V.
[c] A = 0.05 V, B = 0.35 V.
[d] Protected by a (30 × 2.9) guard column, packed with C_{18} corasil (37 to 50 µm).
[e] 0.66 V.
[f] Conditioning cell, + 0.35 V, I = +0.04 V, II = −0.30 V.

L-DOPA (continued)

Extraction — I-1. The sample was mixed with 0.4 mℓ of 2% Na_2EDTA, 30 mg of sodium metabisulfite and 10 μℓ of the internal standard solution (100 ng/mℓ in 0.1 *N* HCl). The pH of the mixture was adjusted to 8.6 with 0.02 *N* NaOH and treated with 300 mg of alumina washed with 5 mℓ of 0.2 *M* phosphate buffer, pH 8.6. After mixing for 1 min, the supernate was discarded, alumina washed three times with 3-mℓ aliquots of water, and finally with 3 mℓ of methanol, and then eluted with 2 mℓ of acetyl chloride-methanol (1:99). The eluate was evaporated at room temperature under vacuum. The residue was reconstituted in 110 μℓ of 0.1 *N* perchloric acid for injection.

I-2. The sample was added to a suspension of 60 ng of activated alumina in 1 mℓ of 1 m*M* HCl containing 0.1 m*M* Na_2 EDTA and 100 μℓ of an aqueous solution of the internal standard (10 μg/mℓ) and 1 mℓ of water, followed by 1 mℓ of 3 *M* Tris buffer, pH 8.6. After 15 to 20 min of mixing, the supernatant was discarded and alumina washed three times with water and finally eluted with 200 μℓ of 0.1 *M* phosphoric acid. Aliquots of 50 μℓ of the supernatant were injected.

I-3. Protein precipitation with 0.45 mℓ of 1 *M* trichloroacetic acid.

I-4. The sample was treated with 50 μℓ of 70% $HClO_4$. After 10 min, the mixture was treated with 500 μℓ of 1 *M* K_2HPO_4 containing 5 m*M* Na_2EDTA and pH adjusted to 8 with 1 *M* KOH (700 μℓ). After centrifugation at 4°C, the pH of the supernatant was adjusted to 5 with HCl.

I-5. The sample was treated with an equal volume of 1.2 *M* perchloric acid, and then the mixture was diluted to 1 mℓ with water. After centrifugation, aliquots of 30 μℓ of clear supernatant were injected.

Elution — E-1. Methanol-0.1 *M* citric acid-0.1 *M* sodium acetate (5:32:37), pH 4.1.

E-2. Methanol-citrate/phosphate buffer, pH 3.1 (14:86) containing 6.5 m*M* 1-octanesulphonic acid and 2 m*M* Na_2EDTA.

E-3. (A) 0.05 *M* phosphate buffer, pH 3.1; (B) methanol. Isocratic 100%(A) for 18 min with linear gradient from 0 to 15%(B) over 8 min.

E-4. 100 m*M* Ammonium phosphate, pH 4.3.

E-5. Acetonitrile-50 m*M* NaH_2PO_4 + 50 m*M* sodium acetate + 0.7 m*M* sodium dodecyl sulfate + 2 m*M* Na_2EDTA (12.5:87.5).

REFERENCES

1. **Shum, A., Van Loon, G. R., and Sole, M. J.,** Measurement of L-dihydroxy-phenylalanine in plasma and other biological fluids by high pressure liquid chromatography with electrochemical detection, *Life Sci.*, 31, 1541, 1982.
2. **Causon, R. C., Brown, M. J., Leenders, K. L., and Wolfson, L.,** High-performance liquid chromatography with amperometric detection of plasma L-3,4-dihydroxyphenylalanine in parkinsonian patients, *J. Chromatogr.*, 277, 115, 1983.
3. **Ishimitsu, T. and Hirose, S.,** Determination of *m*- and *p*-O-methylated products of L-3,4-dihydroxyphenylalanine using high-performance liquid chromatography and electrochemical detection, *Anal. Biochem.*, 150, 300, 1985.
4. **Ishimitsu, T. and Hirose, S.,** Simultaneous assay of 3,4-dihydroxyphenylalanine, catecholamines and O-methylated metabolites in human plasma using high-performance liquid chromatography, *J. Chromatogr.*, 337, 239, 1985.
5. **Beers, M. F., Stern, M., Hurtig, H., Melvin, G., and Scarpa, A.,** Simultaneous determination of L-dopa and 3-O-methyldopa in human serum by high-performance liquid chromatography, *J. Chromatogr.*, 336, 380, 1984.
6. **Baruzzi, A., Contin, M., Albani, F., and Riva, R.,** Simple and rapid micromethod for the determination of levodopa and 3-O-methyldopa in human plasma by high-performance liquid chromatographywith coulometric detection, *J. Chromatogr.*, 375, 165, 1986.

DOTHIEPIN

Gas Chromatography

Specimen (mℓ)	Extraction	Column (m × mm)	Packing (mesh)	Oven temp (°C)	Gas (mℓ/min)	Det.	RT (min)	Internal standard (RT)	Deriv.	Other compounds (RT)	Ref.
Plasma (1)	I-1	1 × 2	3% OV-17 GasChrom Q (100/120)	220	Methane (8)	MS-CI	NA	[$^2H_3$]Dothiepin	—	—	1
Plasma (1—2)	I-2	2 × 2	3% OV-101 GasChrom W (80/100)	T.P.[a]	He (20)	MS-EI	3.6	[$^2H_3$]Dothiepin + Protriptyline (4.1)	Trifluoro-acetyl	Dothiepin-S-oxide (4) Northiaden (5.8)	2

[a] Initial temp = 210°C; rate = 4°C/min; final temp = 230°C.

Extraction — I-1. The sample was mixed with 10 μℓ of an aqueous solution of the internal standard (2 μg/mℓ) 10 μℓ of carrier solution (25 μg/mℓ of imipramine HCl in water) and 200 μℓ of 1 *N* sodium hydroxide. The mixture was extracted with 10 mℓ of hexane. The organic layer was evaporated at 45°C under a stream of nitrogen. The residue was dissolved in 5 μℓ of methyl acetate and injected into the gas chromatograph.
I-2. The sample was diluted to 5 mℓ with water, mixed with ethanolic solutions of the internal standards to give a concentration of 50 ng of dueterodothiepin, 20 ng of protriptyline, and 0.5 mℓ of 5 *M* sodium hydroxide. The mixture was extracted twice with 5-mℓ portions of *n*-hexane. The combined organic layers were back extracted into 1 mℓ of 1 *M* HCl. The aqueous layer was made alkaline with 0.3 mℓ of 5 *M* NaOH and extracted twice with 2-mℓ portions of n-hexane. The combined organic layers were treated with 100 μℓ of trifluoroacetic anhydride. The solvents were removed at 37°C under a stream of air. The residue was reconstituted with 7 μℓ of ethanol and aliquots of 4 to 5 μℓ were injected.

REFERENCES

1. **Crampton, E. L., Glass, R. C., Marchant, B., and Rees, J. A.,** Chemical ionisation mass fragmentographic measurement of dothiepin plasma concentrations following a single oral dose in man, *J. Chromatogr.*, 183, 141, 1980.
2. **Maguire, K. P., Norman, T. R., Burrows, G. D., and Scoggins, B. A.,** Simultaneous measurement of dothiepin and its major metabolites in plasma and whole blood by gas chromatography-mass fragmentography, *J. Chromatogr.*, 222, 399, 1981.

DOXAPRAM

Gas Chromatography

Specimen (mℓ)	Extraction	Column (m × mm)	Packing (mesh)	Oven temp (°C)	Gas (mℓ/min)	Det.	RT (min)	Internal standard (RT)	Deriv.	Other compounds (RT)	Ref.
Blood, plasma (0.1)	I-1	0.75 × 2	1% OV-225 GasChrom Q (100/120)	250	Methane (20)	MS-CI	2.5	Dextromoramide (2)	—	—	1

Extraction — I-1. The plasma sample was mixed with 500 μℓ of the internal standard solution (1 μg/mℓ in 0.1 *M* HCl), water (1.4 mℓ), and 0.5 mℓ of 5 *M* sodium hydroxide. The mixture was extracted with 10 mℓ of diethyl ether. The ether layer was evaporated to about 10 to 15 μℓ. Half of this residual ether extract was analyzed.

REFERENCE

1. **Nichol, H., Vine, J., Thomas, J., and Moore, R. G.,** Quantitation of doxapram in blood, plasma and urine, *J. Chromatogr.*, 182, 191, 1980.

DOXAZOSIN

Liquid Chromatography

Specimen (mℓ)	Extraction	Column (cm × mm)	Packing (μm)	Elution	Flow (mℓ/min)	Det. (nm)	RT (min)	Internal standard (RT)	Other compounds (RT)	Ref.
Plasma (1)	I-1	30 × 4.6	μ-Bondapak-C_{18} (10)	E-1	1.5	Fl (254, 360)	4.7	Prazosin (3.4)	—	1

Extraction — I-1. The sample was mixed with 1 mℓ of the internal standard solution (10 ng/mℓ in water) and 200 μℓ of ammonia. The mixture was extracted with 5 mℓ of dichloromethane-diethyl ether (1:2:5). The organic layer was evaporated under a stream of nitrogen at 75°C. The residue was reconstituted with 20 μℓ of the mobile phase and the entire solution was injected.

Elution — E-1. Methanol containing 0.01 *M* pentane sulfonic acid - water containing 0.01 *M* pentane sulfonic acid + 0.02 *M* tetramethyl ammonium hydroxide, pH 3.4 with acetic acid (700:300).

REFERENCE

1. **Cowlishaw, M. G. and Sharman, J. R.,** Doxazosin determination by high-performance liquid chromatography using fluorescence detection, *J. Chromatogr.*, 344, 403, 1985.

DOXEPIN

Gas Chromatography

Specimen (mℓ)	Extraction	Column (m × mm)	Packing (mesh)	Oven temp (°C)	Gas (mℓ/min)	Det.	RT (min)	Internal standard (RT)	Deriv.	Other compounds (RT)	Ref.
Serum (2)	I-1	1.8 × 2	3% OV-17 GasChrom Q (100/120)	T.P.[a]	He (40)	NPD	4.1	Amitriptyline (3.6)	—	Loxapine (9.5)	1
Plasma (1)	I-2	1.1 × 2	3% OV-17 GasChrom Q (100/120)	220	He (20)	MS-EI	2.1	Nortriptyline (4.9) + Amitriptyline (1.8)	Trifluoro-acetyl	*cis*-Desmethyl doxepine (5.4) *trans*-Desmethyl doxepin (6)	2

Liquid Chromatography

Specimen (mℓ)	Extraction	Column (cm × mm)	Packing (μm)	Elution	Flow (mℓ/min)	Det. (nm)	RT (min)	Internal standard (RT)	Other compounds (RT)	Ref.
Plasma (1)	I-3	15 × 4.6	Spherisorb-ODS (5)	E-1	2.5	ABS (200)	4.6	Desipra-mine (6.3)	Desmethyl-doxepin (3.3)	3
Urine (1)	I-4	12.5 × 3.2	Spherisorb Hexyl (5)	E-2	1.0	ABS (205)	5.5	Imipramine (7.4)	Desmethyl-doxepin (4.5)	4

[a] Initial temp = 235°C; initial time = 6 min; rate = 32°C/min; final temp = 280°C; final time = 4 min.

DOXEPIN (continued)

Extraction — I-1. The sample was made alkaline with 1 mℓ of 0.5 *M* sodium hydroxide and extracted with 30 mℓ of 4% isobutanol in *n*-heptane. The organic layer was back extracted into 5 mℓ of 0.1 *M* HCl. The aqueous layer was washed with 30 mℓ of *n*-heptane and re-extracted with 10 mℓ of ether (presumably after making it alkaline). The ether layer was evaporated and the residue dissolved in 25 μℓ of absolute ethanol containing the internal standard. Aliquots of 1 μℓ of this solution were injected.
I-2. The sample was mixed with 50 μℓ of an aqueous solution of the internal standards (1.2 μg/mℓ of each) and 50 μℓ of 6 *N* NaOH. The mixture was extracted with 5 mℓ of hexane-isoamyl alcohol (95:5). The organic layer was back extracted into 500 μℓ of 0.1 *N* HCl. The aqueous layer was evaporated at 45°C under a stream of nitrogen. The residue was treated with 300 μℓ of a 5% trifluoroacetic anhydride solution at room temperature for 30 min. The reaction mixture was evaporated at room temperature under nitrogen. The residue was dissolved in 25 μℓ of ethyl acetate and 10 μℓ were injected.
I-3. The sample was mixed 50 μℓ of the internal standard solution (2 μg/mℓ) and 0.25 mℓ of saturated sodium carbonate solution. The mixture was extracted with 4.5 mℓ of pentane. The organic layer was back extracted into 0.1 mℓ of 0.1 *N* HCl. Aliquots of the aqueous phase were injected.
I-4. The sample was mixed with 0.1 mℓ of the internal standard solution (2.5 μg/mℓ in water) and 0.2 mℓ of 1 *M* sodium hydroxide. The mixture was extracted with 7 mℓ of 2% *n*-butanol in hexane. The organic layer was back extracted with 0.1 mℓ of 0.1 *M* HCl. An aliquot of 50 μℓ of the acid extract was injected.

Elution — E-1. Acetonitrile-0.01 *M* phosphate buffer containing 600 ppm of *n*-nonylamine, pH 3.1 (45:55).
E-2. Acetonitrile-0.02 *M* phosphate buffer, pH 3.5 (38:62).

REFERENCES

1. **Vasiliades, J., Sahawneh, T. M., and Owens, C.,** Determination of therapeutic and toxic concentrations of doxepin and loxapine using gas-liquid chromatography with a nitrogen-sensitive detector, and gas chromatography-mass spectrometry of loxapine, *J. Chromatogr.*, 164, 457, 1979.
2. **Davis, T. P., Veggeberg, S. K., Hameroff, S. R., and Watts, K. L.,** Sensitive and quantitative determination of plasma doxepin and desmethyldoxepin in chronic pain patients by gas chromatography and mass spectrometry, *J. Chromatogr.*, 273, 436, 1983.
3. **Faulkner, R. D. and Lee, C.,** Comparative assays for doxepin and desmethyldoxepin using high-performance liquid chromatography and high-performance thin-layer chromatography, *J. Pharm. Sci.*, 72, 1165, 1983.
4. **Park, Y. H., Goshorn, C., and Hinsvark, O. N.,** Quantitative determination of doxepin and nordoxepin in urine by high-performance liquid chromatography, *J. Chromatogr.*, 375, 202, 1986.

DOXYCYCLINE

Liquid Chromatography

Specimen (mℓ)	Extraction	Column (cm × mm)	Packing (μm)	Elution	Flow (mℓ/min)	Det. (nm)	RT (min)	Internal standard (RT)	Other compounds (RT)	Ref.
Tissue homogenate	I-1	10 × 2	LiChrosorb RP-8 (5)	E-1	0.5	ABS (350)	6	Demeclo-cycline (2.5)	—	1
Blood, serum, urine (0.1)	I-2	25 × 4	Nucleosil-C_8 (10)	E-2	1.9	ABS (344)	2.8	—	a	2, 3

[a] A new metabolite eluting before doxycycline was isolated.

Extraction — I-1. The weighed amount of tissue was homogenized with 2.5 mℓ of 0.1 *M* HCl and 50 μℓ of the internal standard solution (0.051 m*M* in 0.1 *M* HCl). The homogenate was washed twice with 12-mℓ portions of diethylether after the addition of 3 mg of ascorbic acid. The homogenate was then mixed with 200 μℓ of 1 *M* sodium hydroxide and 1 mℓ of phosphate sulfate buffer (pH 6.1) and then extracted with 10 mℓ of ethylacetate. The organic layer was evaporated at room temperature under reduced pressure. The residue was dissolved in 200 μℓ of the mobile phase and aliquots of 20 μℓ were injected.
I-2. All extraction steps were carried out in the cold (approx. 0°C). The sample was mixed with 150 μℓ of 0.03 *M* H_3PO_4; after 15 min 1 mℓ acetonitrile-buffer (0.01 *M* NaH_2PO_4, pH 2.4) (50:50) was added, allowed to stand for 5 min, and centrifuged. Aliquots of the supernatant were injected.

Elution — E-1. Acetonitrile-0.1 *M* citric acid (25:75).
E-2. Acetonitrile-3.5 m*M* NaH_2PO_4 (30:70), pH 2.7.

REFERENCES

1. **Nelis, H. J. C. F. and De Leenheer, A. P.,** Liquid chromatographic estimation of doxycycline in human tissues, *Clin. Chim. Acta*, 103, 209, 1980.
2. **Bocker, R.,** Analysis and quantitation of a metabolite of doxycycline in mice, rats, and humans by high-performance liquid chromatography, *J. Chromatogr.*, 274, 255, 1983.
3. **Bocker, R.,** Rapid analysis of doxycycline from biological samples by high-performance liquid chromatography, *J. Chromatogr.*, 187, 439, 1980.

DOXYLAMINE

Gas Chromatography

Specimen (mℓ)	Extraction	Column (m × mm)	Packing (mesh)	Oven temp (°C)	Gas (mℓ/min)	Det.	RT (min)	Internal standard (RT)	Deriv.	Other compounds (RT)	Ref.
Plasma, urine (1)	I-1	1.8 × 2	5% Dexsil 300 Chromosorb W (80/100)	220	He (30)	NPD	3.2	—	—	—	1—3

Liquid Chromatography

Specimen (mℓ)	Extraction	Column (cm × mm)	Packing (μm)	Elution	Flow (mℓ/min)	Det. (nm)	RT (min)	Internal standard (RT)	Other compounds (RT)	Ref.
Urine (5)	I-1	25 × 4.6	Ultrasphere-ODS (5)	E-1	1.0	ABS (254)	5.0	—	—	1
Plasma (3)	I-2	30 × 3.9	μ-Porasil (10)	E-2	1.5	ABS (254)	14.6	Amphetamine (12.1)	—	4

Extraction — I-1. The sample was treated with 9 mℓ of 1 *N* HCl. A 5-mℓ aliquot of this solution was washed three times with 5-mℓ portions of dichloromethane. The aqueous phase was made alkaline with 0.5 mℓ of 10 *N* NaOH and 1 mℓ of 1 *M* K_2HPO_4 (pH 9.4) and extracted three times with 5-mℓ portions of dichloromethane. The combined organic extract was drived over anhydrous sodium sulfate and then evaporated under vacuum. The residue was dissolved in 1 mℓ of methanol and aliquots of 2 μℓ were injected for gas chromatography (Aliquots of 10 μℓ were injected for liquid chromatographic analysis of urine).
I-2. The sample was mixed with 10 mℓ of 0.3 *N* NaOH and extracted twice with 5-mℓ portions of dichloromethane. The combined organic extracts were evaporated under nitrogen at 40°C. The residue was dissolved in 200 μℓ of the internal standard solution (30 μg/mℓ in dichloromethane). Aliquots of 80 μℓ of this solution were injected.

Elution — E-1. Methanol-0.01 *M* KH_2PO_4, pH 7 (90:10).
E-2. Chloroform, 8 parts – acetonitrile, 1 part - methanol + ammonium hydroxide + ammonium chloride (57:2:1), 1 part.

REFERENCES

1. **Holder, C. L., Thompson, H. C., Jr., and Slikker, W., Jr.**, Trace level determination of doxylamine in nonhuman primate plasma and urine by GC/NPD and HPLC, *J. Anal. Toxicol.*, 8, 46, 1984.
2. **Holder, C. L., Korfmacher, W. A., Slikker, W., Jr., Thompson, H. C., Jr., and Gosnell, A. B.**, Mass spectral characterization of doxylamine and its rhesus monkey urinary metabolites, *Biomed. Mass Spectrom.*, 12, 151, 1985.
3. **Slikker, W., Jr., Holder, C. L., Lipe, G. W., Korfmacher, W. A., Thompson, H. C., Jr., and Bailey, J. R.**, Metabolism of ^{14}C-labeled doxylamine succinate (Bendectin) in the rhesus monkey *(Macaca mulatta)*, *J. Anal. Toxicol.*, 10, 87, 1986.
4. **Kohlhof, K. J., Stump, D., and Zizzamia, J. A.**, Analysis of doxylamine in plasma by high-performance liquid chromatography, *J. Pharm. Sci.*, 72, 961, 1983.

DROPERIDOL

Liquid Chromatography

Specimen (mℓ)	Extraction	Column (cm × mm)	Packing (μm)	Elution	Flow (mℓ/min)	Det. (nm)	RT (min)	Internal standard (RT)	Other compounds (RT)	Ref.
Dosage	—	25 × 4.6	Ultrasphere-Octyl (5)	E-1	1.5	ABS (230)	14	—	Methylparaben (4.5)[a] Propylparaben (6.5)[a] 2-Benzimidazolinone (4)[b] 4′-Fluoro-4- (4-oxo-piperidine) butyrophenone (6)[b] *p*- Hydroxybenzoic acid (3)[b]	1

[a] Preservatives of droperidol injection solutions.
[b] Possible degradation products of droperidol.

Elution — E-1. Methanol-0.02 *M* phosphate buffer, pH 6.8.

REFERENCE

1. **Dolezalova, M.**, Separation and determination of droperidol, methyl- and propylparaben and their degradation products by high-performance liquid chromatography, *J. Chromatogr.*, 286, 323, 1984.

DROTAVERINE

Liquid Chromatography

Specimen (mℓ)	Extraction	Column (cm × mm)	Packing (μm)	Elution	Flow (mℓ/min)	Det. (nm)	RT (min)	Internal standard (RT)	Other compounds (RT)	Ref.
Plasma (1)	I-1	25 × 3.9	Chromspher-Sil (10)	E-1	1.6	ABS (302)	4.7	Papaverine (6)	Drotaveraldine (3.6)	1

Extraction — I-1. The sample was adjusted to pH 1.5 with 1 *M* HCl and extracted with 10 mℓ of chloroform. The organic layer was acidified with 40 μℓ of glacial acetic acid and then dried with anhydrous sodium sulfate. An aliquot of 8 mℓ of the chloroform extract was evaporated under reduced pressure. The residue was dissolved in 70 μℓ of chloroform and 10 μℓ of the internal standard solution (1 mg/mℓ in chloroform) were added. Aliquots of 20 μℓ of the final solution were injected.

Elution — E-1. *n*-Heptane-dichloromethane-diethylamine (50:25:2).

REFERENCES

1. **Mezei, J., Kuttel, S., Szentmiklosi, P., Marton, S., and Racz, I.,** A new method for high-performance liquid chromatographic determination of drotaverine in plasma, *J. Pharm. Sci.*, 73, 1489, 1984.

DRUG SCREENING

For specific information and tables concerning this subject, the interested reader is directed to the following references.

REFERENCES

1. **Daldrup, T., Susanto, F., and Michalke, P.,** Kombination von DC, GC (OV 1 und OV 17) und HPLC (RP 18) zur schnellen Erkennung von Arzneimitteln, Rauschmitteln und verwandten Verbindungen, *Fresenius Z. Anal. Chem.*, 308, 413, 1981.
2. **Stead, A. H., Gill, R., Wright, T., Gibbs, J. P., and Moffat, A. C.,** Standarised thin-layer chromatographic systems for the identification of drugs and poisons, *Analyst*, 107, 1106, 1982.
3. **Gough, T. A. and Baker, P. B.,** Identification of major drugs of abuse using chromatography: an update, *J. Chromatogr. Sci.*, 21, 145, 1983.
4. **Newton, B. and Foery, R. F.,** Retention indices and dual capillary gas chromatography for rapid identification of sedative hypnotic drugs in emergency toxicology, *J. Anal. Toxicol.*, 4, 129, 1984.

5. **Jarvie, D. R. and Simpson, D.,** Gas chromatographic screening for drugs and metabolites in plasma and urine, *Ann. Clin. Biochem.*, 21, 92, 1984.
6. **Deutsch, D. G. and Bergert, R. J.,** Evaluation of a benchtop capillary gas chromatograph-mass spectrometer for clinical toxicology, *Clin. Chem.*, 31, 741, 1985.
7. **Ehresman, D. J., Price, S. M., and Lakatua, D. J.,** Screening biological samples for underivatized drugs using a splitless injection technique on fused silica capillary column gas chromatography, *J. Anal. Toxicol.*, 9, 55, 1985.
8. **Koves, E. M. and Wells, J.,** An evaluation of fused silica capillary columns for the screening of basic drugs in postmortem blood: qualitative and quantitative analysis, *J. Forensic Sci.*, 30, 692, 1985.
9. **Perrigo, B. J., Peel, H. W., and Ballantyne, D. J.,** Use of dual-column fused-silica capillary gas chromatography in combination with detector response factors for analytical toxicology, *J. Chromatogr.*, 341, 81, 1985.
10. **Fretthold, D., Jones, P., Sebrosky, G., and Sunshine, I.,** Testing for basic drugs in biological fluids by solvent extraction and dual capillary GC/NPD, *J. Anal. Toxicol.*, 10, 10, 1986.
11. **Lora-Tamayo, C., Rams, M. A., and Chacon, J. M. R.,** Gas chromatographic data for 187 nitrogen- or phosphorus-containing drugs and metabolites of toxicological interest analysed on methyl silicone capillary columns, *J. Chromatogr.*, 374, 73, 1986.
12. **Baker, J. K., Skelton, R. E., and Ma, C. Y.,** Identification of drugs by high-pressure liquid chromatography with dual wavelength ultraviolet detection, *J. Chromatogr.*, 168, 417, 1979.

DYPHYLLINE

Liquid Chromatography

Specimen (mℓ)	Extraction	Column (cm × mm)	Packing (μm)	Elution	Flow (mℓ/min)	Det. (nm)	RT (min)	Internal standard (RT)	Other compounds (RT)	Ref.
Plasma (1)	I-1	30 × 3.9	μ-Bondapak-C_{18} (10)[a]	E-1	2.0	ABS (274)	7	β-Hydroxyethyl-theophylline	Theophylline (6) Theobromine (4.2) Caffeine (13.5)	1
Serum (0.25)	I-2	25 × 4	LiChrosorb RP-8 (10)	E-2	2.0	ABS (275)	3.4	Theophylline (4.5)	1,3-Dimethyluric acid (1.4) 3-Methylxanthine (2.4) Caffeine (5.1)	2

DYPHYLLINE (continued)

Serum (0.5)	I-3	25 × 4.6	Ultrasphere-ODS (5)[a]	E-3	1.5	ABS (274)	6.7	8-Chloro-theophylline	Paraxanthine (5.7) Theophylline (6) Theobromine (3.8) Caffeine (12.1) Proxyphylline (15.1)	3

[a] Column temp = 40°C.

Extraction — I-1. The sample was mixed with 100 μℓ of an internal standard solution (0.2 mg/mℓ in pH 4 acetate buffer) and 100 μℓ of a 40% aqueous trichloroacetic acid solution. After mixing and centrifugation a 25-μℓ aliquot of the supernatant was injected.
I-2. The sample was mixed with 50 μℓ of the internal standard solution (30 μg/mℓ in water) and 1.5 g of anhydrous sodium sulfite. The mixture was extracted with 2.5 mℓ of chloroform-methanol (9:1). The organic layer was evaporated at 40°C under nitrogen. The residue was dissolved in 100 μℓ of dichloromethane and back extracted into 100 μℓ of 0.1 *M* ammonium carbonate. Aliquots of 10 μℓ of the aqueous layer were injected.
I-3. The sample was treated with 3 mℓ of 2-propanol containing 1.5 μg of the internal standard. The supernatant was evaporated at 60°C under nitrogen. The residue was dissolved in 50 μℓ of methanol and 10 μℓ were injected.

Elution — E-1. Acetonitrile-0.01 *M* acetate buffer, pH 4 (6:94).
E-2. Methanol-2 m*M* phosphate buffer, pH 3 (250:750).
E-3. Acetonitrile-methanol-0.01 *M* acetate buffer, pH 5.2 (6:3:91).

REFERENCES

1. **Valia, K. H., Hartman, C. A., Kucharczyk, N., and Sofia, R. D.,** Simultaneous determination of dyphylline and theophylline in human plasma by high-performance liquid chromatography, *J. Chromatogr.*, 221, 170, 1980.
2. **Paterson, N.,** High-performance liquid chromatographic method for the determination of diprophylline in human serum, *J. Chromatogr.*, 232, 450, 1982.
3. **Wenk, M., Eggs, B., and Follath, F.,** Simultaneous determination of diprophylline, proxyphylline and theophylline in serum by reversed-phase high-performance liquid chromatography, *J. Chromatogr.*, 276, 341, 1983.

E-0663

Liquid Chromatography

Specimen (mℓ)	Extraction	Column (cm × mm)	Packing (μm)	Elution	Flow (mℓ/min)	Det. (nm)	RT (min)	Internal standard (RT)	Other compounds (RT)	Ref.
Blood, brain homogenate (0.2)	I-1	25 × 4.6	Ultrasphere-ODS (5)[a]	E-1	1.5	Electrochem[b]	6	Promethazine (4)	—	1

[a] Protected by a 10 × 4.5 mm ODS precolumn.
[b] Potential = 0.8 V.

Extraction — I-1. The sample was mixed with 1 mℓ of 1 *M* sodium hydroxide and the internal standard. The mixture was extracted with 5 mℓ of heptane-isoamyl alcohol (99:1). An aliquot of 4 mℓ of the organic phase was back extracted into 100 μℓ of 0.1 *M* HCl. An aliquot of the aqueous layer was injected.

Elution — E-1. Acetonitrile-0.1 *M* acetate buffer, pH 3.5-tetrahydrofuran-pyridine (200:100:3:0.3).

REFERENCE

1. **Shibanoki, S., Kubo, T., and Ishikawa, K.,** Chromatographic assay of 10-[3-(3-hydroxypyrrolidinyl)propyl]-2-trifluoromethyl phenothiazine using electrochemical detection, *J. Chromatogr.*, 377, 436, 1986.

ELLIPTICINE

Liquid Chromatography

Specimen (mℓ)	Extraction	Column (cm × mm)	Packing (μm)	Elution	Flow (mℓ/min)	Det. (nm)	RT (min)	Internal standard (RT)	Other compounds (RT)	Ref.
Blood, tissue (0.59)	I-1	30 × 4	μ-Bondapak-C_{18} (10)	E-1	1.4	Fl (360, 455)	8.4	11-Demethylellipticine (6.4)	9-Hydroxyellipticine[a]	1
Plasma (0.2)	I-2	30 × 3.9	μ-Bondapak-C_{18} (10)	E-2	1.5	Electrochem[b]	—	N-Propyl-9-hydroxyellipticine (5.7)	9-Hydroxyellipticine (3.8)	2

[a] Determined separately by UV detection at 300 nm and a different mobile phase.
[b] Potential = 0.6 V.

Extraction — I-1. The weighed sample was homogenized with 1.5 mℓ of 0.05 *M* sodium phosphate buffer (pH 7.4) containing the internal standard.The homogenate was extracted with 3 mℓ of ethyl acetate. A 2-mℓ aliquot of the organic layer was back extracted into 0.5 mℓ of 0.01 *N* HCl. A 100-μℓ aliquot of the aqueous layer was injected.
I-2. The sample was spiked with 10 μℓ of an aqueous solution of the internal standard (100 ng/mℓ) and extracted twice with 1-mℓ portions of ethyl acetate after the addition of 5 μℓ of tetraphenyl borate (0.5%). The combined organic layers were evaporated under a stream of nitrogen and the residue was dissolved in 200 μℓ of the mobile phase for injection.

Elution — E-1. Acetonitrile-0.01 *M* NaH_2PO_4 (36:64), pH 3.5.
E-2. Methanol-water (60:40) containing 100 m*M* ammonium acetate, pH 6.

REFERENCES

1. **Bykadi, G., Flora, K. P., Cradock, J. C., and Poochikian, G. K.,** Determination of ellipticine in biological sample by high-performance liquid chromatography, *J. Chromatogr.*, 231, 137, 1982.
2. **Bellon, P., Canal, P., Bernadou, J., and Soula, G.,** Use of electrochemical detection in the high-performance liquid chromatographic determination of hydroxylated ellipticine derivatives, *J. Chromatogr.*, 309, 170, 1984.

EMETINE

Liquid Chromatography

Specimen (mℓ)	Extraction	Column (cm × mm)	Packing (μm)	Elution	Flow (mℓ/min)	Det. (nm)	RT (min)	Internal standard (RT)	Other compounds (RT)	Ref.
Plasma (2)	I-1	15 × 4.6	Ultrasphere-ODS (5)	E-1	1.7	Fl (285, 316)	4.1	*N*-Propyl-procain-amide (1.7)	Cephaeline (3)	1

Extraction - I-1. The sample was mixed with 100 μℓ of an aqueous solution of the internal standard (10 μg/mℓ) and 2 mℓ of borate buffer, pH 9. The mixture was extracted with 7 mℓ of *n*-butyl chloride. The organic layer was back extracted into 200 μℓ of 0.01 *M* HCl and a 30- to 50-μℓ aliquot of the aqueous phase was injected.

Elution — E-1. Methanol-0.025 *M* Na_2HPO_4, pH 8 (72:28).

REFERENCE

1. **Crouch, D. J., Moran, D. M., Finkle, B. S., and Peat, M. A.,** Quantative analysis of emetine and cephaeline by reversed-phase high performance liquid chromatography with fluorescence detection, *J. Anal. Toxicol.*, 8, 63, 1984.

ENCAINIDE

Liquid Chromatography

Specimen (mℓ)	Extraction	Column (cm × mm)	Packing (μm)	Elution	Flow (mℓ/min)	Det. (nm)	RT (min)	Internal standard (RT)	Other compounds (RT)	Ref.
Plasma (1)	I-1	30 × 3.9	μ-Porasil (10)	E-1	1.0	ABS (254)	7.5	—	N-Demethylencainide (6.7) N,-O-Didemthylencainide (9.3) O-Desmethylencainide (10.7)	1

Extraction — I-1. The sample was adjusted to pH 8.5 with 0.2 mℓ of 0.5 *M* Tris buffer and extracted with 10 mℓ of *n*-butyl chloride containing 5% isopropyl alcohol. A 9-mℓ aliquot of the organic layer was evaporated under a stream of nitrogen. The residue was dissolved in 100 μℓ of methanol and a 50-μℓ aliquot of this solution was injected.

Elution — E-1. Chloroform-methanol-water-acetic acid (276.5:270:96:24).

REFERENCE

1. **Mayol, R. F. and Gammans, R. E.,** Analysis of encainide in plasma by radioimmunoassay and high pressure liquid chromatography, *Ther. Drug Monit.*, 1, 507, 1979.

ENDRALAZINE

Liquid Chromatography

Specimen (mℓ)	Extraction	Column (cm × mm)	Packing (μm)	Elution	Flow (mℓ/min)	Det. (nm)	RT (min)	Internal standard (RT)	Other compounds (RT)	Ref.
Plasma (1)	I-1	NA	Spectraphysics RP-8 (10)[a]	E-1	2.0	Fl (230, 389)	4.2	Methylen-dralazine (8.7)	Metabolites[b]	1

[a] Column temp = 55°C.
[b] A number of metabolites with the use of a different internal standard have been determined.

Extraction — I-1. The sample was spiked with 100 μℓ of an aqueous solution of the internal standard mixture and extracted with 5 mℓ of chloroform. The organic layer was evaporated at 45°C under a stream of nitrogen. The residue was mixed with 50 μℓ of 90% fomic acid and the mixture was incubated at 90°C for 30 min. The excess reagent was evaporated at 90°C with a stream of nitrogen, 1 drop of ammonia added, and the solution again evaporated. The residue was reconstituted in the mobile phase for injection.

Elution — E-1. Acetonitrile-1.5 m*M* phosphoric acid (18:82).

REFERENCE

1. **Reece, P. A., Cozamanis, I., and Zacest, R.,** Sensitive high-performance liquid chromatographic assay for endralazine and two of its metabolites in human plasma, *J. Chromatogr.*, 225, 151, 1981.

ENOXACIN

Liquid Chromatography

Specimen (mℓ)	Extraction	Column (cm × mm)	Packing (μm)	Elution	Flow (mℓ/min)	Det. (nm)	RT (min)	Internal standard (RT)	Other compounds (RT)	Ref.
Plasma, urine (1)	I-1	30 × 4	μ-Bondapak-C_{18} (10)	E-1	1.5	ABS (340)	5	a (3.5)	M-2[b] (2.5)[c]	1
Plasma, urine (0.1)	I-2	15 × 4.6	LiChrosorb RP-2 (5)	E-2	1.6	ABS (342)	3	—	4-Oxo-enoxacin[d] (5)	2

[a] 8-Ethyl-5,8-dihydro-2-dimethylamino-5-oxopyrido (2,3-d)-pyrimidine-6-carboxylic acid.
[b] 1-Ethyl-6-fluoro-1,4-dihydro-4-oxo-7-(3-oxo-s-piperazinyl)-1,8-naphthypyridine-3-carboxylic acid.
[c] Alternative procedures for the determination of urinary metabolites are described.
[d] The metabolite is monitored at 265 nm.

Extraction — I-1. The sample was mixed with 1 mℓ of the internal standard solution (1 μg/mℓ in 0.2 *M* phosphate buffer, pH 7.4) and extracted with 3 mℓ of chloroform containing 1% ethylchloroformate. The organic layer was evaporated under a stream of air at 50°C. The residue was dissolved in 200 μℓ of methanol and a 20-μℓ aliquot was injected.
I-2. The sample was mixed with 0.3 mℓ of 3% trichloroacetic acid and allowed to stand for 10 min. After centrifugation, a 100-μℓ volume of the clear supernatant was injected.

Elution — E-1. Methanol-0.1 *M* citric acid-acetonitrile (9:5:1).
E-2. Ethanol-dimethylformamide-7 m*M* phosphoric acid (3:20:77).

REFERENCE

1. **Nakamura, R., Yamaguchi, T., Sekine, Y., and Hashimoto, M.,** Determination of a new antibacterial agent (AT-2266) and its metabolites in plasma and urine by high-performance liquid chromatography, *J. Chromatogr.*, 278, 321, 1983.
2. **Vree, T. B., Baars, A. M., and Wijnands, W. J. A.,** High-performance liquid chromatography and preliminary pharmacokinetics of enoxacin and its 4-oxo metabolite in human plasma, urine and saliva, *J. Chromatogr.*, 343, 449, 1985.

ENVIRADENE

Liquid Chromatography

Specimen (mℓ)	Extraction	Column (cm × mm)	Packing (μm)	Elution	Flow (mℓ/min)	Det. (nm)	RT (min)	Internal standard (RT)	Other compounds (RT)	Ref.
Plasma (1)	I-1	25 × 4.6	Zorbax-C_{18} (6)[a]	E-1	1.3	Electrochem[b,c]	14	1-(6-Methoxy-2-benzothiazolyl)-3-phenyl urea (7.5)	—	1

[a] A 3-cm Brownlee, C_{18}-GU column was placed in place of injector loop to avoid late eluting peaks. Analytical column temp = 28°C.
[b] A UV detector (268 nm) was also used.
[c] Potential = 0.9 V.

Extraction — I-1. The sample was spiked with 100 μℓ of the internal standard solution (1 μg/mℓ in benzene) and extracted with 11 mℓ of benzene. The organic layer was evaporated at 37°C under a stream of nitrogen. The residue was reconstituted with 200 μℓ of the mobile phase for injection with an autosampler.

Elution — E-1. Methanol-0.2 *M* sodium acetate (75:25) + 3 mg/ℓ Na_2EDTA, pH 7.8.

REFERENCE

1. **Bopp, R. J., Quay, J. F., Morris, R. M., Stucky, J. F., and Miner, D. J.,** Liquid chromatographic analysis of enviradene, a new antiviral agent, in plasma and its application in bioavailability studies in the dog, *J. Pharm. Sci.*, 74, 846, 1985.

ENVIROXIME

Liquid Chromatography

Specimen (mℓ)	Extraction	Column (cm × mm)	Packing (μm)	Elution	Flow (mℓ/min)	Det. (nm)	RT (min)	Internal standard (RT)	Other compounds (RT)	Ref.
Plasma (1)	I-1	25 × 4.6	Zorbax-C_8 (6)[a]	E-1	0.9	Electrochem[b]	8	Mexestrol (18)	Zinviroxime (9)	1

[a] Protected by a guard column packed with Co:Pell ODS; column temp = 28°C.
[b] Potential = 0.85 V.

Extraction — I-1. The sample was adjusted to pH 7.5 with K_2HPO_4 and extracted with 11.5 mℓ of benzene. The organic layer was washed with 0.5 mℓ of 1 *M* K_3PO_4, pH 11.5. An aliquot of 10 mℓ of benzene layer was mixed with 0.1 mℓ of the internal standard solution (0.5 μg/mℓ in chloroform) and evaporated at 37° under nitrogen. The residue was reconstituted with 200 μℓ of mobile phase for injection with an autosampler.

Elution — E-1. Methanol-0.14 *M* sodium acetate (65:35) with 3 mg/ℓ Na_2EDTA.

REFERENCE

1. **Bopp, R. J. and Miner, D. J.,** Determination of enviroxime in a variety of biological matrixes by liquid chromatography with electrochemical detection, *J. Pharm. Sci.*, 71, 1402, 1982.

EPHEDRINE

Gas Chromatography

Specimen (mℓ)	Extraction	Column (m × mm)	Packing (mesh)	Oven temp (°C)	Gas (mℓ/min)	Det.	RT (min)	Internal standard (RT)	Deriv.	Other compounds (RT)	Ref.
Pure compounds	—	1.2 × 4	3% OV-17 Chromosorb 750 (80/100)	140	He (60)	FID	2	—	Trifluoro-acetyl	a	1

Liquid Chromatography

Specimen (mℓ)	Extraction	Column (cm × mm)	Packing (μm)	Elution	Flow (mℓ/min)	Det. (nm)	RT (min)	Internal standard (RT)	Other compounds (RT)	Ref.
Pure compounds	—	30 × 3.9	μ-Bondapak-C_{18} (10)	E-1	1.0	ABS (210)	18	—	Pseudoephedrine (20)	2
Pure compounds	—	15 × 4.6	Hypersil Phenyl (5)	E-2	NA	ABS[b] (254)	(+) = 8.5 (−) = 10.2	—	c	3

[a] Separation of a number of mono-, di-, and trifluoroacetyl derivatives of stereoisomers of ephedrine and analogs is invertigated.
[b] The detector cell was kept at 22°C while the column was maintained at 25°C.

Elution — E-1. Acetonitrile-0.05 *M* NaH_2PO_4 (1:99).
E-2. Phosphate buffer, pH 6 containing 90 m*M* potassium hexafluorophosphate and saturated with (+)-di-*n*-butyltartarate.

REFERENCE

1. **Coutts, R. T., Dawe, R., Jones, G. R., Liu, S. F., and Midha, K. K.,** Analysis of perfluoroacyl derivatives of ephedrine, pseudoephedrine and analogues by gas chromatography and mass spectrometry, *J. Chromatogr.*, 190, 53, 1980.
2. **Barkan, S., Weber, J. D., and Smith, E.,** Determination of cross-contamination of the diastereomers ephedrine and pseudoephedrine by high-performance liquid chromatography, thin-layer chromatography and carbon-13 nuclear magnetic resonance spectroscopy, *J. Chromatogr.*, 219, 81, 1981.
3. **Patterson, C. and Sturrman, H. W.,** Direct separation of enantiomer of ephedrine and some analogues by reversed-phase liquid chromatography using (+)-di-n-butyltartrate as the liquid stationary phase, *J. Chromatogr. Sci.*, 22, 441, 1984.

4′-EPIDOXORUBICIN

Liquid Chromatography

Specimen (mℓ)	Extraction	Column (cm × mm)	Packing (μm)	Elution	Flow (mℓ/min)	Det. (nm)	RT (min)	Internal standard (RT)	Other compounds (RT)	Ref.
Plasma (2)	I-1	25 × 4	Partisil-ODS (10)[a]	E-1	0.4	Fl (470, 580)	22	13-Dihydro-4′-epidoxarubicin (17.5)	—	1
Plasma (0.6)	I-2	30 × 3.9	μ-Bondapak-C_{18} (10)	E-2; grad	2.5	Fl (470, 585)	20	Daunorubicin (22.2)	4′-Epidoxorubicinol (18.2)	2

[a] Protected by a 70 × 2.1 guard column packed with Co: Pell ODS.

Extraction — I-1. The sample was mixed with an aqueous solution of the internal standard (250 ng/mℓ) containing 125 ng/mℓ of desipramine HCl and 1 mℓ of pH 8.4 phosphate buffer. The mixture was extracted with 10 mℓ of chloroform-1-heptanol (9:1). The organic layer was back extracted into 0.3 mℓ of 0.3 *M* phosphoric acid containing 10 μg/mℓ of desipramine. The aqueous phase was washed with 2 mℓ of hexane and an aliquot of 0.17 mℓ of the aqueous phase was injected.
I-2. The sample was spiked with the internal standard to a final concentration of 150 ng/mℓ and extracted with a mixture of dichloromethane-isopropanol (1:1). The organic layer was evaporated at 30°C under nitrogen, the residue dissolved in 300 μℓ of 25:75 acetonitrile, 0.05 *M* KH_2PO_4, pH 3, and 100 μℓ were injected.

Elution — E-1. Acetonitrile-0.03 *M* phosphoric acid (40:60).
E-2. (A) 0.05 *M* KH_2PO_4, pH 3; (B) acetonitrile-0.05 *M* KH_2PO_4, pH 3 (65:35). Linear gradient from 100% (A) to 40% (A) over 25 min.

REFERENCE

1. **Moro, E., Jannuzzo, M. G., Ranghieri, M., Stegnjaich, S., and Valzelli, G.,** Determination of 4′-epidoxorubicin and its 13-dihydro derivative in human plasma by high-performance liquid chromatography with fluorescence detection, *J. Chromatogr.*, 230, 207, 1982.
2. **Deesen, P. E. and Leyland-Jones, B.,** Sensitive and specific determination of the new anthracycline analog 4′-epidoxorubicin and its metabolites by high pressure liquid chromatography, *Drug Metab. Dispos.*, 12, 9, 1984.

EPOMEDIOL

Gas Chromatography

Specimen (mℓ)	Extraction	Column (m × mm)	Packing (mesh)	Oven temp (°C)	Gas (mℓ/min)	Det.	RT (min)	Internal standard (RT)	Deriv.	Other compounds (RT)	Ref.
Plasma, urine, bile (1—2)	I-1	25 × NA	OV 101	T.P.[a]	He[b]	FID	7.8	Soberol (7)	—	Metabolite MI (7.2)	1,2

[a] Initial temp = 110°C, initial time = 1 min, rate = 5°C/min; final temp. = 155°C
[b] Precolumn flow rate = 40 mℓ/min, splitting ratio = 52:1

Extraction — I-1. The sample was diluted to 3 mℓ with 0.2 *M* acetate buffer (pH 4.6) containing 4 or 40 μg of the internal standard, washed with 3 mℓ of *n*-hexane, saturated with sodium chloride, and extracted three times with 5 mℓ of volumes of chloroform. The combined organic extracts were evaporated *in vacuo* and the residue was dissolved in 0.5 or 1 mℓ of dichloromethane for injection.

REFERENCES

1. **Ventura, P., Serafini, S., and Pria, R.,** Biotransformation of epomediol, *J. Chromatogr.*, 245, 350, 1982.
2. **Ventura, P. and Selva, A.,** Biotransformation of epomediol, *Biomed. Mass Spectrom.*, 9, 18, 1982.

EPRAZINONE

Thin-Layer Chromatography

Specimen (mℓ)	Extraction	Plate (Manufacturer)	Layer (mm)	Solvent	Post-separation treatment	Det. (nm)	Rf	Internal standard (Rf)	Other compounds (Rf)	Ref.
Urine (1.4) (1400)	I-1	20 × 20 cm (Merck)	Silica gel F_{254} (0.25)	S-1	Sp: Dragendorff: reagent	Visual	0.83	—	M_1[a] (0.31)	1

[a] M_1 is the major metabolite. The metabolites were identified by a number of solvent systems, purified by preprative TLC and identified by gas chromatography-mass spectrometry.

Extraction — I-1. Urine was adjusted to pH 1 to 2 and extracted twice with double the quantity of chloroform. The aqueous layer was made alkaline with sodium hydroxide (pH 11 to 12) and again extracted twice with chloroform. The aqueous phase (2 parts) was treated with 37% HCl (1 part) and hydrolyzed for 3 hr and again extracted at acidic and alkaline pH. The chloroform extracts were dried over anhydrous sodium sulfate and evaporated at 40°C *in vacuo*. The residue was dissolved in 10 mℓ of methanol and analyzed by thin-layer and paper chromatography.

Solvent — S-1. Methanol-ammonia (100:1).

REFERENCE

1. **Toffel-Nadolny, P. and Gielsdorf, W.,** Metabolismos von Eprazinon, *Arzneim. Forsch.*, 31, 719, 1981.

ERGOTAMINE

Liquid Chromatography

Specimen (mℓ)	Extraction	Column (cm × mm)	Packing (μm)	Elution	Flow (mℓ/min)	Det. (nm)	RT (min)	Internal standard (RT)	Other compounds (RT)	Ref.
Pure Compounds	—	25 × 2	LiChrosorb NH_2 (5)	E-1	0.67	ABS (310)	7.5	—	Ergosinine (2.5) Ergotaminine (3) Ergosine (5)	1
Dosage	—	25 × 4.6	Nucleosil-C_{18} (5)	E-2	2.5	ABS (NA)	—	Ephedrine (10)	Ergometrine (18) oxytocin (9)	2
Plasma (3)	I-1	25 × 4.6	Hypersil-ODS (5)	E-3	1.5	Fl (328, 389)	4	Ergocris-tine (6)	Ergotaminine (10)	3

Extraction — I-1. The sample was mixed with 0.1 mℓ of an ethanolic solution of the internal standard (100 μg/mℓ) and 3 mℓ of 1 *M* ammonia-HCl buffer, pH 9. The mixture was extracted with 20 mℓ of cyclohexane-1-butanol (9:1). The organic phase was back extracted with 2 mℓ of 0.5 *M* sulfuric acid. The aqueous phase was made alkaline with the ammonia-HCl buffer (2 mℓ) and reextracted with 10 mℓ of the extraction solvent. The organic layer was evaporated at 50°C under nitrogen. The residue was dissolved in 0.2 mℓ of the mobile phase and an aliquot of 50 to 150 μℓ was injected.

Elution — E-1. Diethyl ether-ethanol (93:7).
E-2. Acetonitrile-0.05% sodium tetradecyl sulfate, 0.83 m*M* triethylamine phosphate, pH 5 (35:65).
E-3. Acetonitrile-0.01 *M* ammonium carbonate (50:50).

REFERENCES

1. **Wurst, M., Flieger, M., and Rehacek, Z.,** Analysis of ergot alkaloids by high-performance liquid chromatography, *J. Chromatogr.*, 174, 401, 1979.
2. **Pask-Hughes, R. A., Corran, P. H., and Calam, D. H.,** Assay of the combined formulation of ergometrine and oxytocin by high-performance liquid chromatography., *J. Chromatog.*, 214, 307, 1981.
3. **Edlund, P. O.,** Determination of ergot alkaloids in plasma by high-performance liquid chromatography and fluorescence detection, *J. Chromatogr.*, 226, 107, 1981.

ERYTHROMYCIN

Liquid Chromatography

Specimen (mℓ)	Extraction	Column (cm × mm)	Packing (μm)	Elution	Flow (mℓ/min)	Det. (nm)	RT (min)	Internal standard (RT)	Other compounds (RT)	Ref.
Fermentation broth	I-1	25 × 4	LiChroprep RP-8 (5—20)	E-1	1.5	ABS (215)	A[a] = 6.6 B = 12	—	—	1
Plasma, urine (0.2)	I-2	30 × 3.9	μ-Bondapak-C_{18} (10)	E-2	1.0	Electrochem[b]	A = 9.6 C = 6.9	Erythromycin B (14.2)	Des-N-Methyl erythromycin (7.7) Anhydroerythromycin A (12.4) Anhydroerythromycin C (12.6)	2
Serum (0.25)	I-3	25 × 4.6	Speralyte dip-penyl (5)[c]	E-3[d]	1.0	Electrochem[e]	A = 3 B = 3.6 D = 3.1	Analog (4.5)	f	3
Pure compounds	—	25 × 4.6	TSK-Gel (10)[g]	E-4	1.0	ABS (215)	A = 18	—	f	4
Serum, urine (2)	I-4	15 × 3.9	Novapack-C_{18} (5)[b]	E-5	1.0	ABS (200)	A = 6.2 B = 8.2	Oleandomycin (4.2)	Des-N-Methyl erythromycin (4.5) Anhydroerythromycin (11) Erythralosamine (13) Erythromycin enol ether (22.4)	5

[a] Different components of erythromycin. Erythromycin A is the main component.
[b] Dual electrode coulochem detection. Screening electrode = 0.7 V, sample electrode = 0.9 V.
[c] Protected by a 21 × 3 mm, 40-μm glass head guard column.
[d] Recycled.
[e] Screening electrode = 0.7 V, sample electrode = 0.8 V.

[f] Relative retention times of a number of derivatives of erythromycin A are given.
[g] Column temp = 60°C.
[h] Column temp = 35°C.

Extraction — I-1. An aliquot of the broth was adjusted to pH 9.8 and extracted with isoamyl acetate. Aliquots of the organic extract were injected.
I-2. The sample was mixed with 10 μℓ of the internal standard solution (0.01 to 0.1 mg/mℓ in acetonitrile) and 20 μℓ of saturated sodium carbonate. The mixture was extracted with 1 mℓ of ethyl ether. An aliquot of 0.75 mℓ of the ether extract was allowed to evaporate at room temp at reduced pressure. Just prior to analysis, the residue was reconstituted with 50 μℓ of the mobile phase and an aliquot of 20 μℓ was injected.
I-3. The sample was mixed with an appropriate amount of the internal standard in 1 mℓ of water and 50 μℓ of saturated potassium carbonate solution. The mixture was extracted with 5 mℓ of *t*-butyl ether. The organic layer was evaporated at 40°C under vacuum. The residue was reconstituted in 1 mℓ of acetonitrile/20 m*M* ammonium acetate (1:1) and the solution was washed with 1 mℓ of hexane. Aliquots of the aqueous layer were injected.
I-4. The sample was mixed with 0.25 mℓ of a 12 μg/mℓ aqueous solution of the internal standard and 1 mℓ of acetonitrile. After mixing and centrifugation, the supernatant was diluted with 8 mℓ of water and applied to a prewashed (3 mℓ acetonitrile, 3 mℓ-water) 1 mℓ Baker C_{18} extraction column. The column was washed with 5 mℓ of water, 5 mℓ of acetonitrile-water (1:1), and then eluted successively with two 0.5-mℓ aliquots of acetonitrile-0.05 *M* phosphate buffer (1:1) The eluate was evaporated in a vacuum centrifuge. The residue was reconstituted with 20 μℓ of water and 25 μℓ of acetonitrile. After centrifugation, an aliquot of 15 to 20 μℓ of the acetonitrile phase was injected.

Elution — E-1. Methanol-water-ammonia (80:19.1:0.1).
E-2. Acetonitrile-methanol-0.2 M acetate buffer, pH 6.7 (40:5:55).
E-3. Acetonitrile-20 m*M* sodium perchlorate-20 m*M* ammonium acetate-methanol (50:32:8:10) pH, 7.
E-4. Acetonitrile-methanol-0.2 *M* tetramethyl ammonium hydroxide, pH 8-0.2 *M* phosphate buffer, pH 8-water (30:20:15:5:30).
E-5. Acetonitrile-0.05 *M* phosphate buffer, pH 6.3 (30:70).

REFERENCES

1. **Pellegatta, G., Carugati, G. P., and Coppi, G.,** High-performance liquid chromatographic analysis of erythromycins A and B from fermentation broths, *J. Chromatogr.*, 269, 33, 1983.
2. **Chen, M.-L. and Chiou, W. L.,** Analysis of erythromycin in biological fluids by high-performance liquid chromatography with electrochemical detection, *J. Chromatogr.*, 278, 91, 1983.
3. **Duthu, G. S.,** Assay of erythromycin from human serum by high performance liquid chromatography with electrochemical detection, *J. Liq. Chromatogr.*, 7, 1023, 1984.
4. **Kibwage, I. O., Roets, E., Hoogmartens, J., and Vanderhaeghe, H.,** Separation of erythromycin and related substances by high-performance liquid chromatography on poly(styrene-divinylbenzene) packing materials, *J. Chromatogr.*, 330, 275, 1985.
5. **Stubbs, C., Haigh, J. M., and Kanfer, I.,** Determination of erythromycin in serum and urine by high-performance liquid chromatography with ultraviolet detection, *J. Pharm. Sci.*, 74, 1126, 1985.

ESMOLOL

Gas Chromatography

Specimen (mℓ)	Extraction	Column (m × mm)	Packing (mesh)	Oven temp (°C)	Gas (mℓ/min)	Det.	RT (min)	Internal standard (RT)	Deriv.	Other compounds (RT)	Ref.
Blood (1)	I-1	2 × 2 (steel)	3% SP-2250 Supelcoport (100/200)	230	NA	MS-EI	NA	[$^2H_2$]-Esmolol	Trimethyl-silyl	—	1
Blood[a] (1 g)	I-1	25 × 0.33	CP Sil 8	T.P.[b]	He (2.3)	ECD	19	H 163/37 (24)	Penta-fluoro-propionyl	Flumolol[c] (14)	2

Liquid Chromatography

Specimen (mℓ)	Extraction	Column (cm × mm)	Packing (μm)	Elution	Flow (mℓ/min)	Det. (nm)	RT (min)	Internal standard (RT)	Other compounds (RT)	Ref.
Blood (1)	I-3	30 × 3.9	μ-Bondapak-phenyl (10)	E-1	2.0	ABS (280)	—	3-[1-Amino[3-(4-chloro-phen-oxy)]2-propanol] propionic acid (9.7)	Esmolol acid[d] (7.3)	3
Dosage	—	30 × 3.9	μ-Bondapak CN (10)	E-2	2.0	ABS (280)	7	2-(*p*-chlo-rophenyl)-2-methyl propanol (9.5)	d	4, 5
Urine (1)	I-4	30 × 3.9	μ-Bondapak-C_{18} (10)	E-3	2.0	ABS (229)	6.1	ACC 9038 (8.2)	Esmolol acid[d,e]	6

[a] Blood is collected in the presence of 0.5 mℓ of an aqueous solution of sodium dodecyl sulfate (200 mg/ℓ) to inhibit the hydrolysis of the drug.
[b] Initial temp = 100°C, initial time = 1 min, rate = 10°C/min to 185°C; at 15 min, rate = 30°C/min to 250°C, final time = 7 min.
[c] A different temp program is used for the determination of this drug.
[d] 3-[4-[2-Hydroxy-3-(isopropyl amino)propoxy]phenyl]propionic acid.
[e] Different extraction conditions, internal standard and chromatographic conditions are used for the determination of this metabolite.

Extraction — I-1. Immediately after collection, the sample was spiked with the internal standard and extracted with 10 mℓ of dichloromethane. An aliquot of 9 mℓ of the organic phase was back extracted into 2 mℓ of 0.02 *M* HCl. The aqueous phase was adjusted to pH 8 with 200 μℓ of a 1 *M* potassium phosphate buffer, pH 8 and reextracted with 5 mℓ of dichloromethane. The organic phase was concentrated to 100 mℓ under a stream of nitrogen, treated with 50 μℓ of *bis*(trimethylsilyl) trifluoroacetamide, and the mixture incubated at 60°C for 30 min. Aliquots of 2 μℓ of the reaction mixture were injected.
I-2. The sample was mixed with 100 μℓ of the internal standard solution (5 μmol/ℓ in 0.1 *M* pH 6 phosphate buffer) and 0.5 mℓ of 0.2 *M*, pH 11.8 phosphate buffer. The mixture was extracted with 4 mℓ of toluene. The organic layer was evaporated at 35°C under a stream of nitrogen. The residue was dissolved in 100 μℓ of ethyl acetate and 25 μℓ of pentafluoropropionic anhydride were added. The mixture was allowed to stand at room temp for 20 min and then evaporated at 35°C under a stream of nitrogen. The residue was dissolved in 500 μℓ of toluene and aliquots of 1.5 μℓ were injected.
I-3. The sample was washed with 10 μℓ of dichloromethane. A 0.5-mℓ aliquot of the aqueous layer was mixed with 100 μℓ of the internal standard solution (0.1 mg/mℓ) and 500 μℓ of 14% perchloric acid. After mixing and centrifugation, aliquots of the supernatant were injected.
I-4. The sample was mixed with 1 mℓ of borate buffer (pH 8.4), 0.1 mℓ of an aqueous solution of the internal standard (20 μg/mℓ) and 0.15 mℓ of water. The mixture was extracted with 10 mℓ of dichloromethane. The organic layer was back extracted into 1 mℓ of a 0.25 *M* solution of NaH_2PO_4 (pH 2.8). Aliquots of the aqueous layer were injected with an autosampler. The aqueous phase from the initial extraction was spiked with 0.1 mℓ of an aqueous solution of ACC-8059 (400 μg/mℓ), acidified with dilute acetic acid and washed with 5 mℓ of dichloromethane. Aliquots of the aqueous layer were injected to determine esmolol acid.

Elution — E-1. Acetonitrile-0.01 *M* sodium acetate-acetic acid (12:87:1).
E-2. Acetonitrile-acetic acid-0.068% sodium acetate trihydrate (150:10:840).
E-3.Acetonitrile-10 m*M* phosphate buffer, pH 6.2 (35:65)

REFERENCES

1. **Sum, C. Y. and Yacobi, A.,** Gas chromatographic-mass spectrometric assay for the ultra-short-acting β-blocker esmolol, *J. Pharm. Sci.*, 73, 1177, 1984.
2. **Holm, G., Kylberg-Hanssen, K., and Svensson, L.,** Use of dodecyl sulfate as an esterase inhibitor before gas-chromatographic determination of labile β-adrenoceptor blocking drugs, *Clin. Chem.*, 31, 868, 1985.
3. **Stampfli, H. F., Lai, C. M., Yacobi, A., and Sum, C. Y.,** High-performance liquid chromatographic assay for the major blood metabolite of esmolol-an ultra short acting beta blocker, *J. Chromatogr.*, 309, 203, 1984.
4. **Lee, Y. C., Baaske, D. M., and Alam, A. S.,** High-performance liquid chromatographic method for the determination of esmolol hydrochloride, *J. Pharm. Sci.*, 73, 1660, 1984.
5. **Karnatz, N. N., Baaske, D. M., Herbranson, D. E., and Eliason, M. S.,** High-performance liquid chromatographic method for the determination of esmolol hydrochloride in solutions and parenteral formations, *J. Chromatogr.*, 330, 420, 1985.
6. **Achari, R., Drissel, D., and Hulse, J. D.,** Liquid-chromatographic analysis for esmolol and its major metabolite in urine, *Clin. Chem.*, 32, 374, 1986.

ESTAZOLAM

Liquid Chromatography

Specimen (mℓ)	Extraction	Column (cm × mm)	Packing (μm)	Elution	Flow (mℓ/min)	Det. (nm)	RT (min)	Internal standard (RT)	Other compounds (RT)	Ref.
Serum, urine (0.5)	I-1	25 × 4.6	Ultrasphere-C_{18} (5)	E-1	NA	ABS (240)	5.5	—	—	1

Extraction — I-1. The sample was buffered with 0.1 mℓ of a solution containing 0.1 g/mℓ each of Na_2CO_3 and $NaHCO_3$ and extracted twice with 3-mℓ portions of ethylenechloride-methylenechloride-ethyl acetate (1:1:8). The combined organic extract was dried with Na_2SO_4 and evaporated at 35°C under reduced pressure. The residue was dissolved in 0.5 mℓ of methanol, filtered, and an aliquot of 20 μℓ was injected.

Elution — E-1. Acetonitrile-methanol-0.011 *M* phosphate buffer, pH 7.5 (2:65:33).

REFERENCE

1. **di Tella, A. S., Ricci, P., Di Nunzio, C., and Cassandro, P.,** A new method for the determination in blood and urine of a novel triazolobenzodiazepine (estazolam) by HPLC, *J. Anal. Toxicol.*, 10, 65, 1986.

ESTRAMUSTINE

Liquid Chromatography

Specimen (mℓ)	Extraction	Column (cm × mm)	Packing (μm)	Elution	Flow (mℓ/min)	Det. (nm)	RT (min)	Internal standard (RT)	Other compounds (RT)	Ref.
Plasma (1)	I-1	25 × 4.6	Partisil PXS silica (5)	E-1	1.5	Fl (195, 250)	7.2	—	17-Keto metabolite (5.4)	1

Extraction — I-1. The sample was mixed with 100 μℓ of ethanol and 2 mℓ of 1 *M* borate-KCl-Na_2CO_3 buffer, pH 9. The mixture was extracted with 12 mℓ of hexane. A 10-mℓ aliquot of the hexane extract was evaporated at 50°C under a stream of nitrogen. The residue was dissolved in 100 μℓ of the mobile phase and a 20-μℓ aliquot was injected.

Elution — E-1. Hexane-ethanol (92.5:7.5).

REFERENCE

1. **Brooks, M. A. and Dixon, R.,** Determination of estramustine and its 17-keto metabolite in plasma by high-performance liquid chromatography, *J. Chromatogr.*, 182, 387, 1980.

ESTROGENS

Gas Chromatography

Specimen (mℓ)	Extraction	Column (m × mm)	Packing (mesh)	Oven temp (°C)	Gas (mℓ/min)	Det.	RT (min)	Internal standard (RT)	Deriv.	Other compounds (RT)	Ref.
Dosage	I-1	15 × 0.25	Silar 10C	T.P.[a]	He (0.8)	FID	—	Ethinyl estradiol (20.5)	Oxime, Tri-methyl-silyl	17α-Estradiol (16) 17 β-Estradiol (17.2) 17 α-Dihydro-equilin (18) 17 β-Dihydro-equilin (19.2) Estrone (23) Equilin (24) 17 α-Dihydroequi-lenin (25.5) 17 β-Dihydroequi-lenin (27) Equilenin (29)	1

Liquid Chromatography

Specimen (mℓ)	Extraction	Column (cm × mm)	Packing (μm)	Elution	Flow (mℓ/min)	Det. (nm)	RT (min)	Internal standard (RT)	Other compounds (RT)	Ref.
Dosage	I-2	25 × 32	LiChrosorb Si-60 (5)	E-1	0.98	Fl[b]	—	—	Estrone (5) Equilin (5) Equilenin (5) α-Estradiol (14.5) α-Dihydroequilin (16.5) α-Dihydroequilenin (18)	2

									β-Estradiol (19.5) β-Dihydroequilin (21) β-Dihydroequilenin (24)	
Dosage	I-3	25 × 4.6	Spherisorb ODS[c] (5)	E-2	1.2	ABS (280)	—	Mestranol (19)	Norethisterone (7.5) Ethinylestradiol (9) Ethynodiol diacetate (22.7)[e]	3
Dosage	I-4	25 × 3.2	Express[d] RP-2 (NA)	E-3	1.75	ABS (210, 280)	—	Butylated hydroxy toluene (16.5)[e]	Ethinyl estradiol (3.9)[e] Mestranol (8.1)[e]	4

[a] Initial temp = 170°C, initial time = 7 min; rate = 2.3°C/min, final temp = 220°C; final time = 5 min.
[b] A broad band 7-54 (240 to 420 nm) excitation filter and a 3-72 (440) cut off emission filter were used.
[c] Use of alternative C_{18} column is also described.
[d] From Altex.
[e] Capacity factors.

Extraction — I-1. An amount of powdered tablets corresponding to 1 mg of conjugated estrogens was shaken with 15 mℓ of an acetate buffer (0.02 *M*, pH 5) for 20 min and the incubated with 2000 U of sulfatase for 30 min at 45°C. Then 0.2 mℓ of the internal standard was added in 10 mℓ of chloroform and the mixture was shaken for 30 min. The organic layer was filtered through anhydrous sodium sulfate and evaporated under a stream of nitrogen. The residue was incubated with 200 μℓ of a 2% solution of hydroxylamine hydrochloride in dry pyridine at 70°C for 30 min and evaporated under a stream of nitrogen. The residue was then incubated with 150 μℓ of N,O-*bis*(trimethyl silyl)trifluoroacetamide and 50 μℓ pyridine at 70°C for 10 min. Finally a 2-μℓ aliquot of the resulting solution was injected.
I-2. An amount of powdered tablets corresponding to 3.2 mg of conjugated estrogens was dissolved in 100 mℓ of methanol and filtered discarding the first 20 mℓ of the filtrate. An aliquot of 25 mℓ of the subsequent filtrate was mixed with 1 mℓ of HCl and heated on a steam bath for 3 min. After cooling, the mixture was extracted with 75 mℓ of benzene. The benzene extract was washed successively once with 15 mℓ water, four times with 15 mℓ dilute sodium carbonate (2%), twice with 10 mℓ water, and then passed through a layer of anhydrous sodium sulfate. The sodium sulfate layer was rinsed with additional 25 mℓ of benzene. The combined benzene filtrate was evaporated on a steam bath under a current of air. The residue was treated with 10 mℓ of dansyl chloride solution (0.2 mg/mℓ in acetone) and 30 mℓ of sodium carbonate solution (0.37 g in 300 mℓ water + 150 mℓ acetone). The mixture was allowed to stand at room temp in the dark for 30 min and extracted twice with 50-mℓ portions of ether. The combined ether extract was washed with water, dried over anhydrous sodium sulfate, and evaporated. The residue was dissolved in 5 mℓ of chloroform for injection.
I-3. The contents of 5 capsules were mixed with 1 mℓ of the internal standard solution (0.4 mg/mℓ in acetonitrile), 2 mℓ of acetonitrile and 2 mℓ of tetrahydrofuran. After 3 to 4 min of sonication a 4-mℓ of aliquot of the solution was diluted with 4 mℓ water and filtered through a Millipore FH, 0.5 μm filter. Aliquots of 20 μℓ of the filtrate were injected.

I-4. Tablets (20) were weighed and powdered. An amount of powder equivalent to one tablet was mixed with 2 mℓ of the internal standard solution (50 μg/mℓ in 80% aqueous acetonitrile). After vigorous mixing and centrifugation, aliquots of 20 μℓ of the clear supernate were injected.

Elution — E-1. Chloroform-*n*-heptane (50:50). E-2. Acetonitrile-tetrahydrofuran-water (30:20:50). E-3. Acetonitrile-water (38:62).

REFERENCES

1. **Pillai, G. K. and McErlane, K. M.,** Quantitative determination of conjugated estrogens in formulations by capillary GLC, *J. Pharm. Sci.*, 70, 1072, 1981.
2. **Roos, R. W. and Medwick, T.,** Application of dansyl derivatization to the high pressure liquid chromatographic identification of equine estrogens, *J. Chromatogr. Sci.*, 18, 626, 1980.
3. **Gluck, J. A. P., and Shek, E.,** Determination of ethinylestradiol and norethisterone in an oral contraceptive capsule by reversed-phase high performance liquid chromatography, *J. Chromatogr. Sci.*, 18, 631, 1980.
4. **Carignan, G., Lodge, B. A., and Skakum, W.,** Quantitative analysis of ethynodiol diacetate and ethinyl estradiol/mestranol in oral contraceptive tablets by high-performance liquid chromatography, *J. Pharm. Sci.*, 71, 264, 1982.

Liquid Chromatography

Specimen (mℓ)	Extraction	Column (cm × mm)	Packing (μm)	Elution	Flow (mℓ/min)	Det. (nm)	RT (min)	Internal standard (RT)	Other compounds (RT)	Ref.
Dosage	—	25 × 4.5	Hypersol-ODS (5)[c]	E-1	1.5	ABS (278)	4.5	—	d	2

[a] Initial temp = 200°C; rate = 30°C/min; final temp = 300°C.
[b] Isobutane as the reagent gas.
[c] Column temp = 50°C.
[d] Separation of possible degradation products is shown.

Extraction — I-1. The sample was spiked with a methanolic solution of the internal standard (1 μg) and was extracted with 5 mℓ of diethyl ether after the addition of 1 mℓ of 3 *N* HCl. An aliquot of 4 mℓ of the organic phase was evaporated. The residue was mixed with 2 mℓ of a 2% solution of pentafluorobenzyl bromide in acetonitrile and 10 μg of anhydrous potassium carbonate. The mixture was incubated at 70°C for 45 min and then evaporated *in vacuo*. The residue was dissolved in 50 μℓ of acetonitrile and 5 μℓ of the solution was injected.

Elution — E-1. (A) Methanol; (B) 0.05 *M* phosphate buffer, pH 5.6. Isocratic at 52% (A) for 6 min, then 65% (A) for 7 min.

REFERENCES

1. **Stuber, W., Mutschler, E., and Steinback, D.,** Determination of ethacrynic and tienilic acid in plasma by gas-liquid chromatography-mass spectrometry, *J. Chromatogr.*, 227, 193, 1982.
2. **Yarwood, R. J., Moore, W. D., and Collett, J. H.,** Liquid chromatographic analysis of ethacrynic acid and degradation products in pharmaceutical systems, *J. Pharm. Sci.*, 74, 220, 1985.

ETHACRYNIC ACID

Gas Chromatography

Specimen (mℓ)	Extraction	Column (m × mm)	Packing (mesh)	Oven temp (°C)	Gas (mℓ/min)	Det.	RT (min)	Internal standard (RT)	Deriv.	Other compounds (RT)	Ref.
Plasma (1)	I-1	2 × 2	1% OV-17 Chromosorb W (80/100)	T.P.[a]	He (30)	MS-CI[b]	NA	Tienilic acid	Penta-fluoro-benzyl	—	1

ETHAMBUTOL

Gas Chromatography

Specimen (mℓ)	Extraction	Column (m × mm)	Packing (mesh)	Oven temp (°C)	Gas (mℓ/min)	Det.	RT (min)	Internal standard (RT)	Deriv.	Other compounds (RT)	Ref.
Plasma (0.5—1)	I-1	1 × 3	2% OV-17 Gas Chrom Q (80/100)	150	He (40)	MS-EI	2.5	[$^2H_4$]Ethambutol	Trifluoro-acetyl	—	1
Plasma (0.2)	I-2	1.8 × 2	3% OV-17 Gas Chrom Q (100/200)	160	He (20)	MS-EI	0.8	[$^2H_4$]Ethambutol[a]	Trifluoro-acetyl	—	2
Plasma (0.1)	I-3	1.8 × 3	3% OV-17 Chrom WG (100/120)	157	N_2 (66)	ECD	4.5	(+)-2,2′-(Ethylene-diimino)di-1-prop-anol (3)	Trifluoro-acetyl	—	3

[a] A nondeuterated analog, (+)-2,2′-(ethylenediimino)-di-1-propanol was also used as internal standard for GC-MS as well as for electron capture detection.

Extraction — I-1. The sample was mixed with 0.1 mℓ of an aqueous solution of the internal standard (0.1 mg/mℓ) and 0.5 mℓ of 4 *N* sodium hydroxide. The mixture was extracted with 6 mℓ of chloroform. The organic phase was evaporated after the addition of 3 drops of 1 *M* HCl in methanol. The residue was dissolved with 50 μℓ of benzene-pyridine (4:1) and reacted with 50 μℓ of trifluoroacetic anhydride for 2 hr at room temp. Aliquots of this mixture were injected.
I-2. The sample was diluted with 800 μℓ of water containing 3 μg of the internal standard. After the addition of 1 mℓ of 4 *N* NaOH the sample was extracted with 10 mℓ of chloroform. An aliquot (8 mℓ) of the organic layer was evaporated at room temp under a stream of air. The residue was dissolved in 100 μℓ of benzene-pyridine (7:1) and 25 μℓ of trifluoroacetic anhydride were added. The mixture was allowed to stand at 4°C for 1 hr, and then washed with 400 μℓ of 0.1 *N* HCl. Aliquots of benzene layer were injected within 4 hr after derivatization.
I-3. The sample was mixed with 50 μℓ of the internal standard solution (10 μg/mℓ) and water to make the final volume 0.5 mℓ. The sample was made alkaline and extracted with 5 mℓ of chloroform. Portions of chloroform extract were evaporated under nitrogen. The residue was dissolved in 0.5 mℓ of ethyl acetate and the solution incubated with 20 μℓ of trifluoroacetic anhydride at 50°C for 1 hr and then the reaction mixture was evaporated under nitrogen. The residue was dissolved in 0.5 mℓ of ethyl acetate and aliquots of 1 to 2 μℓ were injected.

REFERENCES

1. **Ohya, K., Shintani, S., and Sano, M.,** Determination of ethambutol in plasma using selected ion monitoring, *J. Chromatogr.*, 221, 293, 1980.
2. **Holdiness, M. R., Israili, Z. H., and Justice, J. B.,** Gas chromatographic-mass spectrometric determination of ethambutol in human plasma, *J. Chromatogr.*, 224, 415, 1981.
3. **Lee, C. S. and Wang, L. H.,** Improved GLC determination of ethambutol, *J. Pharm. Sci.*, 69, 362, 1980.

ETHANOL

Gas Chromatography

Specimen (mℓ)	Extraction	Column (m × mm)	Packing (mesh)	Oven temp (°C)	Gas (mℓ/min)	Det.	RT (min)	Internal standard (RT)	Deriv.	Other compounds (RT)	Ref.
Blood, tissue (0.2)	I-1	1.8 × 2	Porpak Q (100/120)	130	He (37)	FID	3.5	Isopropanol (6.9)	—	Acetaldehyde (6.9)	1
Plasma, blood (0.15)	I-2	1.8 × 2 (Steel)	5% Carbowax 20M Carbopak B (60/80) Haloport F (30/60)	65	N_2 (30)	FID	1.8	Propanol-1 (4.5)	—	2-Propanol (2.8) Acetone (1.4) Acetaldehyde (0.6) Methanol (0.9)	2
Blood (0.1)	I-3	1.8 × 4	5% Carbowax 20M	90	N_2 (30)	FID	1.3	Propanol-1 (2)	—	Acetaldehyde (0.7)	3

Extraction — I-1. The sample was treated with 0.4 mℓ of 40 m*M* thiourea-0.6 *M* perchloric acid and 0.1 mℓ of the internal standard solution (0.78 μg/mℓ in water). The mixture was incubated at 65°C for 30 min in a sealed vial. A 3-mℓ gas aliquot was injected.
I-2. The sample was mixed with 1 mℓ of an aqueous diluent containing 1 *M* ammonium sulfate, 0.1 *M* sodium dithionite and 0.03% 1-propanol. The sealed vial was incubated at 60°C for 30 min. Volumes of head space were injected with an autosampler.
I-3. The sample was mixed with 865 μℓ of a solution of 34 mg/mℓ of perchloric acid + 65 μg/mℓ of NaN_3 in saline, 25 μℓ of an aqueous solution of the internal standard, and 10 μℓ of an aqueous solution of thiourea (76 mg/mℓ). The sealed vial was incubated at 60°C for 15 min. A 1-mℓ volume of the head space was injected.

REFERENCES

1. **Mendenhall, C. L., MacGee, J., and Green, E. S.,** Simple rapid and sensitive method for the simultaneous quantitation of ethanol and acetaldehyde in biological materials using head-space gas chromatography, *J. Chromatogr.*, 190, 197, 1980.
2. **Christmore, D. S., Kelly, R. C., and Doshier, L. A.,** Improved recovery and stability of ethanol in automated headspace analysis, *J. Forensic Sci.*, 29, 1038, 1984.
3. **Steenaart, N. A. E., Clarke, D. W., and Brien, J. F.,** Gas-liquid chromatographic analysis of ethanol and acetaldehyde in blood with minimal artifactual acetaldehyde formation, *J. Pharmacol. Methods*, 14, 199, 1985.

ETHAVERINE

Liquid Chromatography

Specimen (mℓ)	Extraction	Column (cm × mm)	Packing (μm)	Elution	Flow (mℓ/min)	Det. (nm)	RT (min)	Internal standard (RT)	Other compounds (RT)	Ref.
Plasma (3)	I-1	15 × 4.5	LiChrosorb Si (5)	E-1	1.0	ABS (254)	4.6	Papaverine (7.9)	—	1
Plasma (2)	I-2	25 × 4.6	Partisil-ODS (10)	E-2	2.0	ABS (238)	8.5	Papaverine (3.5)	—	2
Plasma (1)	I-3	30 × 3.9	μ-Bondapak-C_{18} (10)	E-3	2.0	ABS (250)	3.2	Chlor-phenir-amine (5.4)		3

Extraction — I-1. The sample was spiked with 30 μℓ of an aqueous solution of the internal standard (50 μg/mℓ), made alkaline with 0.5 mℓ of sodium hydroxide solution, and extracted twice with 10-mℓ portions of *n*-heptane containing 1.6% isoamyl alcohol. An aliquot of 15 mℓ of the combined extract was evaporated at 30°C under nitrogen. The residue was dissolved in 100 μℓ of the mobile phase and 15 μℓ were injected.

I-2. The sample was mixed with 15 μℓ of a methanolic solution of the internal standard (2 μg/mℓ) and 200 μℓ of 4 *M* sodium hydroxide. The mixture was extracted twice with 5- and 2-mℓ volumes of diethyl ether, respectively. The combined ether layers were back extracted into 0.5 mℓ of 1 *M* HCl. The aqueous layer was made alkaline with 0.5 mℓ of 4 *M* NaOH and extracted with 5 mℓ of ether. The ether layer was evaporated at 37°C under nitrogen. The residue was dissolved in 25 μℓ of methanol for injection.

I-3. The sample was mixed with 1 mℓ of an aqueous solution of the internal standard (4 μg/mℓ) and 0.4 mℓ of 9 *N* sodium hydroxide. The mixture was extracted twice with 10-mℓ portions of diethyl ether. The combined ether extracts were evaporated at 40°C under nitrogen. The residue was dissolved in 3 mℓ of ether which was back extracted into 0.3 mℓ of 0.3 *N* HCl. An aliquot of 50 μℓ of the aqueous phase was injected.

Elution — E-1. *n*-Heptane-dichloromethane-methanol-diethylamine (200:30:2.5:0.8).

E-2. Methanol-0.1% KH_2PO_4 (65:35).

E-3. Methanol-water (6:4) containing 0.005 *M* 1-heptanesulfonic acid.

REFERENCES

1. **Renier, E., Kjeldsen, K., and Pays, M.,** Dosage de l'ethaverine dans les milieux biologiques par chromatographie liquide a haute performance, *Feuil. Biol.*, 20, 111, 1979.
2. **Brodie, R. R., Chasseaud, L. F., Walmsley, L. N., Soegtrop, H. H., Darragh, A., and O'Kelly, D. A.,** Determination of the antispasmodic agent ethaverine in human plasma by high performance liquid, *J. Chromatogr.*, 182, 379, 1980.
3. **Meyer, M. C., Raghow, C., and Straughn, A. B.,** Plasma levels of ethaverine after oral administration to humans, *Biopharm. Drug Dispos.*, 4, 401, 1983.

ETHCHLORVYNOL

Gas Chromatography

Specimen (mℓ)	Extraction	Column (m × mm)	Packing (mesh)	Oven temp (°C)	Gas (mℓ/min)	Det.	RT (min)	Internal standard (RT)	Deriv.	Other compounds (RT)	Ref.
Plasma, serum (0.5)	I-1	0.6 × 2	10% SP-1000 Supelcoport (80/100)	110	N_2 (20)	FID	3.5	2-Methylnaphthalene (4.5)	—	—	1

Extraction — I-1. The sample was vortex mixed with 0.5 mℓ of water, 50 μℓ of 3 *N* HCl, and 100 μℓ of the internal standard solution (0.1 mg/mℓ in chloroform) and centrifuged. Aliquots of 2 to 4 μℓ of the chloroform extract were injected.

REFERENCE

1. **Bridges, R. R. and Jennison, T. A.,** Analysis of ethchlorvynol (placidyl): evaluation of a comparison performed in a clinical laboratory, *J. Anal. Toxicol.*, 8, 263, 1984.

ETHIMIZOL

Liquid Chromatography

Specimen (mℓ)	Extraction	Column (cm × mm)	Packing (μm)	Elution	Flow (mℓ/min)	Det. (nm)	RT (min)	Internal standard (RT)	Other compounds (RT)	Ref.
Serum, saliva (2)	I-1	25 × 4.6	LiChrosorb SI-100 (5)	E-2	1.2	ABS (262)	6.3	Antiffeine (7.6)	Desmethylethimizol	1

Extraction — I-1. The sample was diluted with a 2-mℓ aliquot of the aqueous solution of the internal standard (1 μg/mℓ) and applied onto a prewashed (5 mℓ methanol, 5 mℓ water) Sep-Pak C_{18} cartridge. After the sample had passed through, the cartridge was washed with 2 mℓ of water. The cartridge was attached to a micro column packed with silica gel and eluted with 4 mℓ of acetonitrile. The eluate was evaporated at 50°C under a stream of nitrogen. The residue was dissolved in 20 μℓ of chloroform and 10 μℓ of the solution was injected.

Elution — E-1. *n*-Heptane-dichloromethane-methanol-triethylamine (85:10:4.75:0.25).

REFERENCE

1. **Soltes, L., Kallay, Z., Trnovec, T., Durisova, M., and Plskova, M.,** Monitoring of ethimizol and its metabolites in serum or saliva by means of high-performance liquid chromatography, *J. Chromatogr.*, 273, 213, 1983.

ETHIONAMIDE

Liquid Chromatography

Specimen (mℓ)	Extraction	Column (cm × mm)	Packing (μm)	Elution	Flow (mℓ/min)	Det. (nm)	RT (min)	Internal standard (RT)	Other compounds (RT)	Ref.
Plasma, urine (3)	I-1	25 × 5	Hypersil silica (5)[a]	E-1	2.5	ABS (340)	3	2-Methyl-thiosoni-cotina-mide (4)	Ethionamide sulph-oxide (7.5) Prothionamide (2.8)	1, 2

[a] Protected by a 7-cm guard column filled with HC Pellosil/30 to 38 μm.

Extraction — I-1. The sample was spiked with the internal standard and extracted with 7 mℓ of chloroform. The organic layer was back extracted into 1 mℓ of 0.1 *M* HCl. The aqueous phase was neutralized with 10% aqueous ammonia to pH 7 to 8 and extracted with 3 mℓ of chloroform. The organic extract was evaporated at 40°C under nitrogen. The residue was dissolved in 100 μℓ of the mobile phase for injection.

Elution — E-1. Chloroform-propan-2-ol-water (916:8:4).

REFERENCES

1. **Jenner, P. J., Ellard, G. A., Gruer, P. J. K., and Aber, V. R.,** A comparison of the blood levels and urinary excretion of ethionamide and prothionamide in man, *J. Antimicrob. Chemother.*, 13, 267, 1984.
2. **Jenner, P. J. and Ellard, G. A.,** High-performance liquid chromatographic determination of ethionamide and prothionamide in body fluids, *J. Chromatogr.*, 225, 245, 1981.

ETHMOZIN

Liquid Chromatography

Specimen (mℓ)	Extraction	Column (cm × mm)	Packing (μm)	Elution	Flow (mℓ/min)	Det. (nm)	RT (min)	Internal standard (RT)	Other compounds (RT)	Ref.
Plasma (5)	I-1	30 × 3.9	μ-Porasil silica (10)	E-1	1.0	ABS (268)	5	—	a Caffeine (7.5)	1
Plasma, urine (1)	I-2	25 × 3.2	Partisil SCX (10)[b]	E-2	1.0	ABS (254)	6	Nona-chlazin (11)	—	2
Plasma (1)	I-3	25 × 4.6	LiChrosorb CN (10)	E-3	2.0	ABS (268)	11.2	Protrip-tyline (9.2)	—	3

[a] Separation of unidentified metabolites is shown.
[b] Protected by a 40 × 3.2 mm guard column packed with the same material.

Extraction — I-1. The sample was adjusted to pH 9 using a few drops of 1 *N* NaOH, and then diluted with 5 mℓ of 0.1 *M* borate buffer (pH 9) and then extracted with 12 mℓ of dichloromethane. The organic phase was washed with 5 mℓ of water and an aliquot of 8 mℓ of the extract was evaporated. The residue was dissolved in 0.2 mℓ of the mobile phase and aliquots of 50 μℓ were injected.
I-2. The sample was mixed with 0.1 mℓ of an aqueous solution of the internal standard (0.2 μg/mℓ), 1 mℓ of water, and 0.1 mℓ of 0.1 *N* HCl. The mixture was extracted with 10 mℓ of dichloromethane-isopropanol (10:1). The organic layer was evaporated at 35°C under a stream of air. The residue was dissolved in 70 μℓ of acetonitrile, diluted with 70 μℓ of water, and washed with 0.5 mℓ of heptane. An aliquot of 100 mℓ of the lower layer was injected.
I-3. The sample was mixed with 10 μℓ of a methanolic solution of the internal standard (380 μ*M*), 500 μℓ of 2 M N,N-*bis*(2-hydroxyethyl) glycine buffer, pH 9. The mixture was extracted with 6 mL of diethyl ether. The organic phase was evaporated under a stream of nitrogen at 40°C. The residue was dissolved in 200 μℓ of the mobile phase and an aliquot of 150 μℓ was injected.

Elution — E-1. Hexane-tetrahydrofuran-methanol-water (66:27:6.3:0.7).
E-2. Acetonitrile-water-diethylamine-acetic acid (27:73:0.18:0.18).
E-3. Methanol-2-propanol-1.16 *M* perchloric acid (70:30:0.25).

REFERENCE

1. **Whitney, C. C., Weinstein, S. H., and Gaylord, J. C.,** High-performance liquid chromatographic determination of ethmozin in plasma, *J. Pharm. Sci.*, 70, 462, 1981.

2. **Piotrovski, V. K. and Metelitsa, V. I.,** Ion-exchange high-performance liquid chromatography in drug assay in biological fluids, *J. Chromatogr.*, 231, 205, 1982.
3. **Poirier, J. M.,** Sensitive high performance liquid chromatographic analysis of ethmozin in plasma, *Ther. Drug Monit.*, 7, 439, 1985.

ETHOHEPTAZINE

Gas Chromatography

Specimen (mℓ)	Extraction	Column (m × mm)	Packing (mesh)	Oven temp (°C)	Gas (mℓ/min)	Det.	RT (min)	Internal standard (RT)	Deriv.	Other compounds (RT)	Ref.
Plasma, tissue (2)	I-1	2 × 3.2	3% OV-17 Chromosorb W (NA)	225	N_2 (30)	NPD	3.5	Pethidine (2.5)	—	—	1

Extraction — I-1. A weighed amount of the tissue was homogenized with 5 mℓ of the internal standard solution (2.3 mg/mℓ in water) at 0°C and centrifuged. Diluted ammonia (0.25 mℓ, 25%) was added to the supernatant which was then extracted with 6 mℓ of diethyl ether-*n*-hexane-isopropanol (4:1:0.1). An aliquot of the organic layer was evaporated at room temp. The residue was dissolved in 200 μℓ of ethanol and 1 μℓ was injected.

REFERENCES

1. **Drost, R. H., Boelens, M., Maes, R. A. A., and Sunshine, I.,** Determination of ethoheptazine in human post mortem material, *J. Chromatogr.*, 277, 352, 1983.

ETHOXZOLAMIDE

Liquid Chromatography

Specimen (mℓ)	Extraction	Column (cm × mm)	Packing (μm)	Elution	Flow (mℓ/min)	Det. (nm)	RT (min)	Internal standard (RT)	Other compounds (RT)	Ref.
Eye tissue	I-1	30 × 3.9	μ-Bondapak-Phenyl (10)	E-1	1.5	ABS (313)	7	—	—	1

Extraction — I-1. The tissue homogenate was treated with 5 mℓ of 0.025 *M* sodium carbonate solution. The mixture was cooled to room temp, 1 mℓ of 0.5 *M* HCl containing 75 mg NaCl/mℓ was added and the mixture extracted three times with 3-mℓ portions of ethyl acetate. The combined organic extracts were evaporated at 30°C under a stream of nitrogen. The residue was dissolved in 1 mℓ of 50% methanol for injection.

Elution — E-1. Methanol-1% aqueous acetic acid (50:50)

REFERENCE

1. **Eller, M. G. and Schoenwald, R. D.,** Determination of ethoxzolamide in the iris/ciliary body of the rabbit eye by high-performance liquid chromatography: comparison of tissue levels following intravenous and topical administrations, *J. Pharm. Sci.*, 73, 1261, 1984.

ETHOSUXIMIDE

Liquid Chromatography

Specimen (mℓ)	Extraction	Column (cm × mm)	Packing (μm)	Elution	Flow (mℓ/min)	Det. (nm)	RT (min)	Internal standard (RT)	Other compounds (RT)	Ref.
Plasma (0.2)	I-1	15 × 4.1	Hamilton PRP-1 (5)	E-1	1.0	ABS (217)	2.5	α-Methyl-α-propyl succinimide (8.1)	—	1

Extraction — I-1. The sample was mixed with 0.2 mℓ of an aqueous solution of the internal standard (0.1 mg/mℓ) and extracted with 2 mℓ of toluene. The organic layer was back extracted into 0.2 mℓ of 0.5% sodium carbonate solution. An aliquot of 2 μℓ of the aqueous layer was injected.

Elution — E-1. Acetonitrile-water-5% sodium carbonate (50:500:0.5).

REFERENCE

1. **Gupta, R. N. and Stefanec, M.,** Determination of ethosuximide in plasma by liquid chromatography with the use of a resin column and an alkaline mobile phase, unpublished.

ETHYLBISCOUMACETATE

Liquid Chromatography

Specimen (mℓ)	Extraction	Column (cm × mm)	Packing (μm)	Elution	Flow (mℓ/min)	Det. (nm)	RT (min)	Internal standard (RT)	Other compounds (RT)	Ref.
Plasma (0.1—2)	I-1	30 × 3.9	μ-Bondapak-C_{18} (10)	E-1	1.0	ABS (254, 280)	14	Carbama-zepine (17)	—	1

Extraction — I-1. The sample was mixed with 1 mℓ of 0.1 *M* HCl and 1 mℓ of the internal standard solution (40 mg/mℓ in methanol). The mixture was extracted with 10 mℓ of benzene. The benzene layer was evaporated under vacuum and the residue dissolved in 0.25 mℓ of methanol. Aliquots of 25 μℓ of this solution were injected.

Elution — E-1. Methanol-water-acetic acid (56:40:4).

REFERENCE

1. **Arman, M. and Jamali, F.,** High-performance liquid chromatographic determination of ethyl biscoumacetate in human plasma, *J. Chromatogr.*, 272, 406, 1983.

(d,1)-3-ETHYL-2,6-DIMETHYL-4,4α,5,6,7,8,8α,9-OCTAHYDRO-4a,8a-*trans*-IH-PYRROLO[2,3-g] ISOQUINOLIN-4-ONE

Gas Chromatography

Specimen (mℓ)	Extraction	Column (m × mm)	Packing (mesh)	Oven temp (°C)	Gas (mℓ/min)	Det.	RT (min)	Internal standard (RT)	Deriv.	Other compounds (RT)	Ref.
Plasma (2)	I-1	1 × 1	3% OV-17 Gas Chrom Q (120/140)	280	Methane[a]	MS-CI[b]	1.3	[$^2H_3$] analog	—	—	1

[a] Head pressure = 1.2 Kg^{-2}.
[b] Ammonia as a reagent gas.

Extraction — I-1. To the sample were added 10 μℓ of a methanolic solution of the internal standard (4 μg/mℓ) and 2 mℓ of 1 *M* phosphate buffer, pH 11. The mixture was extracted with 6 mℓ of chloroform. The organic layer was evaporated at 50°C under a stream of nitrogen. The residue was reconstituted in 50 μℓ of ethyl acetate and a 5 μℓ aliquot of this solution was injected.

REFERENCE

1. **Min, B. H.,** Quantitation of (d,1)-3-ethyl-2,6-dimethyl-4,4α,5,6,7,8,8α,9-octahydro-4a,8a-*trans*-1H-pyrrolo[2,3-g] isoquinolin-4-one in human plasma by gas chromatography-chemical ionization mass spectrometry, *J. Chromatogr.*, 277, 340, 1983.

ETHYLENEDIAMINEPLATINUM(II)

Liquid Chromatography

Specimen (mℓ)	Extraction	Column (cm × mm)	Packing (μm)	Elution	Flow (mℓ/min)	Det. (nm)	RT (min)	Internal standard (RT)	Other compounds (RT)	Ref.
Plasma	I-1	30 × 3.9	μ-Porasil silica (10)	E-1	2.0	ABS (214)	13	—	—	1

Extraction — I-1. Plasma ultrafiltrates were prepared with Amicon MPS-1 micropartition system provided with YMT filters.

Elution — E-1. Acetonitrile-water (90:10).

REFERENCE

1. **Van Der Vijgh, W. J. F., Elferink, F., Postma, G. J., Vermorken, J. B., and Pinedo, H. M.,** Determination of ethylenediamineplantinum (II) malonate in infusion fluids, human plasma and urine by high-performance liquid chromatography, *J. Chromatogr.*, 310, 335, 1984.

ETHYL LOFLAZEPATE

Gas Chromatography

Specimen (mℓ)	Extraction	Column (m × mm)	Packing (mesh)	Oven temp (°C)	Gas (mℓ/min)	Det.	RT (min)	Internal standard (RT)	Deriv.	Other compounds (RT)	Ref.
Plasma (0.5—2)	I-1	1.8 × 4	3% OV-1-OV-17 (1:3) Gas Chrom Q (80/100)	270	N_2 (60)	ECD	—	CM 7113 (5.5)	—	M_1^a, + M_2^b (3.5)	1

Liquid Chromatography

Specimen (mℓ)	Extraction	Column (cm × mm)	Packing (μm)	Elution	Flow (mℓ/min)	Det. (nm)	RT (min)	Internal standard (RT)	Other compounds (RT)	Ref.
Plasma (0.01)	I-2	25 × 4	μ-Bondapak-C_{18} (10)	E-1	1.0	Radioactivity monitor[c]	—	—	M_1 (2.5) M_2 (6) 3-Hydroxy-M_2 (5)	2, 3

[a] M_1 = Loflazepate.
[b] M_2 = Decarboxylated loflazepate.
[c] The eluate of the column was mixed with the scintillation fluid (3 mℓ/min) just prior to entry in the detector cell.

Extraction — I-1. The sample was spiked with 10 ng of the internal standard, the pH of this mixture adjusted to 2 with 0.25 mℓ of 1 *N* sulfuric acid, and incubated at 37°C for 30 min. The pH was then adjusted to 10 with 1 *N* NaOH/1 *M* phosphate buffer and extracted twice with 8-mℓ portions of diethyl ether. The combined ether extracts were evaporated under vacuum. The residue was dissolved in 2.5 mℓ of 1 *N* sulfuric acid and the solution washed with 10 mℓ of hexane. The aqueous phase was adjusted to pH 10 and extracted twice with 8-mℓ portions of ether. The combined ether extracts were evaporated. The residue was reconstituted with 50 μℓ of toluene and 2- to 3-μℓ aliquots were injected.
I-2. Aliquots of 10 μℓ of plasma sample were injected directly into the HPLC system

Elution — E-1. Acetonitrile-water (50:50).

REFERENCES

1. **Cano, J. P., Sumirtapura, Y. C., Cautreels, W., and Sales, Y.,** Analysis of the metabolites of ethyl loflazepate by gas chromatography with electron-capture detection, *J. Chromatogr.*, 226, 413, 1981.
2. **Davi, H., Guyonnet, J., Necciari, J., and Cautreels, W.,** Determination of circulating ethyl loflazepate metabolites in the baboon by radio-high-performance liquid chromatography with injection of crude plasma samples: comparison with solvent extraction and thin-layer chromatography, *J. Chromatogr.*, 342, 159, 1985.
3. **Davi, H., Guyonnet, J., Sales, Y., and Cautreels, W.,** Metabolism of ethyl loflazepate in the rat, the dog, the baboon and in man, *Arzneim. Forsch.*, 35, 1061, 1985.

ETHYLMORPHINE

Liquid Chromatography

Specimen (mℓ)	Extraction	Column (cm × mm)	Packing (μm)	Elution	Flow (mℓ/min)	Det. (nm)	RT (min)	Internal standard (RT)	Other compounds (RT)	Ref.
Microsomal incubation	I-1	30 × 3.9	μ-Bondapak-C_{18} (10)	E-1	2.0	ABS (254)	9.1	Codeine (5.7)	Morphine (3.4) Norethyl morphine (8.2) Normorphine (2.8) Nicotinamide (3.4)	1

Extraction — I-1. The sample was adjusted to pH 8.7 with 0.5 *M*, pH 8.7 phosphate buffer, and 15 mℓ of the internal standard solution (6 μg/mℓ in isopropanol-dichloromethane, 2:8) were added. After mixing and centrifugation, the organic layer was evaporated at 60°C under a stream of nitrogen. The residue was dissolved in 150 μℓ of the mobile phase and aliquots of this solution were injected with an autosampler.

Elution — E-1. Acetonitrile-1% acetic acid (15:85) containing 0.005 *M* hexanesulfonic acid.

REFERENCE

1. **Jarvi, E. J., Stolzenbach, J. C., and Larson, R. E.,** Simultaneous quantification of ethylmorpine O-deethylase and N-demethylase activity by high-performance liquid chromatography, *J. Chromatogr.*, 377, 261, 1986.

ETODOLAC

Liquid Chromatography

Specimen (mℓ)	Extraction	Column (cm × mm)	Packing (μm)	Elution	Flow (mℓ/min)	Det. (nm)	RT (min)	Internal standard (RT)	Other compounds (RT)	Ref.
Plasma (1)	I-1	25 × 4.6	Spherisorb ODS (5)[a]	E-1	1.8	ABS (226)	5	—	—	1

[a] Column temp = 50°C.

Extraction — I-1. The sample was diluted with 4 mℓ of 1 *N* HCl and extracted with 5 mℓ of isopentyl alcohol-hexane (1:19) An aliquot of 4 mℓ of the organic layer was back extracted into 1 mℓ of 0.1 *M* glycine buffer (pH 11). The aqueous phase was neutralized with 2.5 *M* phosphoric acid for injection.

Elution — Acetonitrile-0.1 *M* phosphate buffer, pH 6 (30:70).

REFERENCE

1. **Cosyns, L., Spain, M., and Kraml, M.,** Sensitive high-performance liquid chromatographic method for the determination of etodolac in serum, *J. Pharm. Sci.*, 72, 275, 1983.

ETOMIDATE

Gas Chromatography

Specimen (mℓ)	Extraction	Column (m × mm)	Packing (mesh)	Oven temp (°C)	Gas (mℓ/min)	Det.	RT (min)	Internal standard (RT)	Deriv.	Other compounds (RT)	Ref.
Serum (1)	I-1	1.5 × 4	3% OV-17 Chromosorb W (80/100)	220	N_2 (70)	NPD	1.3	Propoxate (1.8)	—	—	1

Liquid Chromatography

Specimen (mℓ)	Extraction	Column (cm × mm)	Packing (μm)	Elution	Flow (mℓ/min)	Det. (nm)	RT (min)	Internal standard (RT)	Other compounds (RT)	Ref.
Plasma (2)	I-2	25 × 4.6	Ultrasphere Octyl (8)	E-1	1.2	ABS (248)	5.5	Propoxate (6.8)	—	2, 3
Plasma (0.5)	I-3	10 × 8	Radiao-Pak CN (10)	E-2	1.8	ABS (254)	7.3	Propoxate (9.1)	—	4

Extraction — I-1. The sample was mixed 100 μℓ of an aqueous solution of the internal standard (10 μg/mℓ) and extracted with 10 mℓ of hexane-ether. The organic layer was evaporated at 40°C under nitrogen. The residue was dissolved in 50 μℓ of acetone and 1 to 2 μℓ of the solution were injected.
I-2. The sample was mixed 200 μℓ of an aqueous solution of the internal standard (4 μg/mℓ) and extracted with 10 mℓ of pentane. The organic layer was evaporated at 40 to 45°C. The residue was dissolved in 200 μℓ of the mobile phase and an aliquot of 75 μℓ was injected.
I-3. The sample was mixed with 50 μℓ of the internal standard solution (3.4 μg/mℓ in ethanol) and 100 μℓ of 0.5 *M*, pH 10, borate buffer. This mixture was extracted twice with 3-mℓ portions of hexane-ether (9:1). The combined organic layers were back extracted into 3 mℓ of 0.5 M H_2SO_4. The aqueous layer was washed with 3 mℓ of hexane-ether, made alkaline with 1 mℓ of 3 *M* NH_4OH and extracted with 4 mℓ of dichloromethane. The organic phase was evaporated to dryness at 30°C under reduced pressure. The residue was reconstituted with 100 μℓ of methanol for injection.

Elution — E-1. Acetonitrile-methanol-water (35:32.5:32.5).
E-2. Methanol-water (54:46).

REFERENCES

1. **Haring, C. M. M., Dijkhuis, I. C., and van Dijk, B.,** A rapid method of determining serum levels of etomidate by gas chromatography with the aid of a nitrogen detector, *Acta Anesthesiol. Belg.*, 31, 107, 1980.
2. **Ellis, E. O. and Beck, P. R.,** Determination of etomidate in human plasma by high-performance liquid chromatography, *J. Chromatogr.*, 232, 207, 1982.
3. **Hebron, B. S., Edbrooke, D. L., Newby, D. M., and Mather, S. J.,** Pharmacokinetics of etomidate associatedwith prolonged i.v. infusion, *Br. J. Anaesth.*, 55, 281, 1983.
4. **Avram, M. J., Fragen, R. J., and Linde, H. W.,** High-performance liquid chromatographic assay for etomidate in human plasma: results of preliminary clinical studies using etomidate for hypnosis in total intravenous anesthesia, *J. Pharm. Sci.*, 72, 1424, 1982.

ETOPERIDONE

Gas Chromatography

Specimen (mℓ)	Extraction	Column (m × mm)	Packing (mesh)	Oven temp (°C)	Gas (mℓ/min)	Det.	RT (min)	Internal standard (RT)	Deriv.	Other compounds (RT)	Ref.
Plasma, urine (1)	I-1	1 × 4	3% OV-17 Chromosorb W (100/120)	290	N_2 (40)	NPD	5	Trazodone (9)	—	—	1

Extraction — I-1. The sample was mixed with 50 μℓ of the internal standard solution (10 μg/mℓ in methanol) and 200 μℓ of 5 *M* potassium hydroxide. The mixture was extracted twice with 5-mℓ portions of diethyl ether-petroleum ether (1:1). The combined organic extracts were evaporated under nitrogen. The residue was dissolved in 200 μℓ of methanol for injection.

REFERENCES

1. **Gilmour, W. J. and Leary, J. R.,** Gas-liquid chromatographic determination of etoperidone in plasma, serum, and urine, *J. Chromatogr.*, 233, 381, 1982.

ETOPOSIDE

Liquid Chromatography

Specimen (mℓ)	Extraction	Column (cm × mm)	Packing (μm)	Elution	Flow (mℓ/min)	Det. (nm)	RT (min)	Internal standard (RT)	Other compounds (RT)	Ref.
Plasma (2)	I-1	25 × 3	Partisil ODS (10)[a]	E-1	0.66	ABS (252)	6.6	—	—	1
Plasma (1)	I-2	30 × 3.9	μ-Bondapak C_{18} (10)	E-2	1.0	ABS (254)[b]	5.5	VM 26 (8)	c	2
Plasma (1)	I-3	NA	LiChrosorb RP-8 (5)	E-3	1.0	ABS (254)	5	VM 26 (9)	—	3
Plasma urine (1—2)	I-4	12.5 × 4	Lichrocart RP18 (10)	E-4	1.0	Fl (230, 328)	5	—	Teniposide (7.5)	4

ETOPOSIDE (continued)

Plasma (0.1—1)	I-5	30 × 4.6	μ-Bondapak phenyl (10)	E-5	1.0	Electrochem[d]	5	—	e	5
Plasma, urine (1)	I-6	10 × 8	Radial-Pak C_{18} (10)[f]	E-6	1.0	MS[g]	6	VM 26 (7)	—	6, 7
Dosage	—	30 × 3.9	μ-Bondapak phenyl (10)	E-7	1.0	ABS (254)	17	Methyl-*p*-amino benzoate (8)	h	8
Plasma (1)	I-7	30 × 4.9	Bondapak phenyl (5)	E-8	1.0	Electrochem[i]	5	Teniposide (9)	—	9
Plasma, urine (1)	I-8	10 × 5	Hypersil-ODS (5)	E-9	2.0	ABS (229)	2.1	Phenytoin (2.9)	—	10

[a] Protected by a precolumn packed with the same packing material.
[b] Fluorescene detection (ex = 288 nm, em = 328 nm) was also used.
[c] Conditions for the separation of picro isomer are described.
[d] Potential = 0.5 V.
[e] Conditions for the separation of isomers and possible metabolites are described.
[f] Protected by a Guard-Pak C_{18}.
[g] The drug and the internal standard are monitored by UV absorption to sample the fraction to be introduced into the mass spectrometer.
[h] Separation of impurities in the injectable is shown.
[i] Potential = 0.8 V.

Extraction — I-1. The sample was diluted with 2 mℓ of water and extracted three times with 4-mℓ aliquots of chloroform. The pooled extracts were evaporated at 45°C under vacuum. The residue was dissolved in 0.2 mℓ of dioxane for analysis.

I-2. The sample was mixed with 10 μℓ of the internal standard solution (1 mg/mℓ and extracted with 5 mℓ of chloroform. An aliquot of 4.5 mℓ of the chloroform layer was evaporated at 40°C with a nitrogen stream. The residue was reconstituted with 50 μℓ of methanol and an aliquot of 20 μℓ was injected.

I-3. The sample was washed with 5 mℓ of diisopropyl ether, spiked with the internal standard, and extracted with 8 mℓ of chloroform. The organic extract was

evaporated at room temperature under vacuum. The residue was dissolved in 100 μℓ of the mobile phase and 5 to 20 μℓ of this solution was injected.
I-4. The sample was incubated with an equal volume of an aqueous solution of subtilism (1 mg/mℓ) for 15 min at 50°C. An aliquot of the hydrolyzed sample was injected with an autosampler onto a 10 × 2 mm precolumn packed with Hamilton-PRP-1 (10 μm) packing. The precolumn was backflushed with 4 mℓ of water at a flow rate of 2 mℓ/min. The precolumn was then switched on-line with the analytical column and back flushed for 30 sec using a methanol-water mixture. Sample analysis took 6 min. Meanwhile, the precolumn was backflushed with 6 mℓ of methanol. All operations were carried out automatically.
I-5. The sample was extracted with 1 mℓ of dichloromethane. An aliquot (0.8 mℓ) of the organic extract was evaporated at 30°C under nitrogen. The residue was dissolved in 25 to 100 μℓ of the mobile phase and 5 to 20 μℓ was injected.
I-6. The sample was spiked with the internal standard and extracted with 1 mℓ of chloroform. The organic extract was evaporated by freeze drying and the residue was dissolved in 100 μℓ of the mobile phase and an aliquot of 50 μℓ was injected.

I-7. The sample was spiked with 10 μg of the internal standard and extracted with 4 mℓ of 1,2-dichloroethane. The organic phase was evaporated at 90°. The residue was reconstituted with 0.4 mℓ of the mobile phase and 20-μℓ aliquots were injected.
I-8. The sample was spiked with 50 μℓ of the internal standard solution (200 μg/mℓ) and extracted with 5 mℓ of chloroform. The organic layer was filtered and evaporated at 50°C. The residue was reconstituted with 200 μℓ of the mobile phase.

Elution — E-1. Methanol-5 m*M* phosphate buffer, pH 7.8 (50:50).
E-2. Methanol-water (60:40).
E-3. Methanol-water (55:45).
E-4. Methanol-water-acetic acid (46:64:1).
E-5. Methanol-0.065 M phosphate buffer, pH 7 (60:40 w/w).
E-6. Methanol-acetonitrile-water (2:1:1).
E-7. Acetonitrile-0.02 M sodium acetate buffer, pH 4 (26:74).
E-8. Methanol-0.05 M phosphate buffer, pH 7 (60:40).
E-9. Methanol-water (51:49).

REFERENCES

1. **Allen, L. M.,** Analysis of 4′-demethylepipodophyllotoxin-9-(4,6-0-ethylidene-β-D-glucopyranoside) by high-pressure liquid chromatography, *J. Pharm. Sci.*, 69, 1440, 1980.
2. **Strife, R. J., Jardine, I., and Colvin, M.,** Analysis of the anticancer drugs VP 16-213 and VM 26 and their metabolites by high-performance liquid chromatography., *J. Chromatogr.*, 182, 211, 1980.
3. **Farina, P., Marzillo, G., and D'Incalci, M.,** High-performance liquid chromatography determination of 4′-demethylepipodophyllotoxin-9-(4,6-0-ethylidene β-D-glucopyranoside) (VP 16-213) in human plasma, *J. Chromatogr.*, 222, 141, 1981.
4. **Werkhown-Goewie, C. E., Brinkman, U. A., Frei, R. W., De Ruiter, C., and De Vries, J.,** Automated liquid chromatographic analysis of the anti-tumorigenic drugs etoposide (VP 16-213) and teniposide (VM 26), *J. Chromatogr.*, 276, 349, 1983.
5. **Holthuis, J. J. M., Romkens, F. M. G. M., Pinedo, H. M., and Van Oort, W. J.,** Plasma assay for the antineoplastic agent VP 16-213 (etoposide) using high-performance liquid chromatography with electrochemical detection, *J. Pharm. Biomed. Anal.*, 1, 89, 1983.

6. **Danigel, H., Pfluger, K. H., Jungclas, H., Schmidt, L., and Dellbrugge, J.,** Drug monitoring of etoposide (VP16-213), *Cancer Chemother. Pharmacol.*, 15, 121, 1985.
7. **Danigel, H., Schmidt, L., Jungclas, H., and Pfueger, K. H.,** Combined thin layer chromatography/mass spectrometry: an application of californium-252 plasma desprotion mass spectrometry for drug monitoring. *Biomed. Mass Spectrom.*, 12, 542, 1985.
8. **Floor, B. J., Klein, A. E., Muhammad, N., and Ross, D.,** Stability-indicating liquid chromatographic determination of etoposide and benzyl alcohol in injectable formulations, *J. Pharm. Sci.*, 74, 197, 1985.
9. **Littlewood, T. J., Hutchings, A. L., Bentley, D. P., and Spragg, B. P.,** High-performance liquid chromatographic determination of etoposide in plasma using electrochemical detection, *J. Chromatogr.*, 336, 434, 1984.
10. **Harvey, V. J., Joel, S. P., Johnston, A., and Slevin, M. L.,** High-performance liquid chromatography of etoposide in plasma and urine, *J. Chromatogr.*, 339, 419, 1985.

ETOZOLIN

Liquid Chromatography

Specimen (mℓ)	Extraction	Column (cm × mm)	Packing (μm)	Elution	Flow (mℓ/min)	Det. (nm)	RT (min)	Internal standard (RT)	Other compounds (RT)	Ref.
Plasma (1—2)	I-1	25 × 3	LiChrosorb Si100 (7)	E-1	0.9	ABS (281)	(5)	Piprozolin (9); Go 3284	Ozolinone[a]	1
Plasma (1—2)	I-1	25 × 4	LiChrosorb RP-18 (7)[b]	E-2	NA	ABS (282)	(7)	Piprozolin (9); Go 3284	Ozolinone[a]	2

[a] Separate conditions for the determination of this metabolite are described.
[b] Protected by a 40 × 4 mm precolumn packed with the same material as of the analytical column.

Extraction — I-1. The sample was mixed with 10 μℓ of methanol containing 500 ng of piprozolin and 1 μg of Go 3284 and 2 mℓ of 0.5 *M*, pH 9, glycine buffer. The mixture was extracted twice with 20-mℓ portions of dichloromethane. The combined organic layer was filtered through 5 g basic alumina (activity IV). The filtrate was evaporated. The residue was dissolved in 20 to 50 μℓ of the mobile phase and 10 μℓ was injected. The aqueous phase was made acidic and extracted with dichloromethane to isolate ozolinone and the internal standard Go 3284.
Elution — E-1. Cyclohexane-containing 0.4% acetic acid (68:32). E-2. Methanol-20 m*M* phosphate buffer, pH 2.2 (65:35).

REFERENCES

1. **Hengy, H., Vollmer, K. O., Gladigau, V., and Kolle, E. U.,** Assay of etozolin and its main metabolite, ozolinone, in plasma by high performance liquid chromatography, *Arzneim. Forsch.*, 30, 1788, 1980.
2. **Liddiard, C. and Nau, H.,** Determination of etozolin and ozolinone in human plasma and tissues by reversed-phase high-performance liquid chromatography, *J. Chromatogr.*, 225, 504, 1981.

FAMOTIDINE

Liquid Chromatography

Specimen (mℓ)	Extraction	Column (cm × mm)	Packing (μm)	Elution	Flow (mℓ/min)	Det. (nm)	RT (min)	Internal standard (RT)	Other compounds (RT)	Ref.
Plasma, urine (1)	I-1	25 × 4.6	Ultrasphere RP8 (5)[a]	E-1	NA	ABS (267)[b]	9.5	—	—	1

[a] Protected by a Brownlee 4 cm, RP-8, 10-μm guard column.
[b] Urine extracts were monitored at 254 nm.

Extraction — I-1. The plasma sample was applied to a prewashed (1 mℓ methanol, 1 mℓ water) 2.8 mℓ BondElut silica extraction column. The column was washed with 5 mℓ of water and eluted with 2 mℓ of acetonitrile. The eluate was evaporated at 40°C under a stream of nitrogen and the residue was reconstituted in 0.2 mℓ of 0.017 *M* glacial acetic acid. An aliquot of 150 μℓ of this solution was injected with an autosampler.

Elution — E-1. Acetonitrile-0.019 *M* phosphoric acid (10:90).

REFERENCE

1. **Vincek, W. C., Constanzer, M. L., Hessey, G. A., II, and Bayne, W. F.,** Analytical method for the quantification of famotidine, an H_2-receptor blocker, in plasma and urine, *J. Chromatogr.*, 338, 438, 1985.

FD-1

Liquid Chromatography

Specimen (mℓ)	Extraction	Column (cm × mm)	Packing (μm)	Elution	Flow (mℓ/min)	Det. (nm)	RT (min)	Internal standard (RT)	Other compounds (RT)	Ref.
Visceral tissue	I-1	25 × 6.2	Zorbax Sil (10)	E-1	1.5	ABS (254)	4.5	—	1-(Tetrahydro-2-furanyl)-5-fluoro-2,4,pyrimidine-dione (6.9) 3-(Tetrahydro-2-furanyl)-5-fluoro-2,4-pyrimidine-dione (10.5) 5-Fluoro-2,4-pyrim-idinedione[a]	1

[a] This metabolite was determined by gas chromatography-mass fragmentography.

Extraction — I-1. The weighed sample was homogenized with 2 to 3 volumes of physiological saline and centrifuged. A 1-mℓ aliquot of the supernatant was adjusted to pH 2 with 5 *N* HCl and extracted twice with 20-mℓ portions of chloroform. The chloroform extract was evaporated at a temperature below 25°C under a stream of nitrogen. The residue was dissolved in 100 μℓ of 1,2-dichloroethane and 20 μℓ of this solution was injected.

Elution — E-1. 1,2-Dichloroethane-ethanol (24:1).

REFERENCE

1. **Marunaka, T., Umeno, Y., and Minami, Y.,** Quantitative determination of 1,3-*bis*(tetrahydro-2-furanyl)-5-fluoro-2,4-pyrimidinedione and its metabolites in visceral tissues by high-performance liquid chromatography and gas chromatography-mass fragmentography, *J. Chromatogr.*, 188, 270, 1980.

FEBANTEL

Liquid Chromatography

Specimen (mℓ)	Extraction	Column (cm × mm)	Packing (μm)	Elution	Flow (mℓ/min)	Det. (nm)	RT (min)	Internal standard (RT)	Other compounds (RT)	Ref.
Plasma (4)	I-1	25 × 4	RP-8 (10)[a]	E-1; grad	2.0	ABS (290)	11	—	Fenbendazole (9.4) Metabolite[b] A′ (7.5) Metabolite[b] B (6) Oxfendazole (4.3)	1, 2
Milk (50)	I-2	25 × 4	RP-8 (10)[c]	E-2	3.0	ABS (300)	5.4	PhO-MBC[d] (3)	Fenbendazole (4.2)[e]	3

[a] Column temp = 30°C.
[b] Metabolite A′ = N-2(N′-N′-*bis*-methoxycarbomylguanidino)3-methoxyacetamido-5-phenylsulphinyl benzene. Metabolite B = Methyl-5-phenylsulphonylbenzimidazol-1H-yl carbamate.
[c] Column temp = 50°C.
[d] Methyl(5-phenoxy)-1H-benzimadazol-2-yl carbamate.
[e] Conditions for extraction and chromatography of metabolites A′ and C (oxfendazole) are described.

Extraction — I-1. The sample was mixed with 2 mℓ of pH 7.4 phosphate buffer and extracted twice with 20 mℓ portions of diethyl ether. The combined ether extracts were evaporated at 45°C under a stream of nitrogen. The residue was dissolved in 0.2 mℓ of dimethyl formamide for injection.
I-2. The sample was mixed with the internal standard solution (10 μg/mℓ in dimethyl formamide) and 150 mℓ acetone. The supernatant was adjusted to pH 4.5 and extracted with 300 mℓ of chloroform. The chloroform layer was dried over anhydrous sodium sulfate and evaporated. The residue was dissolved in 50 mℓ of hexane and the solution extracted with 50 mℓ of acetonitrile. The acetonitrile phase was evaporated and the residue dissolved in 2 to 15 mℓ aliquots of hexane-ether (1:1). The solution was extracted with 50 mℓ of 1 *N* NaOH. The aqueous phase was neutralized then buffered to pH 4.5, and extracted twice with 30-mℓ portions of hexane-ether (1:1). The combined extracts were evaporated. The residue was dissolved in 0.5 mℓ of dimethylformamide for injection.

Elution — E-1. (A) Acetonitrile; (B) 1% phosphoric acid. Gradient from 80% to 40% (B) in 10 min.
E-2. Methanol + 1 g/ℓ ammonium carbonate (50:50).

FEBANTEL (continued)

REFERENCES

1. **Delatour, P., Tiberghien, M. P., and Besse, S.,** An HPLC procedure for the quantification of five metabolites of febantel in sheep serum, *J. Vet. Pharmacol. Ther.*, 6, 223, 1983.
2. **Delatour, P., Tiberghien, M. P., Garnier, F., and Benoit, E.,** Comparative pharmacokinetics of febantel and its metabolites in sheep and cattle, *Am. J. Vet. Res.*, 46, 1399, 1985.
3. **Delatour, P., Garnier, F., and Benoit, E.,** Kinetics of four metabolites of febantel in cow's milk, *Vet. Res. Commun.*, 6, 37, 1983.

FEBENDAZOLE

Liquid Chromatography

Specimen (mℓ)	Extraction	Column (cm × mm)	Packing (μm)	Elution	Flow (mℓ/min)	Det. (nm)	RT (min)	Internal standard (RT)	Other compounds (RT)	Ref.
Plasma, urine, tissue (1)	I-1	30 × 4	MicroPak ODS (10)	E-1	a	ABS (290)	9	Mbendazole (7)	Oxfendazole (6.5) Febendazole sulfone (5.5) *p*-Hydroxyfebendazole (7.5) Aminofebendazole[b] (11.5)	1

[a] Flow program 0.6 mℓ/min from 0 to 6 min, 1 mℓ/min from 6 to 7 min, and 2 mℓ from 7 to 13 min.
[b] 5-(Phenylthio)-2-aminobenzimidazole.

Extraction — I-1. The plasma sample was mixed with 20 μℓ of concentrated ammonium hydroxide and applied to a Chem-Elut (1 mℓ) column. The column was eluted twice with 4-mℓ aliquots of dichloromethane. The combined eluates were evaporated at 50°C under nitrogen. The residue was dissolved in 100 μℓ of the mobile phase for analysis.

Elution — E-1. Acetonitrile-water-0.05 *N* phosphoric acid (80:5:15).

REFERENCE

1. **Barker, S. A., Hsieh, L. C., and Short, C. R.,** Methodology for the analysis of fenbendazole and its metabolites in plasma, urine, feces, and tissue homogenates, *Anal. Biochem.*, 155, 112, 1986.

FELODIPINE

Gas Chromatography

Specimen (mℓ)	Extraction	Column (m × mm)	Packing (mesh)	Oven temp (°C)	Gas (mℓ/min)	Det.	RT (min)	Internal standard (RT)	Deriv.	Other compounds (RT)	Ref.
Plasma (1 g)	I-1	25 × 0.32	CP Sil 5	T.P.[a]	He[b]	ECD	14.5	H 165/04 (15.5)	—	H 152/37 (11)	1

[a] Initial temp NA; initial time = 2 min; rate = 8°C/min; final temp = 270°C.
[b] Column head pressure = 17.4 psi.

Extraction — I-1. The sample was mixed with 1 mℓ of water and extracted with 1 mℓ of toluene containing 100 μmol/ℓ of the internal standard. After centrifugation a 3-μℓ aliquot of the organic layer was injected.

REFERENCE

1. **Ahnoff, M.**, Determination of felodipine in plasma by capillary gas chromatography with electron capture detection, *J. Pharm. Biomed. Anal.*, 2, 519, 1984.

FENFLURAMINE

Gas Chromatography

Specimen (mℓ)	Extraction	Column (m × mm)	Packing (mesh)	Oven temp (°C)	Gas (mℓ/min)	Det.	RT (min)	Internal standard (RT)	Deriv.	Other compounds (RT)	Ref.
Plasma (2)	I-1	1.8 × 2	5% Dexsil 410 GC Chromosorb W (100/120)	215	Ar: 95 Methane:5 (80)	ECD	2.4	4-Methoxyphene-thyl-amine (7.2)	Penta-fluoro-benzoyl	Norfenfluramine (3)	1
Plasma (2)	I-2	2 × 2	10% Carbowax 20M + 10%KOH Chromosorb W (80/100)	130	N_2 (60)	NPD	4.1	N,N-Diethylaniline (6)	—	Norfenfluramine (4.8)	2

Plasma (1)	I-3	1.8 × 4	3% OV-1 GasChrom Q (100/120)	155	He (40)	NPD	5.2	N-Propyl-α-methyl-3-(trifluoromethyl)-phenethylamine (6.7)	Chloroformyl	Norfenfluramine (4.1)	3

Extraction — I-1. The sample was mixed 1 mℓ of an aqueous solution of the internal standard (50 ng/mℓ) and 0.5 mℓ of 10 *N* NaOH. The mixture was extracted with 6 mL of *n*-pentane. The organic layer was incubated with 200 μℓ of 0.005% solution of pentafluorobenzoyl chloride in pentane at 65°C for 30 min and then evaporated at 45°C. The residue was treated with 0.5 mℓ of 0.1 *N* NH_4OH and 50 μℓ of amyl acetate. After mixing and centrifugation 2 to 3 μℓ of the organic layer were injected.

I-2. The sample was mixed with 100 μℓ of the internal standard solution (13.4 μmol/ℓ in 0.5 *M* sulfuric acid) and 1 mℓ of 5 *M* sodium hydroxide. The mixture was extracted with 2 mℓ of diethyl ether. The ether layer was back extracted into 1 mℓ of 0.5 *M* sulfuric acid. The aqueous phase was made alkaline and extracted with 0.2 mℓ of *n*-butyl acetate. Aliquots of 5-μℓ of the organic phase were injected.

I-3. The sample was mixed with 0.5 mℓ of an aqueous solution of the internal standard, 0.5 mℓ of absolute ethanol, and 0.5 mℓ of 5% ammonium hydroxide. The mixture was extracted with 8 mℓ of cyclohexane-dichloromethane. The organic layer was back extracted into 1 mℓ of 0.05 *M* sulfuric acid. The aqueous layer was washed with another aliquot (5 mℓ) of this extraction solvent, made alkaline with 0.5 mℓ of 2 *M* sodium hydroxide, and extracted with 0.3 mℓ of 5% ethylchloroformate in cyclohexane-dichloromethane (3:2). A 5-μℓ aliquot of the organic layer was injected.

REFERENCE

1. **Midha, K. K., McGilveray, I. J., and Cooper, J. K.,** A GLC-ECD assay for simultaneous determination of fenfluramine and norfenfluramine in human plasma and urine, *Can. J. Pharm. Sci.*, 14, 18, 1979.
2. **Morris, R. G. and Reece, P. A.,** Improved gas-liquid chromatographic method for measuring fenflura-mine and norfenfluramine in heparinised plasma, *J. Chromatogr.*, 278, 434, 1983.
3. **Krebs, H. A., Cheng, L. K., and Wright, G. J.,** Determination of fenfluramine and norfenfluramine in plasma using a nitrogen-sensitive detector, *J. Chromatogr.*, 310, 412, 1984.

FENOCTIMINE

Gas Chromatography

Specimen (mℓ)	Extraction	Column (m × mm)	Packing (mesh)	Oven temp (°C)	Gas (mℓ/min)	Det.	RT (min)	Internal standard (RT)	Deriv.	Other compounds (RT)	Ref.
Plasma (2)	I-1	15 × 0.32	SE-54 (0.12 μm[a])	T.P.[b]	He (4)	NPD	8.7	Homologue[c] (10.6)	—	—	1

[a] Film thickness.
[b] Initial temp = 210°C, rate = 30°C/min; final temp = 250°C.
[c] 4-(Dimethylphenyl)-1-[(nonylimino)methyl]piperidine.

Extraction — I-1. The sample was spiked with 20 μℓ of a methanolic solution of the internal standard (1 μg/mℓ) and extracted with 10 mℓ of diethyl ether. A 8-mℓ aliquot of the ether layer was evaporated under a stream of nitrogen at room temperature. The residue was reconstituted with 0.5 mℓ of 1% ammonium carbonate-methanol (2:3). The solution was applied to a preconditioned (2 × 3 mℓ methanol, 3 mℓ 1% ammonium carbonate-methanol, 3:2) 3-mℓ BondElute C_{18} column. The column was washed with 2 × 3 mℓ of 1% ammonium carbonate-methanol (3:2) and 3 mℓ of 0.5% ammonium hydroxide in methanol. The column was eluted with 5 mℓ of 0.5% ammonium hydroxide in methanol. The eluate was evaporated under a stream of nitrogen. The residue was dissolved in 30 μℓ of methanol and an aliquot of 5 μℓ was injected.

REFERENCE

1. **Ng, K. T. and Rigney, J. P.,** Determination of fenoctimine in plasma by capillary gas chromatography with nitrogen-phosphorus detection, *J. Chromatogr.*, 377, 373, 1986.

FENOLDOPAM

Liquid Chromatography

Specimen (mℓ)	Extraction	Column (cm × mm)	Packing (μm)	Elution	Flow (mℓ/min)	Det. (nm)	RT (min)	Internal standard (RT)	Other compounds (RT)	Ref.
Plasma (2)	I-1	25 × 4.6	Ultrasphere-ODS (5)	E-1	1.0	Electrochem	8.7	2,3,4,5-Tetrahydro-1-(4-phenyl)-1H-3-benzazepine-7,8 diol (17.5)	b	1, 2

[a] Potential = 0.65 V.
[b] Conditions for the extraction and chromatographic determination of different metabolites are described.

Extraction — I-1. The sample was mixed with 50 μℓ of 0.05 *M* acetic acid containing the internal standard (200 ng/mℓ) and 26 μℓ of 1 *M* sodium hydroxide. The mixture was extracted with 5 mℓ of ethyl acetate. An aliquot of 4.5 mℓ of the organic layer was evaporated under nitrogen at 40°C. The residue was dissolved in 300 μℓ of pH 4 citrate-acetate buffer and the solution was washed with 2 mℓ of diethyl ether. Aliquots of 20 to 100 μℓ of the aqueous layer were injected with an autosampler.

Elution — E-1. Methanol-citrate-acetate buffer, pH 4 (20:80).

REFERENCE

1. **Boppana, V. K., Heineman, F. C., Lynn, R. K., Randolph, W. C., and Ziemniak, J. A.,** Determination of fenoldopam (SK&F 82526) and its metabolites in human plasma and urine by high-performance liquid chromatography with electrochemical detection, *J. Chromatogr.*, 317, 463, 1984.
2. **Boppana, V. K., Fong, K. L. L., Ziemniak, J. A., and Lynn, R. K.,** Use of a post-column immobilized β-glucuronidase enzyme reactor for the determination of diastereomeric glucuronides of fenoldopam in plasma and urine by high-performance liquid chromatography with electrochemical detection, *J. Chromatogr.*, 353, 231, 1986.

FENOTEROL

Liquid Chromatography

Specimen (mℓ)	Extraction	Column (cm × mm)	Packing (μm)	Elution	Flow (mℓ/min)	Det. (nm)	RT (min)	Internal standard (RT)	Other compounds (RT)	Ref.
Plasma (1)	I-1	10 × 5	Hypersil phenyl (5)[a]	E-1	1.2	Electrochem[b]	3.7	Ritodrine (5)	—	1

[a] Protected by a 75 × 2.1 mm guard column packed with pellicular reversed phase.
[b] Potential = 0.9 V.

Extraction — I-1. The sample was mixed with the internal standard solution and 0.5 mℓ of 34 m*M* phosphate buffer, pH 7.4. The sample was washed with 3 mℓ of diethyl ether. An aliquot of the aqueous layer (0.6 mℓ) was extracted with 3 mℓ of ethyl acetate containing 1 m*M* *bis*(2-ethylhexyl)phosphoric acid. An aliquot of the organic phase (2.7 mℓ) was evaporated at 50°C under nitrogen. The residue was dissolved in 0.2 mℓ of acetonitrile-10 m*M* phosphate buffer, pH 6 (35:65) and aliquots of 20 to 150 μℓ were injected.

Elution — E-1. Acetonitrile: 10 m*M* phosphate buffer, pH 6 (35:65) containing 70 μ*M* *bis*(2-ethylhexyl)phosphoric acid.

REFERENCE

1. **Koster, A. S., Hofman, G. A., Frankhuijzen-Sierevogel, A. C., and Noordhoek, J.,** Presystemic and systemic intestinal metabolism of fenoterol in the conscious rat, *Drug Metab. Dispos.*, 13, 464, 1985.

FENPROFEN

Liquid Chromatography

Specimen (mℓ)	Extraction	Column (cm × mm)	Packing (μm)	Elution	Flow (mℓ/min)	Det. (nm)	RT (min)	Internal standard (RT)	Other compounds (RT)	Ref.
Serum (0.05—0.1)	I-1	25 × 2.6	Perkin Elmer ODS (10)[a]	E-1	1.5	ABS (272)	1.5	Valeric acid[b] (2.4)	—	1

Plasma (1)	I-2	30 × 3.9	μ-Bondapak alkylphenyl (10)	E-2	1.0	ABS (272)	8	*dl*-2(4-Phenoxyphenyl) valeric acid (12)	*p*-Hydroxyfenprofen (6)	2
Plasma (0.1)	I-3	15 × 4	Nucleosil C_{18} (5)	E-3	1.0	ABS (240)	4	Diphenylamine (6)	—	3
Serum (0.5)	I-4	15 × 4.6	Ultrasphere-ODS (5)	E-4	2.0	ABS (240)	5.4	Phenolphthalein (2)	Ibuprofen (8) Indomethacin (7.5) Naproxen (3.4) Tolmetin (2.8)	4

[a] Column temp = 40°C.
[b] The identity of this compound appears to be incomplete.

Extraction — I-1. The sample was mixed 100 μℓ of 1 *N* HCl and extracted with 500 μℓ of chloroform containing 20 μg/mℓ of the internal standard. The organic layer was evaporated at 40°C with nitrogen. The residue was dissolved in 20 μℓ of methanol and an aliquot of 5 μℓ was injected.
I-2. The sample was mixed with 0.5 mℓ of the internal standard solution (60 μg/mℓ in 0.01 *N* NaOH) and 1 mℓ of 1 *N* HCl. The mixture was extracted with 10 mℓ of butyl chloride. The organic phase was back extracted into 1 mℓ of 0.1 *N* NaOH. Aliquots of 50 μℓ of the aqueous phase were injected with an autosampler.
I-3. The sample was treated with 0.5 mℓ of methanol containing the internal standard (3 μg/mℓ). After mixing and centrifugation aliquots of 100 μℓ of the supernatant were injected.
I-4. The sample was mixed with 50 μℓ of the internal standard solution (250 μg/mℓ) and 0.5 mℓ of 1 *M* HCl. The mixture was extracted with 10 mℓ of dichloromethane. The organic layer was evaporated under nitrogen, the residue dissolved in 200 μℓ of acetonitrile and aliquots of 20 μℓ of the solution were injected.

Elution — E-1. Acetonitrile-water-acetic acid (50:50:2).
E-2. Acetonitrile-water-acetic acid (50:50:2).
E-3. Acetonitrile-0.35 *M* acetic acid. (60:40).
E-4. Acetonitrile-water-acetic acid (450:550:3.2).

REFERENCE

1. **Miceli, J. N., Ryan, D. M., and Done, A. K.,** High-performance liquid column chromatography of fenoprofen in serum, *J. Chromatogr.*, 183, 250, 1980.
2. **Bopp, R. J., Farid, K. Z., and Nash, J. F.,** High-performance liquid chromatographic assay for fenoprofen in human plasma, *J. Pharm. Sci.*, 70, 507, 1981.
3. **Katogi, Y., Ohmura, T., and Adachi, M.,** Simple and rapid determination of fenoprofen in plasma using high-performance liquid chromatography, *J. Chromatogr.*, 278, 475, 1983.
4. **Levine, B. and Caplan, Y. H.,** Simultaneous liquid-chromatographic determination of five nonsteroidal anti-inflammatory drugs in plasma or blood, *Clin. Chem.*, 31, 346, 1985.

FENQUIZONE

Gas Chromatography

Specimen (mℓ)	Extraction	Column (m × mm)	Packing (mesh)	Oven temp (°C)	Gas (mℓ/min)	Det.	RT (min)	Internal standard (RT)	Deriv.	Other compounds (RT)	Ref.
Plasma, urine (1—2)	I-1	0.35 × 2	3% OV-101 Chromosorb W (80/100)	300	N_2 (60)	ECD	3.7	Penfluridol (2.8)	Methyl	—	1

Extraction — I-1. The sample was extracted with 5 mℓ of methyl isobutyl ketone after the addition of 0.2 g of sodium bicarbonate. An aliquot of the organic layer was extracted into 3 mℓ of 0.1 *N* sodium hydroxide. An aliquot of the aqueous layer was mixed with with 5 mℓ of 0.5 *M* iodomethane in dichloromethane and 50 μℓ of 0.1 *M* tetrahexylammonium acid sulfate in dichloromethane at 50°C for 20 min. After cooling, the organic layer was dried over anhydrous sodium sulfate, evaporated to dryness, and the residue was dissolved in 100 μℓ of acetone containing the internal standard. Aliquots of 1 to 3 μℓ of the final solution were injected.

REFERENCE

1. **Marzo, A., Quadro, G., and Treffner, E.,** Gas-liquid chromatographic evaluation of fenquizone in biological samples for pharmacokinetic investigations, *J. Chromatogr.*, 272, 95, 1983.

FENTANYL

Gas Chromatography

Specimen (mℓ)	Extraction	Column (m × mm)	Packing (mesh)	Oven temp (°C)	Gas (mℓ/min)	Det.	RT (min)	Internal standard (RT)	Deriv.	Other compounds (RT)	Ref.
Plasma (2)	I-1	0.9 × 2	3% SE-30 GasChrom Q (80/100)	235	NA	MS-CI[a]	NA	[$^2H_3$]-Fentanyl	—	—	1
Plasma (2)	I-2	2 × 2	3% OV-17 GasChrom Q (80/100)	280	He (30)	NPD	4	Methoxyfentanyl (8)	—	Sulfentanil (4.7) Alfentanil[b] (7.9)	2

Plasma (1)	I-3	1.2 × 2	3% OV-17 Chromosorb W (80/100)	250	He (35)	NPD	9	Papaverine (19)	Acetyl	c	3
Plasma (1)	I-4	3 × 3	3% OV-17 GasChrom Q (80/100)	290	He (35)	NPD	2.3	Alfentanil (4.6)	—	—	4, 5
Dosage (0.1 mg)	I-5	15 × 0.25	DB-1 (0.25 μm)[d]	T.P.[e]	H_2[f]	ECD	27.9[g] 30.3	Analog (29.7, 31.7)[g]	Hepta-fluoro-butyryl	h	6

[a] Methane as the reagent gas.
[b] Retention times of potential fentanyl metabolites and drugs which may be administered during surgery are given.
[c] Potential metabolites have been identified by GC-MS after acetylation.
[d] Film thickness.
[e] Initial temp = 90°C, initial time = 5 min; rate = 25°C/min; final temp = 160°C; final time = 1 min.
[f] Linear velocity = 60 cm/sec; argon/methane (95/5) was used as a make up gas (30 mℓ/min).
[g] Products after derivatization.
[h] Retention times of derivatization products of a number of analogs of fentanyl are given.

Extraction — I-1. The sample was mixed with 25 μℓ of the internal standard solution (1 μg/mℓ in ethyl acetate) and saturated with ammonium carbonate. The mixture was extracted twice with 3-mℓ portions of ethyl acetate. The combined organic extracts were evaporated. The residue was dissolved in 30 μℓ of ethyl acetate and aliquots of 10 μℓ were injected.
I-2. The sample was mixed with 5 μℓ of the internal standard solution (1 μg/mℓ in ethanol) and 0.5 mℓ of 2 *N* sodium hydroxide. The mixture was extracted with 5 mℓ of hexane-ethanol (19:1). The organic layer was back extracted into 5 mℓ of 0.1 *N* HCl. The aqueous layer was made alkaline with 0.5 mℓ of 2 *N* sodium hydroxide and extracted with 5 mℓ of the extraction solvent. The organic layer was evaporated, the residue dissolved in 30 μℓ of ethanol and aliquots of 8 μℓ were injected.
I-3. The sample was made alkaline with 0.1 mℓ of 4 *N* sodium hydroxide and extracted three times with 5 mℓ portions of benzene. The pooled organic extracts were treated with 0.5 mℓ of acetic annhydride and 10 μℓ of pyridine. The mixture was incubated at 75°C for 2 hr and then evaporated at 75°C under a stream of nitrogen. The residue was dissolved in 50 μℓ of a solution of the internal standard (5 μg/mℓ in benzene). Aliquots of 10 μℓ of this solution were injected.
I-4. The sample was mixed with 0.1 mℓ of 4 *M* sodium hydroxide and 1 mℓ of the internal standard solution (2 μg/mℓ) in benzene. The mixture was extracted with 5 mℓ of benzene. The organic layer was evaporated at 40°C. The residue was reconstituted with 10 μℓ of benzene and 1 μℓ was injected.
I-5. The sample is dissolved in 2 mℓ of water and the solution mixed with 100 μℓ of an aqueous solution of the internal standard (10 μg/mℓ), a small amount of sodium carbonate, and 3 g of acid washed Celite 545. The mixture is packed in a column which is eluted with 5 mℓ of diethyl ether. The eluate is evaporated under a stream of nitrogen. The residue is dissolved in 1 mℓ of acetonitrile, the solution treated with 50 mg of 4-(dimethylamino)pyridine and 50 μℓ of heptafluorobutyric anhydride, and the mixture incubated at 75°C for 1 hr. After cooling this mixture is treated with 5 mℓ of 1 *N* sodium carbonate and extracted with 5 mℓ of isooctane containing 200 pg/μℓ of *p,p'*DDT and 10 ng/mℓ of dioctylphthalate. A 1-mℓ aliquot of the isooctane layer is diluted to 10 mℓ with isooctane containing p, p'DDT-dioctylphthalate. An aliquot (5 mℓ) of diluted isooctane solution is washed with 5 mℓ of 1 *N* sulfuric acid and an aliquot of 2 μℓ of the organic layer is injected.

FENTANYL (continued)

REFERENCES

1. **Lin, S. N., Wang, T. P. F., Caprioli, R. M., and Mo, B. P. N.**, Determination of plasma fentanyl by GC-mass spectrometry and pharmacokinetic analysis, *J. Pharm. Sci.*, 70, 1276, 1981.
2. **Gillespie, T. J., Gandolfi, A. J., Maiorino, R. M., and Vaughan, R. W.**, Gas chromatographic determinatino of fentanyl and its analogues in human plasma, *J. Anal. Toxicol.*, 5, 133, 1981.
3. **Van Rooy, H. H., Vermeulen, N. P. E., and Bovill, J. G.**, The assay of fentanyl and its metabolites in plasma of patients using gas chromatography with alkali flame ionisation detection and gas chromatography-mass spectrometry, *J. Chromatogr.*, 223, 85, 1981.
4. **Phipps, J. A., Sabourin, M. A., Buckingham, W., and Strunin, L.**, Detection of picogram concentrations of fentanyl in plasma by gas-liquid chromatography, *J. Chromatogr.*, 272, 392, 1983.
5. **Phipps, J. A., Sabourin, M. A., Buckingham, W., and Strunin, L.**, Measurement of plasma fentanyl concentration: Comparison of three methods, *Can. Anesth. Soc. J.*, 30, 162, 1983.
6. **Moore, J. M., Allen, A. C., Cooper, D. A., and Carr, S. M.**, Determination of fentanyl and related compounds by capillary gas chromatography with electron capture detection, *Anal. Chem.*, 58, 1656, 1986.

FENTIAZAC

Gas Chromatography

Specimen (mℓ)	Extraction	Column (m × mm)	Packing (mesh)	Oven temp (°C)	Gas (mℓ/min)	Det.	RT (min)	Internal standard (RT)	Deriv.	Other compounds (RT)	Ref.
Plasma, urine (1—2)	I-1	30 × 0.32	OV-101	230	N_2 (3)	ECD	NA	Fentiazac butyl ester	Methyl	—	1
Serum (0.05—1)	I-2	2 × 4	1% OV 1 GasChrom Q (100/200)	T.P.[a]	N_2 (35)	FID	NA	Cholesteryl butyrate	Methyl	—	2

Liquid Chromatography

Specimen (mℓ)	Extraction	Column (cm × mm)	Packing (μm)	Elution	Flow (mℓ/min)	Det. (nm)	RT (min)	Internal standard (RT)	Other compounds (RT)	Ref.
Plasma (0.5)	I-3	25 × 5	μ-Bondapak-C_{18} (10)[b]	E-1	2.0	ABS (310)	7.7	[2,4-Di-(*p*-methoxy-phenyl) thiazol-5-yl] acetic acid (6)	*p*-Hydroxyfentiazac (5.1)	3

[a] Initial temp = 240°C; final temp = 290°C.
[b] Protected by 50 × 5 mm guard column packed with the same material.

Extraction — I-1. The sample was mixed with 100 μℓ of 3 *N* HCl and 50 μℓ of the internal standard solution (10 μg/mℓ in ethanol). The mixture was extracted twice with 5-mℓ portions of ethyl ether. The combined extracts were evaporated under nitrogen. The residue was dissolved in 0.1 mℓ of diazomethane in ether. The excess reagent was removed with nitrogen after 5 min. The residue was dissolved in 50 μℓ of *n*-hexane and 1 μℓ of this solution was injected.
I-2. The sample was spiked with 0.1 mℓ of the internal standard solution (250 μg/mℓ in 3 *N* HCl) and extracted twice with 5-mℓ portions of diethyl ether. The combined extracts were evaporated and the residue treated with 0.1 mℓ of diazomethane in ether. The excess reagent was removed with nitrogen after 5 min. The residue was reconstituted with 50 μℓ of ethyl ether. An aliquot 5 μℓ of this solution was spread on a Ni-Cr spiral for injection.
I-3. The sample was mixed with 25 μℓ of a methanolic solution of the internal standard (10 μg/mℓ) and 25 μℓ of 5 *N* HCl. The mixture was extracted with 3.5 mℓ of dichloromethane. The organic layer was evaporated at 45°C under a stream of nitrogen. The residue was reconstituted with 150 μℓ of the mobile phase for injection.

Elution — E-1. Methanol-1% formic acid (77:23).

REFERENCES

1. **Quattrini, M., Zanolo, G., Mondino, A., Giachetti, C., and Silvestri, S.,** Serum and urinary levels of fentiazac after a single oral and epicutaneous administration in human subjects, *Arzneim. Forsch.*, 31, 1046, 1981.
2. **Zanolo, G., Giachetti, C., Mondino, A., Silvestri, S., Bianchi, E., Segre, G., Gomarasca, P., and De Marchi, G.,** Pharmacokinetics of fentiazac in rats and monkeys, *Arzneim. Forsch.*, 31, 1098, 1981.
3. **Dowell, P. S.,** Simultaneous determination of fentiazac and *p*-hydroxyfentiazac in plasma by high-performance liquid chromatography with ultraviolet detection, *Analyst*, 108, 1535, 1983.

FEPRAZONE

Liquid Chromatography

Specimen (mℓ)	Extraction	Column (cm × mm)	Packing (μm)	Elution	Flow (mℓ/min)	Det. (nm)	RT (min)	Internal standard (RT)	Other compounds (RT)	Ref.
Plasma (0.5)	I-1	25 × 4.5	LiChrosorb Si100 (7)[a]	E-1	3.0	ABS (240)	1.7	b (2.8)	DA 3305[c] (7.5)	1

[a] Column temp = 50°C.
[b] 2,4-Dinitrophenyl hydrazone of 3,4-dimethoxybenzaldehyde.
[c] Alcoholic metabolite of feprazone.

Extraction — I-1. The sample was mixed with 0.5 mℓ of 1 *N* HCl and extracted with 0.3 mℓ of chloroform-diisopropyl ether (1:3) containing 25 μg/mℓ of the internal standard. Aliquots of 20 μℓ of the organic layer were injected.

Elution E-1. *n*-Hexane-tetrahydrofuran-acetic acid (780:220:0.5).

REFERENCE

1. **Spahn, H. and Mutschler, E.,** Die Quantitative Bestimmung von Feprazon und einem Feprazon-Metaboliten in Humanplasma nach Hochleistungsflussigkeitschromatographischer oder Dunnschichtchromatographischer trennung, *J. Chromatogr.*, 232, 145, 1982.

FEZOLAMINE

Liquid Chromatography

Specimen (mℓ)	Extraction	Column (cm × mm)	Packing (μm)	Elution	Flow (mℓ/min)	Det. (nm)	RT (min)	Internal standard (RT)	Other compounds (RT)	Ref.
Plasma, urine (1)	I-1	30 × 3.9	μ-Bondapak C_{18} (10)	NA	NA	ABS (254)	13.2	Bupivacaine (7.4)	Desmethylfezolamine (10.8) Didesmethylfezolamine (11.7)	1

Extraction — I-1. The sample was mixed with 50 μℓ of an aqueous solution of the internal standard (100 μg/mℓ), 4 mℓ of water and 100 μℓ of 1 *M* HCl. The mixture was washed with 10 mℓ of diethyl ether. The aqueous layer was made alkaline with 100 μℓ of 10 *M* sodium hydroxide and extracted twice with 5-mℓ volumes of diethyl ether. The combined ether extracts were evaporated at 40°C under a stream of nitrogen. The residue was dissolved in 100 μℓ of the mobile phase and aliquots of 100 μℓ were injected.

REFERENCE

1. **McCoy, L., Skee, D., and Edelson, J.,** Determination of fezolamine and its desmethyl metabolite in human plasma and urine by high-performance liquid chromatography Intravenous pharmacokinetics in the beagle hound, *J. Chromatogr.*, 344, 211, 1985.

FK-027

Liquid Chromatography

Specimen (mℓ)	Extraction	Column (cm × mm)	Packing (μm)	Elution	Flow (mℓ/min)	Det. (nm)	RT (min)	Internal standard (RT)	Other compounds (RT)	Ref.
Serum, urine (0.3)	I-1	15 × 4.6	TSK-LS410 ODS (5)[a]	E-1[b]	1.0	ABS (295)	12.5	—	—	1

[a] The analytical column is preceded by two columns for sample purification. Column I (10 × 4 nm) is an anion exchange column packed with TSK-IEX540 DEAE (5 um); Column II (10 × 4 mm) is packed with the material of the analytical column,

[b] There are three mobile phases. Mobile phase 3 (E-1) is being pumped through this analytical column. In addition a different mobile phase 3 is used for the analysis of urine extracts.

Extraction — I.1 The sample was treated with 30 μℓ of 1/15 *M* phosphate buffer, pH 7 and 600 μℓ of ethanol. After mixing, the mixture was allowed to stand at room temperature for 5 min and centrifuged. An 80-μℓ of the supernatant was injected.

Elution — E-1. Methanol-0.03 *M* ammonium dihydrogen phosphate-phosphoric acid, pH 2.5 (27:73).

REFERENCE

1. **Tokuma, Y., Shiozaki, Y., and Noguchi, H.,** Determination of a new orally active cephalosporin in human plasma and urine by high-performance liquid chromatography using automated column switching, *J. Chromatogr.*, 311, 339, 1984.

FLAVODATE

Liquid Chromatography

Specimen (mℓ)	Extraction	Column (cm × mm)	Packing (μm)	Elution	Flow (mℓ/min)	Det. (nm)	RT (min)	Internal standard (RT)	Other compounds (RT)	Ref.
Plasma, urine (1)	I-1	25 × 4	LiChrosorb RP-18 (10)	E-1[a]	1.3	ABS (268)	6	1-Naphthyl-acetic acid (7.5)	—	1

[a] A different mobile phase (containing 57% methanol) is used for the analysis of urine extracts.

Extraction — I-1. The sample was mixed with 300 μℓ of 18.5% HCl and 100 μℓ of an aqueous solution of the internal standard (100 μg/mℓ). The mixture was extracted with 7 mℓ of diethyl ether. An aliquot of 6 mℓ of the ether layer was evaporated under a stream of nitrogen at room temperature. The residue was reconstituted with 100 μℓ of the mobile phase and aliquots of 10 to 20 μℓ were injected.

Elution — E-1. Methanol-0.1% phosphoric acid (60:40).

REFERENCE

1. **Zecca, L., Guadagni, L., and Bareggi, S. R.,** Determination of sodium flavodate in body fluids by high-performance liquid chromatography. Application to clinical pharmacokinetic studies, *J. Chromatogr.*, 230, 168, 1982.

FLECAINIDE

Gas Chromatography

Specimen (mℓ)	Extraction	Column (m × mm)	Packing (mesh)	Oven temp (°C)	Gas (mℓ/min)	Det.	RT (min)	Internal standard (RT)	Deriv.	Other compounds (RT)	Ref.
Plasma (1)	I-1	1.8 × 2	3% SP-2250 Supelcoport (100/120)	268	Ar:95-Methane:5 (20)	ECD	16.2	Positional isomer[a] (13.8)	Penta-fluoro-benzoyl	—	1

FLECAINIDE (continued)

Liquid Chromatography

Specimen (mℓ)	Extraction	Column (cm × mm)	Packing (μm)	Elution	Flow (mℓ/min)	Det. (nm)	RT (min)	Internal standard (RT)	Other compounds (RT)	Ref.
Plasma (1)	I-2	15 × 4	μ-Bondapak-C_{18} (10)	E-1	2.0	Fl (300, 370)	6	—	b	2
Plasma (1)	I-3	15 × 4.6	Zorbac TMS (5)[c]	E-2	2.0	ABS (308)	5	N-(2-Piperidylmethyl)-2,5-diethoxybenzamide (7)	—	3
Plasma (1)	I-4	30 × 3.9	μ-Bondapak phenyl (10)	E-3	2.0	Fl (300, 370)	5.6	Positional isomer[a] (5)	—	4
Plasma, urine, dialysate (1)	I-5	30 × 4.6[d]	μ-Bondapak phenyl (10)	E-4	2.0	Fl (230, 340)	—	—	*meta*-O-Dealkylated flecainide (4)	5
Plasma (1)	I-4	10 × 8	Radial-Pak C_{18} (5)[e]	E-5	0.7	Fl (293, 340)	7.2	Positional isomer[a] (8.3)	—	6

[a] N-(2-Piperidylmethyl)-2,3-*bis*(2,2,2-trifluoroethoxy)benzamide.
[b] Retention times of a number of cardiac drugs are given to check possible interferences.
[c] Protected by a 2-cm guard column packed with LiChrosorb RP-2 (10 μm).
[d] This system is used for the analysis of plasma samples. Alternative systems are described for the analysis of other fluids.
[e] Protected by a guard column packed with Corasil C_{18} (37 to 50 μm).

Extraction — I-1. The sample was mixed with 0.5 mℓ of water, 0.5 mℓ of an aqueous solution of the internal standard (0.2 μg/mℓ), 0.2 mℓ of 0.2 *M* trimethylamine in benzene, and 1 mℓ of 1 *M* NaOH. The mixture was extracted with 10 mℓ of diethyl ether. The ether layer was back extracted into 2 mℓ of 0.5 *M* HCl. Pentafluorobenzoyl chloride (2 μℓ) was added to the aqueous extract followed by the addition of 1 mℓ of 1 *M* sodium hydroxide. The reaction mixture was vortexed for 15 sec and extracted with 1 mℓ of hexane. The hexane layer was washed with 3 mℓ of 0.1 *M* sodium hydroxide. Aliquots of 5 μℓ of the hexane layer were injected.

I-2. The sample was treated with an equal volume of 0.8 *M* perchloric acid at 0°C. An aliquot of the supernatant was adjusted to pH 5.7 with 6 *M* KOH-1 *M* K_2CO_3 solution. After centrifugation a 200 μℓ aliquot of the supernatant was injected.
I-3. The sample was mixed with 0.1 mℓ of an aqueous solution of the internal standard (5 μg/ℓ), 0.5 mℓ of water and 2 mℓ of 0.1 *M* carbonate buffer. The mixture was applied to a prewashed (twice with methanol, once with water) BondElut 1-mℓC_{18} column. The column was washed once with carbonate buffer, once with water, twice with aqueous methanol, and twice with aqueous acetonitrile. The column was eluted with 0.5 mℓ of methanol. The eluate was evaporated at 60°C under nitrogen. The residue was dissolved in 0.25 mℓ of the mobile phase and 100 μℓ of the solution were injected.
I-4. The sample, 0.5 mℓ of water, 0.5 mℓ of an aqueous solution of the internal standard (10 μg/mℓ, and 0.2 mℓ of 0.2 *M* Na_1CO_3 were applied to a prewashed (twice with methanol, twice with water) Baker 3-mℓ C_8 column. The column was washed twice with water and two 1-mℓ volumes of acetonitrile. The column was eluted with 0.5 mℓ of methanol. An aliquot of 25 to 50 μℓ of the eluate was injected.
I-5. The extraction procedures is the same as described in I-4 except that no internal standard is used.

Elution — E-1. Methanol-50 m*M* $NH_4H_2PO_4$, pH 3 (60:40).
E-2. Acetonitrile-water-acetic acid (450:544.5:4.5) containing 0.1 *M* pentane-1-sulfonic acid.
E-3. Acetonitrile-0.06% phosphoric acid (40:60).
E-4. Acetonitrile-0.06% phosphoric acid (24:76).
E-5. Methanol-25% ammonia (99.9:0.1).

REFERENCE

1. **Johnson, J. D., Carlson, G. L., Fox, J. M., Miller, A. M., Chang, S. F., and Conard, G. J.,** Quantitation of flecainide acetate, a new antiarrhythmic agent, in biological fluids by gas chromatography with electron-capture detection, *J. Pharm. Sci.*, 73, 1469, 1984.
2. **DeJong, J. W., Hegge, J. A. J., Harmsen, E., and DeTombe, P. Ph.,** Fluorometric liquid chromatographic assay of the antiarrhythmic agent flecainide in blood plasma, *J. Chromatogr.*, 229, 498, 1982.
3. **Becker, J. U.,** Bestimmung der Konzentration eines neuen Antiarrhythmicums, Flecainid, im Plasma durch Hochleistungsflussigkeitschromatographie (HPLC): Probenvorbereitung Durch Extraktionssaulen, *J. Clin. Chem. Clin. Biochem.*, 22, 389, 1984.
4. **Chang, S. F., Miller, A. M., Fox, J. M., and Welscher, T. M.,** Application of a bonded-phase extraction column for rapid sample preparation of flecainide from human plasma for high-performance liquid chromatographic analysis-fluorescence or ultraviolet detection, *Ther. Drug Monit.*, 6, 105, 1984.
5. **Chang, S. F., Welscher, T. M., Miller, A. M., McQuinn, R. L., and Fox, J. M.,** High-performance liquid chromatographic method for the quantitation of a *meta*-O-dealkylated metabolite of flecainide acetate, a new antiarrhythmic, *J. Chromatogr.*, 343, 119, 1985.
6. **Plomp, T. A., Boom, H. T., and Maes, R. A. A.,** Measurement of flecainide plasma concentrations by high performance liquid chromatography with fluorescence detection, *J. Anal. Toxicol.*, 10, 102, 1986.

FLESTOLOL

Liquid Chromatography

Specimen (mℓ)	Extraction	Column (cm × mm)	Packing (μm)	Elution	Flow (mℓ/min)	Det. (nm)	RT (min)	Internal standard (RT)	Other compounds (RT)	Ref.
Blood (1)	I-1	30 × 3.9	μ-Bondapak-C_{18} (10)	E-1	2.0	ABS (229)	3.9	*p*-Ethoxy-phenyl-ethyl alcohol (16.9)	2-Fluorobenzoic acid (5.2) 2-Fluorohippuric acid (3.2)	1

Extraction — I-1. The sample was mixed with 7 mℓ of acetonitrile-dichloromethane (2:5) containing 1.9 μg/mℓ of the internal standard. After mixing and centrifugation the upper aqueous layer was discarded. The organic layer was extracted with 250 μℓ of 0.05 *M* phosphate buffer, pH 3.4. After centrifugation, a 200-μℓ aliquot of the top aqueous layer was injected.

Elution — E-1. Acetonitrile-0.05 *M* phosphate buffer, pH 3 to 4 (65:300).

REFERENCE

1. **Moore, P., Mai, K., and Lai, C. M.,** Quantitation of the ultra short acting β-adrenergic antagonist flestolol in blood by liquid chromatography, *J. Pharm. Sci.*, 75, 424, 1986.

FLORIDIPINE

Liquid Chromatography

Specimen (mℓ)	Extraction	Column (cm × mm)	Packing (μm)	Elution	Flow (mℓ/min)	Det. (nm)	RT (min)	Internal standard (RT)	Other compounds (RT)	Ref.
Plasma (1)	I-1	25 × 4.6	Zorbax ODS (5)[a]	E-1	1.5	ABS (238)	9.1	—	Metabolite[b] (7.6)	1

[a] Column temp = 55°C.
[b] Unidentified metabolite.

Extraction — I-1. The sample was mixed with 100 μℓ of distilled triethylamine and extracted with 10 mℓ of hexane. The organic layer was evaporated with a stream of nitrogen. The residue was reconstituted with 75 μℓ of mobile phase and a 50-gmℓ aliquot was injected.

Elution — E-1. Acetonitrile-0.005 *M* potassium phosphate buffer, pH 7 (60:40).

REFERENCE

1. **Rosenberg, M. and Choi, R. L.,** High-performance liquid chromatographic analysis of flordipine in human plasma, *J. Chromatogr.*, 308, 382, 1984.

FLUCLOXACILLIN

Liquid Chromatography

Specimen (mℓ)	Extraction	Column (cm × mm)	Packing (μm)	Elution	Flow (mℓ/min)	Det. (nm)	RT (min)	Internal standard (RT)	Other compounds (RT)	Ref.
Urine	I-1	25 × 4.6	LiChrosorb RP-18 (NA)[a]	E-1	3.0	ABS (254)	40	—	Penicilloic acid of 5-hydroxymethyl derivative of flucloxaccilin (11) Penicilloic acid of flucloxacillin (17) 5-Hydroxymethyl derivative of flucloxacillin (20)	1

[a] Protected by a 50 × 4.6 mm guard column packed with RP-2.

Extraction — I-1. Aliquots of urine samples were filtered through 0.45-μm pore size membrane filters and 5-μℓ portions of the filtrate were injected.

Elution — E-1. Acetonitrile-5 m*M* tetrabutylammonium bromide + 1/120 *M* Na_2HPO_4 + 1/120 *M* KH_2PO_4 (1:3), pH 7.48.

REFERENCE

1. **Murai, Y., Nakagawa, T., Yamaoka, K., and Uno, T.,** High-performance liquid chromatographic determination and moment analysis of urinary excretion of flucloxacillin and its metabolites in man, *Int. J. Pharm.*, 15, 309, 1983.

FLUCYTOSINE

Liquid Chromatography

Specimen (mℓ)	Extraction	Column (cm × mm)	Packing (μm)	Elution	Flow (mℓ/min)	Det. (nm)	RT (min)	Internal standard (RT)	Other compounds (RT)	Ref.
Plasma (1)	I-1	25 × 4.6	Spherisorb ODS (10)[a]	E-1	1.0	ABS (280)	5	—	—	1
Plasma CSF (0.1)	I-2	30 × 3.9	μ-BondapakC_{18} (10)	E-2	1.5	ABS (276)	4	5-Methyl-cytosine (7)	Salicylic acid (8)	2
Serum (0.5)	I-3	10 × 8	Radial-Pak A (10)	E-3	1.0	ABS (254)	4.5	—	—	3
Serum (0.5)	I-4	15 × 4.6	LiChrosorb RP-18 (5)[b]	E-4	1.1	ABS (280)	9.5	*p*-Amino-benzoic acid (11.5)	Cefoxitin (15)	4
Plasma (1)	I-5	30 × 3.9	μ-Bondapak-C_{18} (10)	E-5	1.0	ABS (280)	5.4	5-Fluorou-racid (6.5)	—	5
Plasma (0.1)	I-6	15 × 4.6	Speralyte SCX (NA)	E-6	1.0	ABS (254)	4	5-Iodocyto-sine (6.2)	—	6
Serum (0.01)	I-7	25 × 4.1	Hamilton PRP-1 (10)	E-7	0.8	Fl (300, 370)	4.5	—	—	7

[a] Protected by a 50 × 4.6 mm guard column packed with reversed phase packing.
[b] Protected by a 40 × 4.6mm guard column packed with Corasil C_{18} (37 to 50 μm).

Extraction — I-1. The sample was treated with 50 μℓ of 4 *M* trichloroacetic acid. After mixing and centrifugation at 0°C, the supernatant was diluted 1:4 with water and 100-μℓ aliquots were injected.
I-2. The sample was mixed with an equal volume of 100 mg/ℓ trichloroacetic acid solution containing 100 mg/ℓ of the internal standard. After mixing and centrifugation, aliquots of 10 μℓ of the supernate were injected.
I-3. The sample was mixed with an equal volume of acetonitrile. After mixing and centrifugation aliquots of 10 μℓ of the supernate were injected.
I-4. The sample was diluted with an equal volume of water and filtered through Amicon Centriflo CF 25 membrane. The filtrate was diluted 1:2 with water containing 5 μg/mℓ of the internal standard. An aliquot of 50 μℓ of this solution was injected.

I-5. The sample was mixed with 50 μℓ of an aqueous solution of the internal standard (5 μg/mℓ) and 1 mℓ of pH 7 phosphate buffer. The mixture was extracted with 6 mℓ of ethyl acetate. The organic layer was evaporated under a stream of nitrogen 40°C. The residue was dissolved in 200 μℓ of the mobile phase and an aliquot of 20 μℓ was injected.
I-6. The sample was mixed with 100 μℓ of an aqueous solution of the internal standard (500 μg/mℓ) and 100 μℓ of 10% trichloroacetic acid. A 100-μℓ aliquot of the supernatant was diluted with 100 μℓ of 0.1 *M* ammonium phosphate. A 10-μℓ aliquot of the resulting solution was injected.
I-7. The sample was diluted with 1 mℓ of the mobile phase and an aliquot of 50 μℓ was injected.

Elution — E-1. Methanol-0.025 *M* KH_2PO_4, pH 2.5 (5:95).
E-2. 10 m*M* KH_2PO_4, pH 7.
E-3. Methanol-water (40:60) containing one vial of PIC B7 reagent/ℓ.
E-4. Acetonitrile-water-acetic acid (7:88:5) containing 15 m*M* octane sulfonic acid.
E-5. Methanol-5 m*M* acetate buffer, pH 4.8 (1:99).
E-6. Acetonitrile-methanol-water (30:90:90) containing 40 μℓ of phosphoric acid and 0.23 g of ammonium phosphate.
E-7. 0.09 *M* Sodium carbonate.

REFERENCES

1. **Bury, R. W., Mashford, M. L., and Miles, H. M.,** Assay of flucytosine (5-fluorocytosine) in human plasma by high-pressure liquid chromatography, *Antimicrob. Agents Chemother.*, 16, 529, 1979.
2. **Miners, J. O., Foenander, T., and Birkett, D. J.,** Liquid-chromatographic determination of 5-fluoro-cytosine, *Clin. Chem.*, 26, 117, 1980.
3. **Warnock, D. W. and Turner, A.,** High performance liquid chromatographic determination of 5-fluoro-cytosine in human serum, *J. Antimicrob Chemother.*, 7, 363, 1981.
4. **Essers, L., Hantschke, D., and Leisse, K. H.,** Serumspiegeluberwachung von 5-fluorocytosin durch hochleistungsflussigkeits-chromatographie, *Mykosen*, 25, 183, 1982.
5. **Bouquet, S., Quehen, S., Brisson, A. M., Courtois, Ph., and Fourtillan, J. B.,** High performance liquid chromatographic determination of 5-fluorocytosine in human plasma, *J. Liq. Chromatogr.*, 7, 743, 1984.
6. **Schwertschlag, U., Nakata, L. M., and Gal, J.,** Improved procedure for determination of flucytosine in human blood plasma by high-pressure liquid chromatography, *Antimicrob. Agents Chemother.*, 26, 303, 1984.
7. **Lacroix, C., Levert, P., Laine, G., Goulle, J. P., and Gringore, A.,** Microdosage de la 5-fluorocytosine par chromatographie en phase liquide et detection fluorimetrique, *J. Chromatogr.*, 345, 436, 1985.

FLUDALANINE

Liquid Chromatography

Specimen (mℓ)	Extraction	Column (cm × mm)	Packing (μm)	Elution	Flow (mℓ/min)	Det. (nm)	RT (min)	Internal standard (RT)	Other compounds (RT)	Ref.
Plasma, urine (1)	I-1	10 × 8	Radial-Pak C_{18} (5)[a]	E-1	2.0	Fl (340, 455)	7	3,3-Difluoroalanine (4)	—	1

[a] Protected by a guard column packed with LiChrosorb C_{18} (10 μm).

Extraction — I-1. The sample was mixed with 50 μℓ of an aqueous solution of the internal standard (0.9 mg/mℓ) and filtered through a centriflo ultrafilter (CF 50A from Amicon). Aliquots of the filtrate were injected. The eluate of the analytical column was mixed with the *o*-phthalaldehyde-2-mercaptoethanol reagent in borate buffer and passed through a 25 cm × 4.6 mm column packed with 40 μm glass beads maintained at 40°C.

Elution — E-1. Methanol-water (100:900) containing sodium dodecyl sulfate (50 mg) + 85% phosphoric acid (2 mℓ), pH 2.5 with 1 *M* KOH.

REFERENCE

1. **Musson, D. G., Maglietto, S. M., and Bayne, W. F.,** Determination of the antibiotic fludalanine in plasma and urine by high-performance liquid chromatography using a packed-bed, post-column reactor with *o*-phthalaldehyde and 2-mercaptoethanol, *J. Chromatogr.*, 338, 357, 1985.

FLUFENAMIC ACID

Liquid Chromatography

Specimen (mℓ)	Extraction	Column (cm × mm)	Packing (μm)	Elution	Flow (mℓ/min)	Det. (nm)	RT (min)	Internal standard (RT)	Other compounds (RT)	Ref.
Plasma (1)	I-1	30 × 4	μ-Bondapak CN (10)	E-1	1.0	ABS (254)	10.4	Mefenamic acid (9.2)	—	1

Extraction — I-1. The sample was spiked with 4 μg of the internal standard, acidified with 0.9 *M* sulfuric acid, and extracted with 8 mℓ of carbon tetrachloride. The organic layer was evaporated under nitrogen, the residue dissolved in 0.5 mℓ of methanol, and a 40-μℓ aliquot was injected.

Elution — E-1. Acetonitrile-water-acetic acid (30:60:10).

REFERENCE

1. **Lin, C. K., Lee, C. S., and Perrin, J. H.,** Determination of two fenamates in plasma by high-performance liquid chromatography, *J. Pharm. Sci.*, 69, 95, 1980.

FLUMECINOL

Gas Chromatography

Specimen (mℓ)	Extraction	Column (m × mm)	Packing (mesh)	Oven temp (°C)	Gas (mℓ/min)	Det.	RT (min)	Internal standard (RT)	Deriv.	Other compounds (RT)	Ref.
Plasma, saliva (1)	I-1	25 × 0.20	SP2100-Carbowax 20M	160	N_2 (1.6)	FID	3.1	3-Trifluoromethyl-benzhydrol (4.1)	—	—	1, 2

Extraction — I-1. The sample was spiked with 600 ng of the internal standard, mixed with 0.6 mℓ of 2 *N* KOH, and extracted with 4 mℓ of diethyl ether. The ether layer was evaporated at room temperature. The dry residue was dissolved in 20 μℓ of chloroform and 1 to 2 μℓ was injected.

REFERENCE

1. **Klebovich, I. and Vereczkey, L.,** Gas chromatographic method for the determination of flumecinol in biological fluids, *J. Chromatogr.*, 221, 403, 1980.
2. **Klebovich, I., Kapas, M., and Vereczkey, L.,** Capillary gas chromatographic method for determination of flumecinol in plasma and saliva, *J. Chromatogr.*, 273, 207, 1983.

FLUNARIZINE

Gas Chromatography

Specimen (mℓ)	Extraction	Column (m × mm)	Packing (mesh)	Oven temp (°C)	Gas (mℓ/min)	Det.	RT (min)	Internal standard (RT)	Deriv.	Other compounds (RT)	Ref.
Plasma, urine (1)	I-1	1.8 × 2	3% OV-17 GasChrom Q (80/100)	300	He (30)	NPD	2.5	Cinnarizine (3)	—	—	1

Extraction — I-1. The samples was mixed with 20 μℓ of a methanolic solution of the internal standard (9 μg/mℓ), 0.5 mℓ of concentrated ammonium hydroxide and 1 mℓ of 1 *M* carbonate buffer, pH 10.8. The mixture was extracted twice with 4-mℓ portions of hexane-ethyl acetate (3:1). The combined organic layers were back extracted into 6 mℓ of 1 *N* HCl. The aqueous layer was made alkaline with 1 mℓ of 10 *M* sodium hydroxide and extracted twice with the above extraction solvent. The combined organic layers were evaporated, the residue reconstituted in 100 μℓ of toluene, and 2 to 5 μℓ were injected.

REFERENCE

1. **Flor, S. C.,** Determination of the calcium antagonist flunarizine in biological fluids by gas-liquid chromatography, *J. Chromatogr.*, 272, 315, 1983.

FLUNITRAZEPAM

Gas Chromatography

Specimen (mℓ)	Extraction	Column (m × mm)	Packing (mesh)	Oven temp (°C)	Gas (mℓ/min)	Det.	RT (min)	Internal standard (RT)	Deriv.	Other compounds (RT)	Ref.
Plasma (0.2 — 2)	I-1	2.5 × 4	3% OV-101 Gas Chrom Q (80/100)	262	Ar:90-Methane: 10 (60)	ECD	7.9	Methyclonazepam (11.4)	—	—	1
Plasma (0.1 — 1)	I-2	1.8 × 2	3% SP-2250 Supelcoport (80/100)	265	Ar:95-Methane: 5 (30)	ECD	8	Methylnitrazepam (10)	—	Desmethylfluni-trazepam (11.5)	2
Plasma (1)	I-3	10 × 0.4	3% OV-17 Tutlanox (10 μm)	215	He (10)	ECD	2.2	Nordiazepam (1.7)	—	a	3

FLUNITRAZEPAM (continued)

Liquid Chromatography

Specimen (mℓ)	Extraction	Column (cm × mm)	Packing (μm)	Elution	Flow (mℓ/min)	Det. (nm)	RT (min)	Internal standard (RT)	Other compounds (RT)	Ref.
Plasma (0.5 — 4)	I-4	30 × 4	μ-Bondapak-C_{18} (10)	E-1	2.5	Fl (390,470)	—	7-Amino-methyl clona-zepam (14.6)	7-Aminodesmethyl flunitrazepam (4.5) 7-Aminoflunitraze-pam (9.2) 7-Aminoclonaze-pam (6)	4
Urine (0.5)	I-5	10 × 4.6	Hypersil ODS (5)	E-2	NA	Fl (396, 445)	—	—	Acridine derivative of 7-aminofluni-trazepam (3) Acidine derivative of 7-aminodesme-thyl flunitrazepam (2)	5,6

a. Conditions for the determination of desalkylflurazepam and temazepam are also described.

Extraction — I-1. The sample was mixed 5 to 20 ng of the internal standard and 2 mℓ of pH 9 borate buffer. The mixture was extracted with 10 mℓ of hexane. A 9-mℓ aliquot of hexane layer was back extracted into 3 mℓ of 2 *N* sulfuric acid. The aqueous layer was adjusted to pH 2 with 1 *M* K_2HPO_4 and extracted twice with 8 mℓ of toluene-hexane (1:4). The combined organic extracts were evaporated to dryness at 40°C under a stream of nitrogen. The residue was dissolved in 100 μℓ of acetone-hexane (1:4). Aliquots of 2 to 3 μℓ of this solution were injected.

I-2. The sample was mixed with the residue after evaporation of 25 μℓ of the internal standard solution (1 μg/mℓ in benzene) and extracted with 2 mℓ of benzene. The organic layer was evaporated, the residue dissolved in 150 μℓ of toluene (containing 15% isoamyl alcohol). Aliquots, 6 μℓ, of this solution were injected with an auto sampler.

I-3. The sample was mixed with 50 μℓ of ethanol containin the internal standard and 1 mℓ of 0.2 *M* borate buffer (pH 9). The mixture was extracted twice with 5 mℓ of pentane-dichloromethane (1:1). The combined extracts were evaporated at 50°C under a stream of nitrogen. The residue was dissolved in 40 μℓ of ethyl acetate and 2 to 3 μℓ of this solution were applied to the needle of solid injection system.

I-4. The sample was mixed with 50 μℓ of the internal standard solution (0.01 to 10 μg/mℓ in acetone-hexane, 1:4) and 2 mℓ of pH 10 buffer. The mixture was extracted with 10 mℓ of diethyl ether containing 1% isoamyl alcohol. The organic layer was back extracted into 2 mℓ of 0.5 *M* sulfuric acid. The aqueous layer was adjusted to pH 9 to 10 with 1 *M* sodium hydroxide solution containing 1 *M* dipotassium hydrogen phosphate and extracted with 10 mℓ of diethyl ether. The organic phase was evaporated at 45°C under vacuum. To the residue, 100 μℓ of the mobile phase and 20 μℓ of 0.5% fluorescamine solution in acetone were added. After mixing, aliquots of this solution were injected.

I-5. The samples was mixed with an equal volume of 10 *M* HCl. The solution was heated for 15 min at 100°C. After washing, the pH was adjusted to 9 with 10 *M* sodium hydroxide at 0.05 *M* borate buffer, pH 9 and extracted twice with 1.2-mℓ portions of ethyl acetate. The combined organic extracts were evapoated on a water bath at 50°C under nitrogen. The residue was dissolved in 0.5 mℓ of a saturated solution of sodium nitrite in dimethylformamide. The vials were closed with aluminum caps and heated at 180°C for 2 hr. After cooling, the mixture was diluted with 0.5 mℓ of water and aliquots of 20 μℓ of this solution were injected.

Elution — E-1. Acetonitrile-20 m*M* phosphate buffer, pH 8 (25:75).
E-2. Methanol-water (55:45) containing 0.05 *M* acetate buffer, pH 4.7 + 0.05 m*M* tetramethylammonium hydroxide.

REFERENCES

1. **Sumirtapura, Y. C., Aubert, C., and Cano, J. P.,** Highly specific and sensitive method for the determination of flunitrazepam in plasma by electron capture gas-liquid chromatography, *Arzneim. Forsch.*, 32, 252, 1982.
2. **Greenblatt, D. J., Ochs, H. ., Locniskar, A., and Lauven, P. M.,** Automated electron-capture gas chromatographic analysis of flunitrazepam in plasma, *Pharmacology,* 2, 82, 1982.
3. **Jochemsen, R. and Breimer, D. D.,** Assay of flunitrazepam, temazepam and desalkylflurazepam in plasma by capillary gas chromatography with electron-capture detection, *J. Chromatogr.*, 227, 199, 1982.
4. **Sumirtapura, Y.C., Aubert, C., Coassolo, P., and Cano, J. P.,** Determination of 7-amino-flunitrazepam (Ro 20-1815) and 7-amino-desmethylflunitrazepam (Ro 5-4650) in plasma by high-performance liquid chromatography and fluorescence detection, *J. Chromatogr.*, 232, 111, 1982.
5. **Weijers-Everhard, J.P., Wijker, J., Verrijk, R., Van Rooij, H. H., and Soudijn, W.,** Improved qualitative method for establishing flunitrazepam abuse using urine samples and column liquid chromatography with fluorimetric detection, *J. Chromatogr.*, 374, 339, 1986.
6. **Van Rooij, H. H., Fakiera, A., Verrijk, R., and Soudijn, W.,** The identification of flunitrazepam and its metabolites in urine samples, *Anal. Chim. Acta,* 170, 153, 1985.

FLUNIXIN

Liquid Chromatography

Specimen (mℓ)	Extraction	Column (cm × mm)	Packing (μm)	Elution	Flow (mℓ/min)	Det. (nm)	RT (min)	Internal standard (RT)	Other compounds (RT)	Ref.
Plasma, serum (1)	I-1	25 × 4.6	Spherisorb ODS1 (5)[a]	E-1	1.2	ABS (254)	9	Naproxen (7.7)	γ-Hydroxyphenyl-butazone (3.5) Oxyphenbutazone (5.8) Phenylbutazone (14.2)	1

FLUNIXIN (continued)

Liquid Chromatography

Specimen (mℓ)	Extraction	Column (cm × mm)	Packing (μm)	Elution	Flow (mℓ/min)	Det. (nm)	RT (min)	Internal standard (RT)	Other compounds (RT)	Ref.
Plasma (0.1)	I-2	15 × 4.5	LiChrosorb RP-18 (10)[b]	E-2	0.8	ABS (284)	6	—	—	2
Plasma (1)	I-3	10 × 8	Radial-Pak-C_{18} (10)[c]	E-3	2.0	ABS (280)	3.2	Sch 13476 (4.8)	—	3

[a] Protected by a 5-cm guard column packed with (30 μm) pellicular ODS packing.
[b] Protected by a 50 × 3 mm guard column packed with the same materials as of the analytical column.
[c] Protected by a guard column packed with Co:Pell ODS (40 μm).

Extraction — I-1. The samples was mixed with 4 mℓ of acetonitrile containing 250 ng/mℓ of the internal standard. An aliquot of the supernatant (4 mℓ) was evaporated under a streamof nitrogen at 37°C. The residue was dissolved in 500 μℓ of the mobile phase and aliquots of 50 μℓ of this solution were injected.
I-2. The sample was treated with 200 μℓ of methanol. After mixing and centrifugation aliquots of 50 μℓ of the supernatant were injected.
I-3. The plasma sample was mixed with 10 μℓ of the internal standard solution (6 μg/mℓ), 1 mℓ of 1 *M* HCl and 3 mℓ of water. The mixture was applied to a ClinElut column. The column was eluted with 20 to 8-mℓ aliquots of dichloromethane. The combined eluates were evaporated at 50°C unde a steam of nitrogen, the residue dissolved in 200 μℓ of methanol, and aliquots of 50 μℓ were injected.

Elution — E-1. Acetonitrile-methanol-1% acetate buffer, pH 3 (30:20:50).
E-2. Methanol-phosphate buffer,pH 3.1 (7:3).
E-3. Acetonitrile-0.025 *M* phosphate buffer, pH 2.5 (50:50).

REFERENCES

1. **Hardee, G. E., Lai, J. W., and Moore, J. N.,** Simultaneous determination of flunixin, phenlybutazone, oxyphenbutazone and γ-hydroxyphenylbutazone in equine plasma by high-performance liquid chromatography: with application to pharmacokinetics, *J. Liq. Chromatogr.*, 5, 1991, 1982.
2. **Johansson, I. M. and Schubert, B.,** Determination of flunixin in equine plasma by reversed-phase liquid chromatography, *J. Pharm. Biomed. Anal.*, 2, 501, 1984.
3. **Neff-Davis, C. A. and Bosch, K.,** An HPLC method for the determination of flunixin in bovine plasma and milk, *J. Vet. Pharmacol. Ther.*, 8, 331, 1985.

1-(2-FLUORO-2-DEOXYARABINOFLURANOSYL)-5-IODOCYTOSINE

Liquid Chromatography

Specimen (mℓ)	Extraction	Column (cm × mm)	Packing (μm)	Elution	Flow (mℓ/min)	Det. (nm)	RT (min)	Internal standard (RT)	Other compounds (RT)	Ref.
Pure compounds	—	25 × 4.6	Partisil ODS-1 (10)[a,b]	E-1; grad	1.0	ABS (254)	18.5	—	R[c]-Cytosine (5.8) R-Uracil (7.5) R-5-Methylcytosine (9) R-5-Methyluracil (14.5) R-5-Iodouracil (20)	1

[a] Protected by a guard column packed with Co:Pell ODS.
[b] Column temp = 40°C.
[c] R = 10(2-Fluor-2-deoxy-β-D-arabinosyl).

Elution — E-1. (A) 0.01 *m* Phosphate buffer, pH 5.3; (B) methanol. Isocratic 96% (A) from 0 to 4 min, then isocratic 80% (A).

REFERENCE

1. **Feinberg, A.,** Separation of 2-fluoro-2-deoxyarabinofuranosylpyrimidine nucleosides by high-performance liquid chromatography, *J. Chromatogr.*, 210, 527, 1981.

FLUORESCEIN

Liquid Chromatography

Specimen (mℓ)	Extraction	Column (cm × mm)	Packing (μm)	Elution	Flow (mℓ/min)	Det. (nm)	RT (min)	Internal standard (RT)	Other compounds (RT)	Ref.
Serum (0.5)	I-1	15 × 3.9	Nova-Pak C_{18} (5)	E-1	1.0	Fl (450, 500)	3	—	Fluorescein monoglucuronide	1

Extraction — I-1. The samples was filtered through Amicon YMT membrane filter. An aliquot of 100 μℓ of the ultrafiltrate was injected.

Elution — E-1. Methanol-0.005 *M* tetrabutylammonium phosphate (47:53).

REFERENCE

1. **Selan, F., Blair, N., and Evans, M. A.,** High-performance liquid chromatographic analysis for fluorescein and fluorescein monoglucuronide in plasma, *J. Chromatogr.*, 338, 213, 1985.

5-FLUORO-2′-DEOXYURIDINE

Gas Chromatography

Specimen (mℓ)	Extraction	Column (m × mm)	Packing (mesh)	Oven temp (°C)	Gas (mℓ/min)	Det.	RT (min)	Internal standard (RT)	Deriv.	Other compounds (RT)	Ref.
Urine (1)	I-1	12 × 0.25	OV-17	195	He:9, H_2:1 (3)	NPD	6.9	5-Chloro-2′-deoxyuridine (12.5)	Methyl	—	1, 2
Plasma, urine (0.72)	I-2	1.8 × 2	3% SP-2100[a] Supelcoport (100/200)	205	He (30)	NPD	6.7	5-Chloro-2′-deoxyuridine (11.3)		5-Fluorouridine (7.9) 5-Bromo-2′-deoxy uridine (14.4) 2′-Deoxyuridine (8.3) Uridine (9.5)	3

Liquid Chromatography

Specimen (mℓ)	Extraction	Column (cm × mm)	Packing (μm)	Elution	Flow (mℓ/min)	Det. (nm)	RT (min)	Internal standard (RT)	Other compounds (RT)	Ref.
Plasma (2 — 10)	I-3	30 × 7.8	μ-Bondpak-C_{18} (10)	E-1	b	ABS (280)	27	5[^{3}H]-Fluorodeoxyuridine	5-Fluorouracil (12.5)	4

[a] Alternative column packings were also used.
[b] Flow programming.

Extraction — I-1. The sample was treated with 100 μℓ of an aqueous solution of the internal standard (20 μg/mℓ) and 1 mℓ of 0.1 *M* barium hydroxide. After centrifugation, the supernatant was applied to a 20 × 6 mm column of AG 1X-4 (Cl) resin. The column was washed with 10 mℓ of water and 10 mℓ of methanol and eluted with 10 mℓ of 0.3 *M* acetic acid in methanol. The eluate was evaporated at 50°C under a stream of nitrogen. The residue was dissolved in 200 μℓ of dichloromethane-methanol (93:7) and applied to a 60 × 6 mm column of Sephadex LH-20. The column was washed with 10 mℓ of dichloromethane-methanol (97:3) and eluted with 10 mℓ of dichloromethane-methanol (94:6). The eluate was evaporated under a stream of nitrogen. The residue was dissolved in 200 μℓ of dry dimethyl sulfoxide, 50 μℓ of potassium *tert*-butoxide reagent and after 10 sec, 100 μℓ of iodomethane were added. The mixture was allowed to stand for 1 hr diluted with 5 mℓ of 0.05 *N* sulfuric acid and extracted twice with 5-mℓ portions of cyclohexane-dichloromethane (9:1). The combined extracts were evaporated under a stream of nitrogen. The residue was dissolved in 50 μℓ of ethyl acetate and an aliquot of 1 μℓ was applied to the falling needle injection system.
I-2. The sample was mixed with 80 μℓ of the internal standard solution dissolved in 0.3 *M* ammonium formate buffer, pH 5 and applied to a cation exchange column (Bio-Rad AG 50W-X4, 200 to 400 mesh, H^+ form, 4.5 × 0.7 cm). The column was eluted with 0.03 *m* ammonium formate buffer, pH 5. After the initial 0.8 mℓ of the eluate was discarded,a 4-mℓ fraction was collected. This was combined with 1 mℓ of 0.5 *m* carbonate-bicarbonate buffer, pH 10.7 and applied to an anion exchange column (Bio-Rad AG 1-X4, 100 to 200 mesh, Cl form, 2 × 0.7cm). The column was washed sequentially with 2 mℓ of 0.1 *M* carbonate-bicarbonate buffer, pH 10, 10 mℓ of water, 10 mℓ of methanol, and then eluted with 10 mℓ of 0.3 *m* acetic acid in methanol. The eluate was evaporated under nitrogen at 50°C. The residue was dissolved in 100 μℓ of 0.025 *M* phenyltrimethylammonium hydroxide in methanol under nitrogen at 50°C. A 0.5 - to 1-μℓ aliquot was injected.
I-3. The sample was spiked with the tritiated internal standard (10,000 dpm), brought to pH 10 with dropwise addition of 5 *N* KOH and passed through an anion exchange column (12 × 0.8cm, Dowex AG 1-X2, 200 to 400 mesh, acetate form). The column was washed with 30 mℓ of water and 4 mℓ of 1 *N* acetic acid and then eluted with another 4 mℓ aliquot of 1 *N* acetic acid. The eluate was lyophilized and the residue dissolved in 500 μℓ of 1 M KH_2PO_4 buffer, pH 6.8. This solution was extracted with 7 mℓ of ethyl acetate. The organic layer was evaporated and the residue dissolved in 100 μℓ of the mobile phase for injection. After chromatography, fractions containing 5-fluorodeoxy-uridine or 5-fluorouracil are collected to determine radioactivity for monitoring extraction recovery.

Elution — E-1. 0.01 M KH_2PO_4 pH 4.

5-FLUORO-2′-DEOXYURIDINE (continued)

REFERENCES

1. **Gelijkens, C. F. and De Leenheer, A. P.,** Gas chromatographic measurement of urinary 5-fluoro-2′-deoxyuridine levels after barium salt precipitation and sephadex LH-20 cleanup, *Anal. Biochem.*, 105, 106, 1980.
2. **Gelijkens, C. F., De Leenheer, A. P., and Sandra, P.,** Measurement of 5-fluoro-2′-deoxyuridine serum levels by gas chromatography chemical ionization mass spectrometry, *Biomed. Mass Spectrom.*, 7, 572, 1980.
3. **Williams, W. M., Warren, B. S., and Lin, F. H.,** Gas-liquid chromatographic analysis of fluoropyrimidine nucleosides and fluorouracil in plasma and urine, *Anal. Biochem.*, 147, 478, 1985.
4. **Buckpitt, A. R. and Boyd, M. R.,** A sensitive method for determination of 5-fluorouracil and 5-fluoro-2′-deoxyuridine in human by high-pressure liquid chromatography, *Anal. Biochem.*, 106, 432, 1980.

α-FLUOROMETHYLHISTIDINE

Liquid Chromatography

Specimen (mℓ)	Extraction	Column (cm × mm)	Packing (μm)	Elution	Flow (mℓ/min)	Det. (nm)	RT (min)	Internal standard (RT)	Other compounds (RT)	Ref.
Plasma (1)	I-1	10 × 8	Nova-Pak-C_{18} (5)[a]	E-1	2.0	ABS[b] (214)	6.8	α-Hydroxy-methyl histidine (10.4)	—	1

[a] Protected by a Guard-Pak C_{18} cartridge packed with the same material as in the analytical column.
[b] For the analysis of urine a fluorescence detector (ex = 350 nm, em = 450 nm) was used coupled with a postcolumn reactor using o-phthalaldehyde reagent.

Extraction — I-1. The sample was mixed with 50 μℓ of an aqueous solution of the internal standard (0.25 mg/mℓ) and the pH adjusted to 6.5 with 3 *M* HCl. The sample was applied to a BioRex 70 resin column (200 to 400 mesh) as H^+ form. The column was washed twice with 1-mℓ aliquots of water and eluted with 1 mℓ of 0.18 *M* sulfuric acid. An aliquot of 50 μℓ of the eluant was injected.

Elution — E-1. Isopropanol-water (9:91) containing 0.2 mℓ phosphoric acid and 0.002 *M* sodium 1-decanesulfonate.

REFERENCE

1. **August, T. F., Musson, D. G., Hwang, S. S., Duggan, D. E., Hooke, K. F., Roman, I. J., Ferguson, R. J., and Bayne, W. F.,** Bioanalysis and disposition of α-fluoromethylhistidine, a new histidine decarboxylase inhibitor, *J. Pharm. Sci.*, 74, 871, 1985.

5-FLUOROURACIL

Gas Chromatography

Specimen (mℓ)	Extraction	Column (m × mm)	Packing (mesh)	Oven temp (°C)	Gas (mℓ/min)	Det.	RT (min)	Internal standard (RT)	Deriv.	Other compounds (RT)	Ref.
Plasma (0.5)	I-1	1 × 2	3% OV-17 Chromosorb W (80/100)	T.P.	He (30)	MS-CI[b]	1.5	[1,3-$^{15}N_2$]-5-Fluorouracil [1,3-$^{15}N_2$] Uracil	Trimethyl-silyl	Uracil (1.7) Thymine (2.1) Cytosine (3.7)	1, 2, 3
Plasma (1)	I-2	1.8 × 2	3% OV-1 GasChrom Q (80/100)	190	He (30)	MS-EI	5.5	[4-^{18}O]-5-Fluorouracil	*n*-Butyl	—	4
Plasma (0.1 — 0.5)	I-3	1.2 × 2	3% Poly I-110 GasChrom Q (100/120)	170	Isobutane (30)	MS-CI	2	[1,3-$^{15}N_2$]-5-Fluorouracil	Methyl	—	5
Plasma (0.2)	I-4	7 × NA[c]	OV-275	215	He (12)	NPD	3	5-Chlorouracil (6)	—	5,6-Dihydrofluorouracil (1.5)[d]	6
Plasma (0.9)	I-5	25 × 0.22	CP Sil 19 (0.12 μm)	290		MS-CI	4.1	5-Chlorouracil	Penta-fluoro-benzyl (5.1)	5-Bromouracil (5.8)	7

Liquid Chromatography

Specimen (mℓ)	Extraction	Column (cm × mm)	Packing (μm)	Elution	Flow (mℓ/min)	Det. (nm)	RT (min)	Internal standard (RT)	Other compounds (RT)	Ref.
Pure compounds	—	30 × 4	μ-Bondapak-C_{18} (10)	E-1[e]	NA	ABS (254)	4	—	5-Fluororidine (6) 5-Fluorodeoxyuridine (10) 5-Fluorouridine monophosphate (15)	8

5-FLUOROURACIL (continued)

Liquid Chromatography

Specimen (mℓ)	Extraction	Column (cm × mm)	Packing (μm)	Elution	Flow (mℓ/min)	Det. (nm)	RT (min)	Internal standard (RT)	Other compounds (RT)	Ref.
									5′-Deoxy-5-fluorouridine (18) 5-Fluorodeoxyuridine monophosphate (24)	
Plasma (1)	I-6	30 × 4	μ-Bondapak-C_{18} (10)	E-2	0.8	ABS (266)	6.5	5-Bromouracil (12)	—	9
Plasma	I-7	15 × 4.1	Hamilton PRP-1 (10)	E-3	1.0	ABS (254)	6.8	—	Uracil (1.8) Uridine (1.8) Uric acid (4.5)	10
Serum (0.5)	I-8	20 × 4	Nucleosil-C_{18} (5)	E-4	0.8	Fl (346, 395)	6.5	—	Ftorafur (4.5)	11
Plasma (1)	I-9	25 × 4.6	Alltech-C_{18} (10)[f]	E-5	g	ABS (280)	5.4	—	—	12
Plasma (1)	I-10	25 × 4.6	Zorbax-C_8 (6)	E-6		ABS (254)	4.4	5-Chlorouracil (7.6)	—	13

[a] Initial temp = 150°C; rate = 10°C/min.
[b] Ammonia as the reagent gas.
[c] A SCOT column.
[d] Chromatographed at 195°C, using diphenylsuccinimide as the internal standard.
[e] The mobile phase is changed at 30 min to elute nonfluorinated nucleotides.
[f] Protected by a 3-cm pellicular guard column.
[g] Flow gradient; 1 mℓ/min for first 10 min; 3 mℓ/min from 10 to 35 min.

Extraction — I-1. The sample was diluted with 2 mℓ of water containing 0.1 μg of ^{15}N-F-Uracil and 0.5 μg of ^{15}N-Uracil as the internal standards, pH adjusted to 4 with 5 *n* HCl, and washed twice with 20 mℓ portions of chloroform. The aqueous layer was neutralized with sodium hydroxide solution and adjusted to pH 6 with 0.2 mℓ of 0.5 *M* NaH_2PO_4 and extracted with 40 mℓ of ethyl acetate. The organic layer was evaporated at 40°C under a stream of nitrogen. The residue was dried over phosphorus pentoxide and incubated with 100 μℓ of a 20% solution of N,O-*bis*(trimethylsilyl) trifluoroacetaide in pyridine at 70°C for 20 min. Aliquots of 1 to 3 μℓ of this reaction mixture were injected. The ethyl acetate extract of plasma obtained after the administration of ethyl-6-*n*-butoxy-5-fluoro-5,6-dihydrouracil-γ-carboxylate was purified by TLC prior to derivatization.
I-2. The sample was spiked with 75 pg of the internal standard, diluted with 4 mℓ of carbonate buffer, pH 10 and applied to a 20 × 6 mm column packed with AG 1-X4 (100 to 200 mesh) anion exchange resin prewashed with 10 mℓ of 0.3 *m* acetic acid in methanol followed with 10 mℓ of carbonate buffer, pH 10. After the

sample had passed through, the column was washed with 10 mℓ water and then with 10 mℓ of methanol. The column was eluted with 10 mℓ of 0.3 *M* acetic acid in methanol. The eluate was evaporated at 50°C under a stream of nitrogen. The residue was dissolved in 50 μℓ of N,N-dimethylacetamide, 15 μℓ of 2% tetramethyl ammonium hydroxide in methanol, and 20 μℓ of *n*-butyl iodide were added. The reaction mixture was allowed to stand at room temperature for 10 min, diluted with 1 mℓ of water, and extracted into 2 mℓ of cyclohexane-dichloromethane (95:5). The organic phase was evaporated at 40°C under a stream of nitrogen. The residue was dissolved in 100 μℓ of methanol and 1 μℓ was injected.

I-3. The sample was spiked with the internal standard, treated with an equal volume of saturated ammonium sulfate, and washed with benzene and extracted with 20% *n*-propanol in ether. The organic phase was evaported, the residue dissolved in methanol, and treated with etherealdiazomethane. The methylating solvents were removed, the residue dissolved in acetone and aliquots of this solution were injected.

I-4. The sample was spiked with 100 ng of the internal standard and washed twice with 3-mℓ portions of chloroform. The aqueous layer was then extracted twice with 3-mℓ portions of ethyl acetate. The combined organic layer was spiked with 250 ng of diphenylsuccinimide and evaporated at room temperature under a stream of nitrogen. The residue was dissolved in 100 μℓ of ethyl acetate and aliquots of 10 μℓ were brought onto the needle of the solid sample injection system.

I-5. The sample was mixed with 30 μℓ of the internal standard solution (final concentration 10^{-7} *M*) and 0.1 mℓ of a 2 *M* Tris-HCl buffer, pH 6. The mixture was extracted twice with 4-mℓ portions of 2-propanol-diethyl ether (22:78). The combined organic extract was back extracted into 0.5 mℓ of 0.2 *m* phosphate buffer, pH 10.5. An aqueous solution (0.5 mℓ) containing 0.5 *m* tetrabutylammonium hydrogen sulfate and 0.2 *M* phosphate buffer, pH 10.5 was added to the aqeuous layer. The mixture was then shaken for 1 hr at room temperature with 5 mℓ of dichloromethane and 10 μℓ of pentafluorobenzyl bromide. The organic layer was washed with 1 mℓ of 0.1 *M* HCl and 1 mℓ of 0.1 *M* phosphate buffer (pH 8). An aliquot of 4 mℓ of the organic layer was evaporated under a stream of nitrogen at 50°C. The residue was dissolved in 0.5 mℓ of hexane-chloroform (3:1) for injection.

I-6. The sample was mixed with 100 μℓ of an aqueous solution of the internal standard (100 μmol/ℓ) and 100 μℓ of 1 *M* HCl. The mixture was extracted with 10 mℓ of *n*-propanol-diethylether (16:84). The organic layer was dried at 50°C under a stream of nitrogen at 50°C. The residue was dissolved in 200 μℓ of 50 m*M* phosphate buffer, pH 11, neutralized with 2 μℓ of 1 *M* H_2SO_4 and an aliquot of 10 μℓ was injected.

I-7. The sample was deproteinized with perchloric acid (final concentration 0.4 *M*). One volume of the supernatant was mixed with 2 volumes of triocytlamine-Freon (1:4). After centrifugation aliquots of this solution were injected.

I-8. The sample was mixed with 0.5 mℓ of physiological saline and 0.1 mℓ of 0.5 *M* sodium dihydrogen phosphate buffer. The mixture was extracted with 8 mℓ of ethyl acetate. The organic layer was evaporated under vacuum. The residue was treated with 1 mℓ acetone-acetonitrile (1:2) containing 0.5 mg/mℓ of 4-bromomethyl-7-methoxycoumarin and 0.1 mg/mℓ of 18-crown-6, and 1 mg of potassium carbonate. The mixture was refluxed for 45 min. After cooling 0.2 mℓ of *n*-valeric acid was added to the mixture for the esterification of excess reagent. The reagent was again refluxed for 5 min. The mixture was diluted with acetone for injection.

I-9. The sample was acidified with 2 drops of 6 *N* HCl and extracted twice with 5 mℓ aliquots of methyl isobutyl ketone. The combined organic phase was evaporated at 40 to 45°C under a stream of nitrogen. The residue was dissolved in 200 μℓ of the mobile phase and an aliquot of 25 μℓ of the clear supernatant was injected.

I-10. The sample was mixed with 100 μℓ of the internal standard solution (10 μg/mℓ), 2 mℓ of saturated ammonium sulfate solution, and 100 μℓ of ammonium phosphate buffer, pH 6.7. The mixture was extracted with 8 mℓ of ethyl acetate. The organic layer was concentrated to approximately 1 mℓ and back extracted into 400 μℓ of 0.5 *M* KOH. Aliquots of 3 to 30 μℓ of the alkaline aqueous phase were injected.

Elution — E-1. Methanol-0.1 m*M* tetrabutyl ammonium hydrogen sulfate in 2 m*M* sodium acetate + 1.5 m*M* phosphate buffer, pH 6 (2:98).

E-2. 50 mM KH_2PO_4 buffer, pH 3.

E-3. 0.05 *M* Tris-HCl + 0.025 *M* cetrimide, pH 8.

E-4. Methanol-water (70:30).

E-5. 0.05 *M* NaH_2PO_4 containing 3.2 g/ℓ tetrabutylammonium hydroxide, pH 6.2.

E-6. Methanol-0.05 *M* ammonium phosphate buffer, pH 6.8 (2:98).

5-FLUOROURACIL (continued)

REFERENCES

1. **Marunaka, T. and Umeno, Y.,** Determination of 5-fluorouracil and pyrimidine bases in plasma by gas chromatography-chemical ionization-mass fragmentography, *J. Chromatogr.*, 221, 382, 1980.
2. **Marunaka, T., Umeno, Y., Yoshida, K., Nagamachi, M., Minami, Y., and Fujii, S.,** High-pressure liquid chromatographic determination of ftorafur[1-(Tetrahydro-2-furanyl)-5-fluorouracil] and GLC-mass spectrometric determination of 5-fluorouracil and uracil in biological materials after oral administration of uracil plus ftorafur, *J. Pharm. Sci.*, 69, 1296, 1980.
3. **Marunaka, T. and Umeno, Y.,** Gas chromatographic-mass fragmentographic determination of 5-fluorouracil as a metabolite of new 5-fluorouracil derivatives, *Chem. Pharm. Bull.*, 30, 1868, 1982.
4. **Cosyns-Duyck, M. C., Cruyl, A. A. M., De Leenheer, A. P., De Schryver, A., Huys, J. V., and Belpaire, F. M.,** Analysis of fluorouracil in human plasma by combined gas-liquid chromatography mass spectrometry, *Biomed. Mass Spectrom.*, 7, 61, 1980.
5. **Min, B. H., Garland, W. A., Lewinson, T. M., and Mehta, B. M.,** Comparison of gas chromatographic/mass spectrometric and microbiological assays for 5-fluorouracil in plasma, *Biomed. Mass Spectrom.*, 12, 238, 1985.
6. **De Bruijn, E. A., Driessen, O., Van Den Bosch, N., Van Strijen, E., Slee, P. H. T., Van Oosterom, A. T., and Tjaden, V.,** A gas chromatographic assay for the determination of 5,6-dihydrofluorouracil and 5-fluorouracil in human plasma, *J. Chromatogr.*, 278, 283, 1983.
7. **Kok, R. M., De Jong, A. P. J. M., Van Groeningen, C. J., Peters, G. J., and Lankelma, J.,** Highly sensitive determination of 5-fluorouracil in human plasma by capillary gas chromatography and negative ion chemical ionization mass spectrometry, *J. Chromatogr.*, 343, 59, 1985.
8. **Au, J. L. S., Wientjes, M. G., Luccioni, C. M., and Rustum, Y. M.,** Reversed-phase ion-pair high-performance liquid chromatographic assay of 5-fluorouracil, 5′-deoxy-5-fluorouridine, their nucleosides, mono-, di-, and triphosphate nucleotides with a mixture of quaternary ammonium ions, *J. Chromatogr.*, 228, 245, 1982.
9. **Sampson, D. C., Fox, R. M., Tattersall, M. H. N., and Hensley, W. J.,** A rapid high-performance liquid chromatographic method for quantitation of 5-fluorouracil in plasma after continuous intravenous infusion, *Ann. Clin. Biochem.*, 19, 125, 1982.
10. **Peters, G. J., Kraal, I., Laurensse, E., Leyva, A., and Pinedo, H. M.,** Separation of 5-fluorouracil and uracil by ion-pair reversed-phase high-performance liquid chromatography on a column with porous polymeric packing, *J. Chromatogr.*, 307, 464, 1984.
11. **Iwamoto, M., Yoshida, S., and Hirose, S.,** Fluorescence determination of 5-fluorouracil and 1-(tetrahydro-2-furanyl)-5-fluorouracil in blood serum by high-performance liquid chromatography, *J. Chromatogr.*, 310, 151, 1984.
12. **Quebbeman, E. J., Hoffman, N. E., Hamid, A. A. R., and Ausman, R. K.,** An HPLC method for measuring 5-fluorouracil in plasma, *J. Liq. Chromatogr.*, 7, 1489, 1984.
13. **Stetson, P. L., Shukla, U. A., and Ensminger, W. D.,** Sensitive high-performance liquid chromatographic method for the determination of 5-fluorouracil in plasma, *J. Chromatogr.*, 344, 385, 1985.

FLUORPROQUAZONE

Liquid Chromatography

Specimen (mℓ)	Extraction	Column (cm × mm)	Packing (μm)	Elution	Flow (mℓ/min)	Det. (nm)	RT (min)	Internal standard (RT)	Other compounds (RT)	Ref.
Animal feed (50 g)	I-1	25 × 4.6	LiChrosorb RP-8 (5)	E-1	1.5	ABS (240)	7	—	—	1

[a] A 50 × 4.6mm column packed with LiChrosorb RP-18 (10 μm) was used for sample clean up.

Extraction — I-1. The sample was mixed with 100 mℓ of methanol for 30 min. After centrifugation an aliquot of the supernatant was injected.

Elution — E-1. Acetonitrile-water (55:45).

REFERENCE

1. **Gfeller, J. C. and Stockmeyer, M.,** High-performance liquid chromatographic column switching technique in the analysis of medicated feed for an automated clean-up procedure, *J. Chromatogr.*, 198, 162, 1980.

FLUOXETINE

Gas Chromatography

Specimen (mℓ)	Extraction	Column (m × mm)	Packing (mesh)	Oven temp (°C)	Gas (mℓ/min)	Det	RT (min)	Internal standard (RT)	Deriv	Other compounds (RT)	Ref.
Plasma (1)	I-1	1.2 × 3	3% SP 2100 Supelcoport (80/100)	190	Ar:95-Methane: 5 (40)	ECD	3.4	4,4-*bis*(*p*-fluoro-phenyl)-N-ethyl butylamine (5.2)	Penta-fluoro-propionyl	Norfluoxetine (2.7)	1

FLUOXETINE (continued)

Extraction — I-1. The sample was mixed with 0.5 mℓ of the internal standard solution (1 mgℓ in 10 m*M* HCl) and 2 mℓ of 2.7 *M* carbonate buffer, pH 9.8. The mixture was extracted with 11 mℓ of butyl choride. An aliquot of 10 mℓ of the organic layer was back extracted into 5 mℓ of 0.2 *M* HCl. The aqueous phase was made basic with 2 mℓ of carbonate buffer and extracted with 6 mℓ of butyl chloride. One drop of 0.3 *M* methanolic HCl was added to an aliquot of 5 mℓ of the organic layer which was then evaporated at 40°C under a stream of nitrogen. The residue was dissolved in 0.2 mℓ of benzene and the solution was incubated after the addition of 50 μℓ of pentafluoropropionic anhydride at 90°C for 30 min. The reaction mixture was evaporated at room temperature under nitrogen, the residue dissolved in 0.7 mℓ of hexane and an aliquot of 3 μℓ of this solution was injected.

REFERENCE

1. **Nash, J. F., Bopp, R. J., Carmichael, R. H., Farid, K. Z., and Lemberger, L.,** Determination of fluoxetine and norfluoxetine in plasma by gas chromatography with electron-capture detection, *Clin. Chem.*, 28, 2100, 1982.

FLUOXYMESTERONE

Liquid Chromatography

Specimen (mℓ)	Extraction	Column (cm × mm)	Packing (μm)	Elution	Flow (mℓ/min)	Det. (nm)	RT (min)	Internal standard (RT)	Other compounds (RT)	Ref.
Serum (1)	I-1	25 × 4.6	Zorbax Silica (6)[a]	E-1	2.0	ABS (236)	13	6-α-Methyl-prednisolone (22)	—	1

[a] Protected by a 3-cm guard column packed with Spherisorb silica (5 μm).

Extraction — I-1. The sample was mixed with 10 μℓ of a 20-μg/mℓ solution of the internal standard in methanol and then extracted with 10 mℓ of dichloromethane. The organic layer was washed with 1 mℓ of 0.1 *M* NaOH, then with 1 mℓ of water, and finally evaporated under nitrogen at 40 to 45°C. The residue was reconstituted with 200 μℓ of the mobile phase. Aliquots of 50 μℓ were injected with an autosampler.

Elution — E-1. Butyl chloride (50% water saturated)-tetrahydrofuran-methanol-phosphoric acid (880:100:15:0.5).

REFERENCE

1. **Capponi, V. J., Cox, S. R., Harrington, E. L., Wright, C. E., Antal, E. J., and Albert, K. S.,** Liquid chromatographic assay for fluoxymesterone in human serum with application to a preliminary bioavailability study, *J. Pharm. Sci.*, 74, 308, 1985.

FLUPENTIXOL

Gas Chromatography

Specimen (mℓ)	Extraction	Column (m × mm)	Packing (mesh)	Oven temp (°C)	Gas (mℓ/min)	Det	RT (min)	Internal standard (RT)	Deriv.	Other compounds (RT)	Ref.
Plasma (2)	I-1	0.9 × 2	2% OV-101 Supelcoport (80/100)	250	He (40)	NPD	1.5	Perphenazine[a] (3)	Acetyl	—	1, 2

[a] An alternative compound (Lu-9) was also used as the internal standard.

Extraction — I-1. The sample was mixed with 50 μℓ of a methanolic solution of the internal standard (1 μg/mℓ) and 200 μℓ of 4 *N* NaOH. The mixture was extracted with 5 mℓ of heptane-isopropanol (9:1). The organic layer was back extracted into 1 mℓ of 0.1 *N* HCl. The aqueous layer was made alkaline with 0.1 mℓ of 6 *N* NaOH and extracted with 0.5 mℓ of the extraction solvent. The organic layer was evaporated at 40°C with a gentle stream of nitrogen. The residue was treated with 50 μℓ of ethyl acetate-acetic anhydride (9:1) for 40 min at 40°C and aliquots of 5 μℓ of this solution were injected.

REFERENCES

1. **Balant-Gorgia, A. E., Balant, L. P., Genet, Ch., and Eisele, R.,** Comparative determination of flupentixol in plasma by gas chromatography and radioimmunoassay in schizophrenic patients, *Ther. Drug Monit.*, 7, 229, 1985.
2. **Balant-Gorgia, A. E., Eisele, R., Aeschlimann, J. M., Balant, L. P., and Garrone, G.,** Plasma flupentixol concentrations and clinical response in acute schizophrenia, *Ther. Drug Monit.*, 7, 411, 1985.

FLUPHENAZINE

Gas Chromatography

Specimen (mℓ)	Extraction	Column (m × mm)	Packing (mesh)	Oven temp (°C)	Gas (mℓ/min)	Det.	RT (min)	Internal standard (RT)	Deriv.	Other compounds (RT)	Ref.
Plasma (5)	I-1	0.9 × 2	2% OV-101 Chromosorb W (80/100)	240	He (30)	NPD	1.3	Perphenazine (2.5)	Acetyl	—	1

FLUPHENAZINE (continued)

Gas Chromatography

Specimen (mℓ)	Extraction	Column (m × mm)	Packing (mesh)	Oven temp (°C)	Gas (mℓ/min)	Det.	RT (min)	Internal standard (RT)	Deriv.	Other compounds (RT)	Ref.
Plasma (2)	I-2	1.8 × 2	3% OV-1 GasChrom W (NA)	300	N_2 (20)	MS-EI	1.3	Perphenazine (2.4)	Trimethyl-silyl	—	2

Liquid Chromatography

Specimen (mℓ)	Extraction	Column (cm × mm)	Packing (μm)	Elution	Flow (mℓ/min)	Det. (nm)	RT (min)	Internal standard (RT)	Other compounds (RT)	Ref.
Urine (10)	I-3	20 × 4.6	Hypersil-ODS (NA)	E-1	NA	ABS (NA)	—	—	Fluphenazine sulf-oxide (2.1) Fluphenazine sul-fone (2.8) 7-Hydroxylfluphen-azine (3.5) Fluphenazine di N-oxide (5.5) Dealkylated flu-phenazine (8.2)	3
Serum (2)	I-4	30 × 4	LiChrosorb RP18 (7)	E-2	1.6	ABS[a] (254)	7.5	—	—	4

Thin-Layer Chromatography

Specimen (mℓ)	Extraction	Plate (Manufacturer)	Layer (mm)	Solvent	Post-separation treatment	Det. (nm)	Rf	Internal standard (Rf)	Other compounds (Rf)	Ref.
Plasma (4)	I-5	10 × 10 cm (Merck)	Silica gel[b] (HPTLC) (0.25)	S-1[c]	D: 5% Par-affin oil in toluene E: UV light 20 min	Fl (Reflectance) (254, 400)	0.2	Trifluopro-mazine (0.5)	—	5

[a] A fluorescence detector (em = 375 nm, em = 470 nm) was also used.
[b] The plates were prewashed.
[c] The plates were developed twice in two different solvents.

Extraction — I-1. The sample was spiked with 20 ng of the internal standard, mixed with 0.2 mℓ of 4 *N* NaOH, and extracted with 5 mℓ of heptane-isopropanol (9:1). The organic layer was back extracted into 1 mℓ of 0.1 *N* HCl. The aqueous layer was made alkaline with 0.1 mℓ of 5 *N* NaOH and extracted with 0.5 mℓ of the extraction solvent. The organic layer was evaporated in a stream of nitrogen and the residue treated with 0.1 mℓ of ethyl acetate-acetic anhydride (9:1) at 65°C for 30 min. Aliquots of 3 to 5 μℓ of this solution were injected.
I-2. The sample was mixed with 1 mℓ of an aqeuous solution of the internal standard (100 ng/mℓ) and 0.5 mℓ of saturated sodium carbonate. The mixture was extracted twice with 5-mℓ portions of isopropanol-*n*-pentane (5:95). The combined extracts were evaporated at 65°C. The residue was dissolved in 1 mℓ of 0.1 *N* HCl and the solution washed twice with 5-mℓ portions of *n*-hexane. The aqueous layer was made alkaline with 1 mℓ of 1 *N* NaOH and extracted twice with 5-mℓ portions of the extraction solvent. The combined extracts were evaporated. The residue was dissolved in 50 μℓ of ethyl acetate and 50 μℓ of N,O-*bis*(trimethylsilyl)acetamide were added. The mixture was incubated at 65°C for 1 hr and then evaporated in a stream of nitrogen. The residue was dissolved in 30 μℓ of ethyl acetate and aliquots of 1 to 2 μℓ were injected.
I-3. The sample was made alkaline with 1 *M* sodium hydroxide, saturated with potassium chloride and extracted with diethyl ether. The organic layer was evaporated in a stream of nitrogen. The residue dissolved in 1 mℓ of the mobile phase and 50-μℓ aliquots were injected.
I-4. The sample was mixed with 0.2 mℓ of 30% sodium carbonate and heated on a boiling water bath for 3 min. After cooling the mixture was extracted twice with 5-mℓ portions of chloroform. The combined extracts were evaporated. The residue was dissolved in 100 μℓ for injection.
I-5. The sample was mixed with 100 μℓ of the internal standard solution (120 ng/mℓ in heptane) and 0.6 mℓ of saturated sodium carbonate solution. The mixture was extracted with 20 mℓ of heptane containing 0.05% isoamyl alcohol. The organic layer was back extracted into 2 mℓ of 0.05 *N* HCl. The aqueous layer was made alkaline with 0.3 mℓ of saturated sodium carbonate and extracted with 4.5 mℓ of pentane. The organic layer was evaporated under nitrogen at 40°C. The residue was dissolved in 50 μℓ of heptane-ethanol-*n*-dodecane (75:25:0.05) and the entire solution was spotted on the TLC plate with the contact spotter at low temperature.

Elution — E-1. Methanol-1% potassium chloride containing 0.01% phosphoric acid (2:1).
E-2. Methanol-water-diethylamine (80:20:0.05).

Solvent — S-1. (A) Toluene-acetone (60:40); (B) Toluene-acetone-ammonium hydroxide (60:40:2).

REFERENCES

1. **Javaid, J. I., Dekirmenjian, H., Liskevych, U., Lin, R. L., and Davis, J. M.,** Fluphenazine determination in human plasma by a sensitive gas chromatographic method using nitrogen detector, *J. Chromatogr. Sci.*, 19, 439, 1981.
2. **McKay, G., Hall, K., Edom, R., Hawes, E. M., and Midha, K. K.,** Subnanogram determination of fluphenazine in human plasma by gas chromatography mass spectrometry, *Biomed. Mass Spectrom.*, 10, 550, 1983.
3. **Heyes, W. F. and Robinson, M. L.,** HPLC procedure for the determination of some potential fluphenazine metabolites in urine, *J. Pharm. Biomed. Anal.*, 3, 477, 1985.
4. **Poey, J., Puig, Ph., Bourbon, Ch., and Bourbon, P.,** Dosage de la fluphenazine dans le serum humain par chromatographie en phase liquide, *Analysis*, 13, 65, 1985.
5. **Davis, C. M. and Fenimore, D. C.,** Determination of fluphenazine in plasma by high-performance thin-layer chromatography, *J. Chromatogr.*, 272, 157, 1983.

FLURAZEPAM

Gas Chromatography

Specimen (mℓ)	Extraction	Column (m × mm)	Packing (mesh)	Oven temp (°C)	Gas (mℓ/min)	Det.	RT (min)	Internal standard (RT)	Deriv.	Other compounds (RT)	Ref.
Plasma (2)	I-1	1.2 × 1	OV-17 μ-Partisorb[a]	290	Methane (15)	MS-NCI	NA	[$^2H_{10}$]Flurazepam	—	—	1
Plasma (3)	I-2	1.8 × 2	3% OV-17 Gas Chrom Q (100/120)	255	Ar:95 Methane:5 (22)	ECD	9.8	Diazepam (4.9)	*tert*-Butyl-dimethyl-silyl	N-1-Desalkyl flura-zepam (5.7) N-1-Hydroxyethyl-flurazepam (11.9)	2
Plasma (0.5 — 1)	I-3	25 × 0.5[b]	CPSil5	265	He	ECD	—	Delorazepam (3.3)	c	N-1 Desalkyl flura-zepam (2.4)	3

Liquid Chromatography

Specimen (mℓ)	Extraction	Column (cm × mm)	Packing (μm)	Elution	Flow (mℓ/min)	Det. (nm)	RT (min)	Internal standard (RT)	Other compounds (RT)	Ref.
Urine (1)	I-4	25 × 4.6	Partisil silica (10)	E-1	1.5	ABS (254)	4	d (7)	N-1-Hydroxy ethyl-flurazepam (4.5)	4
Plasma (1)	I-5	15 × 4.6	LiChrosorb RP18 (5)	E-2	1.0	ABS (254)	3.6	N-Des-methyl-diazepam (2.6)	e	5

[a] Chemically bound stationary phase on solid support (Whatman).
[b] A packed column was also used.
[c] A separate extraction procedure, a different internal standard was used and trimethyl silyl derivative was prepared for the determination of N-1-hydroxyethyl-flurazepam.
[d] 7-Chloro-5-(2′-Chlorophenyl)-1,3-dihydro-1-2-dimethylaminoethyl-2H-1,4-benzodiazepin-2-one.
[e] The metabolites were separated with a different mobile phase.

Extraction — I-1. The sample was mixed with 50 μℓ of an aqueous solution of the internal standard (0.1 μg/mℓ) and 1 mℓ of 1 *M* borate buffer, pH 10. The mixture was extracted with 6 mℓ of benzene-dichloromethane (7:3). The organic layer was evaporated at 50°C under a stream of nitrogen. The residue was dissolved in 1 mℓ of 0.1 *M* HCl and the solution was washed with 6 mℓ of benzene. The aqueous layer was made alkaline with 0.2 mℓ of 0.5 *M* NaOH, 0.5 mℓ of 1 *M* pH 10 borate buffer saturated with sodium chloride, and extracted with 6 mℓ of the extraction solvent. The organic layer was evaporated at 50°C under a stream of nitrogen. The residue was dissolved in 40 μℓ of hexane-acetone (9:1) and 1 to 5 μℓ of the solution were injected.
I-2. The sample was mixed with 100 μℓ of the methanolic solution of the internal standard (400 ng/mℓ) and 3 mℓ of 1 *M* borate buffer. The mixture was extracted with 6 mℓ of benzene-dichloromethane (9:1). The organic layer was back extracted with 3 mℓ of 4 *N* HCl. The aqueous layer was washed with 5 mℓ of diethyl ether. The combined ether extracts were dried over sodium sulfate and evaporated at 40°C under a stream of nitrogen., The residue was dissolved in 100 μℓ of benzene-acetone-methanol (85:10:5) and an aliquot of 5 μℓ was injected. The remaining solution was evaporated at 60°C with a stream of nitrogen. The residue was treated with 5 μℓ of *tert*-butyldimethyl chlorosilane-imidazole at 60°C for 10 min and then diluted with 200 μℓ of toluene. A 5-μℓ aliquot of this solution was injected.
I-3. The sample was mixed with an appropriate volume of methanolic solution of the internal standard and 1 to 2 mℓ of pH 9 buffer. The mixture was extracted with 10 mℓ of ether. The organic layer was evaporated under a stream of nitrogen. The residue was dissolved in 1 mℓ of 1 *N* HCl and the solution was washed with 3 mℓ of hexane. The aqueous phase was adjusted to pH 7.5 to 7.8 with phosphate buffer and re-extracted with 10 mℓ of ether. The organic layer was evaporated and the residue dissolved in 50 to 100 μℓ of toluene for injection.
I-4. The sample was adjusted to pH 5.4 with 0.25 *N* HCl, mixed with 2 mℓ of 1 *M* phosphate buffer, pH 5.4 and incubated with 0.1 of Glusulase overnight at 37°C. After cooling to room temp, the sample was adjusted to pH 9 with 6 *N* sodium hydroxide and extracted twice with 12-mℓ portions of anhydrous diethyl ether. The combined extracts were evaporated at 40°C under a stream of nitrogen. The residue was reconstituted in 1.5 to 5 mℓ of the internal standard solution (5 μg/mℓ in the mobile phase). Aliquots of 50 μℓ were injected with an autosampler.
I-5. The sample was diluted with an equal volume of water and then treated with 0.1 *M* NaOH containing 0.05 *M* ammonium sulfate. After centrifugation, the pH of the supernatant was adjusted to 9 and extracted twice with 2-mℓ portions of ethyl acetate. The combined extracts were evaporated at 55°C under nitrogen. The residue was dissolved in 0.5 mℓ of the mobile phase for injection.

Elution — E-1. Dichloromethane-X (500:25); X = Methanol-water-ammonium hydroxide (150:9:1).
E-2. Methanol-0.66 *M* phosphate buffer, pH 7.6 (85:15).

REFERENCES

1. **Miwa, B. J., Garland, W. A., and Blumenthal, P.,** Determination of flurazepam in human plasma by gas chromatography-electron capture negative chemical ionization mass spectrometry, *Anal. Chem.*, 53, 793, 1981.
2. **Cooper, S. F. and Drolet, D.,** Gas-liquid chromatographic determination of flurazepam and its major metabolites in plasma with electron-capture detection, *J. Chromatogr.*, 231, 321, 1982.
3. **Coassolo, P., Aubert, C., and Cano, J. P.,** Determination of flurazepam and its hydroxyethyl and dealkyl metabolites by some packed and capillary ECD-GLC methods: application or pharmacokinetic studies, *J. High Resol. Chromatogr. Chromatogr. Commun.*, 7, 258, 1984.
4. **Weinfeld, R. E. and Miller, K. F.,** Determination of the major urinary metabolites of flurazepam in man by high-performance liquid chromatography, *J. Chromatogr.*, 223, 123, 1981.
5. **Dadgar, D., Smyth, W. F., and Hojabri, H.,** High-performance liquid chromatographic determination of flurazepam and its metabolites in human blood plasma, *Anal. Chim. Acta*, 147, 381, 1983.

FLURBIPROFEN

Gas Chromatography

Specimen (mℓ)	Extraction	Column (m × mm)	Packing (mesh)	Oven temp (°C)	Gas (mℓ/min)	Det.	RT (min)	Internal standard (RT)	Deriv.	Other compounds (RT)	Ref.
Plasma (0.5 — 2)	I-1	1 × 3	1.5% OV-17 Chromosorb W (80/100)	175	He (50)	MS-EI	4	[$^2H_3$]-Flurbiprofen	Methyl	—	1
Plasma (0.05)	I-2	1.4 × 1.9	3% SP-100V Chromosorb W (100/120)	255	N_2 (52)	NPD	3.8	Naproxen (6.2)	Dipropyl-amino	—	2

Liquid Chromatography

Specimen (mℓ)	Extraction	Column (cm × mm)	Packing (μm)	Elution	Flow (mℓ/min)	Det. (nm)	RT (min)	Internal standard (RT)	Other compounds (RT)	Ref.
Plasma (1)	I-3	25 × 4.6	Rheodyne RP-8 (10)[a]	E-1	2.0	ABS (254)	17	Ibuprofen (22)	4′-Hydroxyflurbi-profen (8.5) 2′-Hydroxyflurbi-profen (9.5) 3′-Hydroxyflurbi-profen (9.1)	3
Serum (0.5)	I-4	25 × 4.6	Zorbax ODS (6)	E-2	1.0	Fl (250,315)	6.1	Biphenyl-acetic acid (5)	—	4

[a] Protected by a 42 × 3 mm guard column packed with Permaphase-ODS (30 μm).

Extraction — I-1. The sample was mixed with 1 mℓ of the internal standard solution (50 ng/mℓ in 0.001 *N* sodium hydroxide) and 1 mℓ of 3 *N* HCl. The mixture was extracted with 15 mℓ of benzene. The benzene layer was back extracted into 5 mℓ of carbonate buffer (pH 9.5). A 0.5-mℓ aliquot of 1 *N* NaOH was added to the aqueous layer which was then washed with 5 mℓ of benzene, acidified (pH 1) with 1 mℓ of 3 *N* HCl, and extracted ith 10 mℓ of diethyl ether. The organic layer was evaporated under reduced pressure, the residue treated with 0.5 mℓ of diazomethane-ether solution, and the mixture allowed to stand at room temperature for 5 min. The solvents were removed at 15°C under reduced pressure, the residue dissolved in 50 μℓ of methyl alcohol, and aliquots of 1 to 6-μℓ were injected.
I-2. The sample was spiked with 0.5 μg of the internal standard, acidified with 10 μℓ of 4 *M* HCl and extracted with 200 μℓ of dichloromethane. The organic layer was evaporated. The residue was treated with 30 μℓ of a solution of 2-bromo-1-methylpyridinium iodide (10 μg/mℓ in acetonitrile) and 40 μℓ of a solution of dipropylamine (10 μg/μℓ in dichloromethane). The mixture was allowed to stand at room temperature for 15 min. The reaction mixture was diluted with 500 μℓ of dichloromethane and extracted three times with 500-μℓ aliquots of 2 *M* sulfuric acid. The organic layer was evaporated under a stream of nitrogen. The residue was dissolved in 20 μℓ of methanol and aliquots of 1 μℓ of the resulting solution were injected.
I-3. The sample was applied to an extraction cartridge packed with styrene-divinyl resin along with 0.2 mℓ aliquot of 0.5 *M* sulfuric acid. The cartridges were washed (1 mℓ of water) and eluted (2 mℓ of methanol) in a DuPont Prep 1 automated sample processor. The residue obtained after evaporation of the eluate was reconstituted with 1 mℓ of the mobile phase containing 10 μg/mℓ of ibuprofen.
I-4. The sample was mixed with 50 μℓ of the internal standard solution (10 μg/mℓ in methanol) and 0.5 mℓ of 1 *M* HCl. The mixture was extracted with 10 mℓ of pentane-ether (8:2). The organic layer was evaporated at 30°C under nitrogen. The residue was dissolved in 5 mℓ of acetonitrile. Aliquots of this solution were injected with an autosampler.

Elution — E-1. Acetonitrile-O. 1 *M* acetic acid (50:50).
E-2. Acetonitrile-water-phosphoric acid (650:350:0.5).

REFERENCES

1. **Kawahara, K., Matsumura, M., and Kimura, K.,** Determination of flurbiprofen in human plasma using gas chromatography-mass spectrometry with selected ion monitoring, *J. Chromatogr.*, 223, 202, 1981.
2. **Lingeman, H., Haan, H. B. P., and Hulshoff, A.,** Rapid and selective derivatization method for the nitrogen-sensitive detection of carboxylic acids in biological fluids prior to gas chromatographic analysis, *J. Chromatogr.*, 336, 241, 1984.
3. **Snider, B. G., Beaubien, L. J., Sears, D. J., and Rahn, P. D.,** Determination of flurbiprofen and ibuprofen in dog serum with automated sample preparation, *J. Pharm. Sci.*, 70,, 1347, 1981.
4. **Albert, K.S., Gillespie, W. R., Raabe, A., and Garry, M.,** Determination of flurbiprofen in human serum by reverse-phase high-performance liquid chromatography with fluorescence detection, *J. Pharm. Sci.*, 73, 1823, 1984.

FLUTROLINE

Gas Chromatography

Specimen (mℓ)	Extraction	Column (m × mm)	Packing (mesh)	Oven temp (°C)	Gas (mℓ/min)	Det.	RT (min)	Internal standard (RT)	Deriv.	Other compounds (RT)	Ref.
Plasma (1)	I-1	5 × 0.32	DB-1 (1 μm)[a]	250	He (2)	MS-EI	2.4	Homologue (1.8)	Trimethyl-silyl	—	1

[a] Film thickness.

Extraction — I-1. The sample was spiked with 200 ng of the internal standard and treated with 5 mℓ of acetonitrile. Following centrifugation, the supernatant was mixed with 1 mℓ of 1 *N* sodium hydroxide and washed with 4 mℓ of hexane. The top hexane layer and the bottom aqueous layer were discarded. The middle (aqueous acetonitrile) layer was concentrated to approximately 0.1 mℓ, mixed with 1 mℓ of 0.2 *N* sodium hydroxide, and extracted with 5 mℓ of benzene. The benzene layer was evaporated under nitrogen. The residue was dissolved in 20 μℓ of ethyl acetate. Aliquots of 1 to 4 μℓ were loaded onto the tip of the moving needle injector and allowed to dry. Then 1 μℓ of N-methyl-N-trimethylsilyltrifluoroacetamide-pyridine (1:0.2) was applied to the needle, allowed to dry, and injected.

REFERENCE

1. **Falkner, F. C., Fouda, H. G., and Mullins, F. G.,** A gas chromatographic/mass spectrometric assay for flutroline, a α-carboline antipsychotic agent, with direct derivatization on a moving needle injector, *Biomed. Mass Spectrom.* 11, 482, 1984.

FLUZINAMIDE

Liquid Chromatography

Specimen (mℓ)	Extraction	Column (cm × mm)	Packing (μm)	Elution	Flow (mℓ/min)	Det. (nm)	RT (min)	Internal standard (RT)	Other compounds (RT)	Ref.
Plasma (0.5)	I-1	30 × 3.9	μ-Bondapak-C_{18} (10)	D-1	1.5	ABS (220)	9.5	Analog (11.5)	N-Hydroxymethyl-fluzinamide (6.2) 3-[3-(Trifluoromethyl)phenoxy]1-acetidine carboxamide (7.5) N-Formylfluzinamide (10.5)	1

Extraction — I-1. The sample was spiked with 80 μℓ of the internal standard solution (50 μg/mℓ in acetonitrile-water, 40:60) and extracted with 5 mℓ of hexane-dichloromethane-butanol (50:40:5). The organic layer was evaporated to dryness under a stream of nitrogen at 25°C. The residue was reconstituted in 200 μℓ of the mobile phase and 100 μℓ of this solution were injected.

Elution — E-1. Acetonitrile-tetrahydrofuran-0.025 *M* phosphate buffer, pH 4.2 (30:5).

REFERENCE

1. **Osman, M.A., Pinchbekc, F. M., Cheng, L. K., and Wright, G. J.,** Simultaneous determination of fluzinamide and three of its active metabolites in plasma by high-performance liquid chromatography, *J. Chromatogr.*, 336, 329, 1984.

FM24

Gas Chromatography

Specimen (mℓ)	Extraction	Column (m × mm)	Packing (mesh)	Oven temp (°C)	Gas (mℓ/min)	Det.	RT (min)	Internal standard (RT)	Deriv.	Other compounds (RT)	Ref.
Plasma (1)	I-1	2 × 3	3% OV-17 Chromosorb W (100/120)	215	Ar:90-Methane: 10 (40)	ECD	4.5	FM 25 (6.8)	Penta-fluoro-propionyl	—	1

Liquid Chromatography

Specimen (mℓ)	Extraction	Column (cm × mm)	Packing (μm)	Elution	Flow (mℓ/min)	Det. (nm)	RT (min)	Internal standard (RT)	Other compounds (RT)	Ref.
Plasma (2)	I-2	30 × 3.9	μ-Bondapak-C_{18} (10)	E-1	1.4	Fl (230)[a]	5.8	Imipramine (4.5)	—	2

[a] No emission filter was used.

Extraction — I-1. An aliquot of 0.5 mℓ of a methanolic solution of the internal standard (1 μg/mℓ) was evaporated under a stream of nitrogen. The residue was mixed with 1 mℓ of the sample, 0.5 mℓ of 1 *N* NaOH, and 7 mℓ of toluene. The organic phase (6.5 mℓ) was back extracted into 5 mℓ of 0.1 *N* HCl. An aliquot (4.5 mℓ) of the aqueous layer was alkalinized with 1 mℓ of 1 *N* sodium hydroxide under a stream of nitrogen. The residue was incubated with 50 μℓ of pentafluoropropionic anhydride-ethyl acetate (1:1) at 65 to 70°C for 1 hr. The excess reagent was removed at 50 to 60°C under a stream of nitrogen. The residue was dried thoroughly under an increased nitrogen flow. The residue was dissolved in 150 μℓ of hexane and aliquots of 5 μℓ of this solution were injected.
I-2. The sample was spiked with 50 μℓ of an aqueous solution of the internal standard (20 μg/mℓ) and applied to a column with a 4-cm bed of Carboxymethyl Sephadex. The column was washed twice with 5-mℓ aliquots of water. The column was eluted with 2 mℓ of 0.1 *N* NaOH in 0.1 *M* sodium sulfate solution and 2 × 2 mℓ of borate buffer, pH 9. The combined eluates were extracted with 9 mℓ of ethyl acetate. The organic layer was evaporated under nitrogen. The residue was dissolved in 200 μℓ of the mobile phase and aliquots of 100 μℓ of this solution were injected.

Elution — E-1. Acetonitrile-water-acetic acid (47:48:5) containing 1-heptane sulfonic acid.

REFERENCE

1. **Bernard, N., Cuisinaud, G., Jozefczak, Seccia, M., Ferry, N., and Sassard, J.,** Sensitive gas chromatographic method for the determination in plasma of FM 24, 1-(2-exo-bicyclo[2,2,1]hept-2-ylphenoxyl)-3-[(1-methylethyl)amino]-2-propanol, a new β-adrenoceptor blocking agent, *J. Chromatogr.*, 183, 99, 1980.
2. **Lefebvre, M. A., Julian, B., and Fourtillan, J. B.,** High-performance liquid chromatographic assay for determination of a new β-blocking agent FM 24, *J. Chromatogr.*, 230, 199, 1982.

FOLIC ACID

Liquid Chromatography

Specimen (mℓ)	Extraction	Column (cm × mm)	Packing (μm)	Elution	Flow (mℓ/min)	Det. (nm)	RT (min)	Internal standard (RT)	Other compounds (RT)	Ref.
Dosage	I-1	30 × 4	μ-Bondapak-C_{18} (10)	E-1	1.0	ABS (280)	11.5	—	—	1
Dosage	I-2	30 × 4	μ-Bondapak-C_{18} (10)[a]	E-2	1.5	ABS (280)	12	Methyl-paraben (17)	—	2
Plasma (0.3)	I-3	25 × 4	LiChrosorb RP-18 (5)	E-3	1.0	Fl (295, 350)	25[b]	—	5-Methyltetrahydro-folate (13)	3
Pure compounds	—	30 × 3.9	μ-Bondapak-C_{18} (5)[c]	E-4; grad	1.0	ABS (280)	55	—	*p*-Aminobenzoly-glutamic acid (11) 10-Formyltetrahy-drofolic acid (21) Tetrahydrofolic acid (28) 5-Formyltetrahydro-folic acid (32) Dihydrofolic acid (40) 5-Methyltetrahydro-folic acid (48)	4

FOLIC ACID (continued)

Liquid Chromatography

Specimen (mℓ)	Extraction	Column (cm × mm)	Packing (μm)	Elution	Flow (mℓ/min)	Det. (nm)	RT (min)	Internal standard (RT)	Other compounds (RT)	Ref.
Tissue	I-4	30 × 3.9	μ-Bondapak-C_{18} (10)[c,d]	E-5	1.0	ABS (284)	33		*p*-Aminobenzoly-glutamic acid (10) 10-Formyltetrahy-drofolic acid (13.5) 10-Formyldihydro-folic acid (15) Tetrahydrofolic acid (17) 10-Formylfolic acid (18) 5-Formyltetrahydro-folic acid (20) Dihydrofolic acid (27) 5-Methyltetrahydro-folic acid (37) 10-Methyltetrahy-drofolic acid (48)	5
Total nutritional diet	I-5	30 × 3.9	μ-Bondapak-C_{18} (10)	E-6	2.0	ABS (365)	5.5	—	—	6
Pure compounds	—	25 × 4.6	LiChrosorb-RP-18 (10)[e]	E-7; grad	1.0	ABS (280)	47	—	*p*-Aminobenzoly-glutamic acid (15) 5-Formyltetrahydro-folic acid (25) 5-Methyltetrahydro-folic acid (45)	7
Plasma (0.01—0.05)	I-6	25 × 4.6	Spherisorb ODS-2 (5)	E-8	1.0	ABS (280)	18	—	Tetrahydrofolic acid (4.5) 5-Formyltetrahydro-folic acid (7) 10-Formyltetrahy-drofolic acid (12) 5-Methyltetrahydro-folic acid (15) Dihydrofolic acid	8

[a] Column temp = 35°C.
[b] Not detected by fluorescence detector. Detected by absorbance detector at 288 nm.
[c] Protected by a 40 × 3.9 guard column packed with Corasil-C_{18} (37 to 50 μm).
[d] Alternatively a Radial-Pak-C_{18} cartridge was also used.
[e] Protected by a 30 × 4.6 mm guard column packed with the material of the analytical column.

Extraction — I-1. A sample weight of ground tablets equivalent to 1 mg of folic acid was mixed with an appropriate amount of calcium carbonate (depending upon the amount of calcium present in the tablet), 5 mℓ of 10% thioglycerol solution, and lactate-phosphoric acid extraction solution (prepared by dissolving 75 g of calcium lactate trihydrate and 67 mℓ of phosphoric acid to a volume of 500 with water). The mixture was stirred for 30 min at 50°C, diluted with water, adjusted to pH 7 with 1 *N* NaOH, mixed with 40 mℓ of acetonitrile, and made up to 250 mℓ with water. After mixing, the solution was filtered, the first 10 mℓ were discarded and the next 8 mℓ collected for chromatography.
I-2. A quantity of capsule fillings equivalent to 0.3 mg of folic acid was washed with 30 mℓ of hexane. The residue was dried at 60°C, treated with 25 mℓ of the internal standard solution (40 mg of methylparaben, 240 mℓ of methanol, 650 mℓ of water, 12 mℓ of a 40% solution of tetrabutylammonium hydroxide, 2.04 g of KH_2PO_4, and 30 mℓ of a solution containing 100 mg/mℓ of pentetic acid in 0.75 *N* ammonium hydroxide, dilute to 1 ℓ with water). The head space of the tube was immediately flushed with nitrogen, the contents vigorously shaken and centrifuged. Aliquots of the supernatant were injected.
I-3. The sample, stabilized with ascorbic acid, was mixed with 4 μℓ of 1 *M* sodium carbonate and treated in a boiling water bath for 5 min. After cooling and centrifugation, aliquots of 10 to 50 μℓ were injected.
I-4. The sample was washed with ice-cold phosphate-buffered saline, minced and then treated for 5 min at 95°C in 3 vol of 1% sodium ascorbate containing 10 m*M* mercaptoethanol. After cooling in ice, the extracts were homogenized and centrifuged. The supernatant was adjusted to pH 4.5 with glacial acetic acid. a 4-mℓ aliquot of the supernatant was incubated with 100 μℓ of the conjugase preparation at 37°C for 1.5 hr. After centrifugation, the reaction mixture was passed through a 0.5 × 2.5 cm column of Dowex 50X-4, 200 to 400 mesh, (NH_4^+) which had been washed with 2 mℓ of 1 *M* ammonium acetate, pH 4.5 containing 50 m*M* mercaptoethanol and 0.2% ammonium ascorbate. The column was then eluted with 3 mℓ of 0.05 *M* ammonium acetate, pH 7.3 containing 50 m*M* mercaptoethanol. The eluate was lyophilized and then dissolved in 200 μℓ of 0.1 *M* diethyloglycine containing 50 m*M* mercaptoethanol and 0.01% ascorbate, pH 8.3, kept at 4°C for 3 hr prior to injection.
I-5. An aliquot of 10 g of diet was shaken with 100 mℓ of 0.01 *M* pH 7.4 phosphate buffer, and filtered through Whtman GF/A glass fiber paper. An aliquot of 10 mℓ of the filtrate was applied to a prewashed (one column-volume each of hexane, methanol, and water respectively) Bond-Elut SAX column. The column was washed with two column volumes of water and then eluted with 4.5 mℓ of 10% sodium chloride in 0.1 *M* sodium acetate solution. Aliquots of this solution, after appropriate dilution, were injected.
I-6. Aliquots (10 to 50 μℓ) of the plasma sample were injected directly onto the precolumn (40 × 4.6 mm packed with Nucleosol C_{18}, 5 μm) which was flushed with 50 m*M* phosphate buffer, pH 7 at 1 mℓ/min. After 50 sec, the column was swtched automatically to be in line with the analytical column and the mobile phase. In the meantime, another similar precolumn was rinsed with phosphate buffer to receive the next sample.

Elution — E-1. Acetonitrile-water (160:840) containing 13 mℓ of 10% tetrabutylammonium hydroxide + 50 mg Na_2 EDTA.
E-2. Methanol-water (240:760) containing 7.5 mℓ of a 40% tetrabutylammonium hydroxide, 2.04 g of KH_2PO_4 and 7 mℓ of 3 *N* phosphoric acid, pH 7.
E-3. Methanol-0.1 *M* phosphate buffer, pH 4.5 (200:800) containing 5 mmol of tetrabutylammonium phosphate.
E-4. (A). 10 m*M* Tetrabutylammonium phosphate, pH 7.55; (B) Water-95% ethanol (50:50) containing 10 m*M* tetrabutylammonium phosphate. A concave gradient #7 for 50 min from 15% (B) to 25% (B).

FOLIC ACID (continued)

E-5. Methanol-0.01 *M* ammonium dihydrogen phosphate + 5 m*M* tetrabutylammonium phosphate (20:80).
E-6. Acetonitrile-0.1 *M* acetate buffer, pH 5.7 (6:94).
E-7. (A) 5 m*M* Tetrabutylammonium phosphate, pH 7.4; (B) methanol-water (30:70) containing 5 m*M* tetrabutylammonium phosphate. A linear gradient over 45 min from 50% (A) to 0% (A).
E-8. Methanol-25 m*M* phosphate buffer, pH 7 (5:95).

REFERENCES

1. **Holcomb. I. J. and Fusari, S. A.,** Liquid chromatographic determination of folic acid in multivitamin-mineral preparations, *Anal. Chem.*, 53, 607, 1981.
2. **Tafolla, W. H., Sarapu, A. C., and Dukes, G. R.,** Rapid and specific high-pressure liquid chromatographic assay for folic acid in multivitamin-mineral pharmaceutical preparations, *J. Pharm.Sci.*, 70, 1273, 1981.
3. **Giulidori, P., Galli-Kienle, M., and Stramentionoli, G.,** Liquid-chromatographic monitoring of 5-methyltetrahydrofolate in plasma, *Clin. Chem.*, 27, 2041, 1981.
4. **Horne, D. W., Briggs, W. T., and Wagner, C.,** High-pressure liquid chromatographic separation of the naturally occurring folic acid monoglutamate derivatives, *Anal. Biochem.*, 116, 393, 1981.
5. **Duch, D. S., Bowers, S. W., and Nichol, C. A.,** Analysis of folate cofactor levels in tissues using high-performance liquid chromatography, *Anal. Biochem.*, 130, 385, 1983.
6. **Schieffer, G. W., Wheeler, G. P., and Cimino, C. O.,** Determination of folic acid in commercial diets by anion-exchange solid-phase extraction and subsequent reversed-phase HPLC, *J. Liq. Chromatogr.*, 7, 2659, 1984.
7. **Schulz, A., Wiedemann, K., and Bitsch, I.,** Stabilization of 5-methyltetrahydrofolic acid and subsequent analysis by reversed-phase high-performance liquid chromatography, *J. Chromatogr.*, 328, 417, 1985.
8. **Wegner, C., Trotz, M., and Nau, H.,** Direct determination of folate monoglutamates in plasma by high-performance liquid chromatography using an automatic precolumn-switching system as sample clean-up procedure, *J. Chromatogr.*, 378, 55, 1986.

FOLINIC ACID

Liquid Chromatography

Specimen (mℓ)	Extraction	Column (cm × mm)	Packing (μm)	Elution	Flow (mℓ/min)	Det. (nm)	RT (min)	Internal standard (RT)	Other compounds (RT)	Ref.
Serum (1)	I-1	10 × 8	Radial-Pak-C_{18} (10)[a]	E-1	3.0	Electrochem[b]	8	—	—	1

[a] Column temp = 40°C.
[b] Potential = 0.7 V.

Extraction — I-1. The sample was applied to a prewashed (10 mℓ methanol, 10 mℓ phosphate buffer, pH 5.5) Sep-Pak C_{18} cartridge which was then washed with 3 mℓ of 5 m*M* phosphate buffer, pH 5.5, and finally eluted with 2 mℓ of methanol. The eluate was evaporated at 50°C under a stream of nitrogen. The residue was reconstituted with 500 μℓ of the mobile phase for injection.

Elution — E-1. Acetonitrile-methanol-0.5 *M* phosphate buffer, pH 3.5 (3:4:95) containing 10 m*M* EDTA.

REFERENCE

1. **Birmingham, B. K. and Greene, D. S.,** Analysis of folinic acid in human serum using high-performance liquid chromatography with amperometric detection, *J. Pharm. Sci.*, 72, 1306, 1983.

FORMOTEROL

Gas Chromatography

Specimen (mℓ)	Extraction	Column (m × mm)	Packing (mesh)	Oven temp (°C)	Gas (mℓ/min)	Det.	RT (min)	Internal standard (RT)	Deriv.	Other compounds (RT)	Ref.
Urine (2)	I-1	0.5 × 3	3% OV-1 Chromosorb-W (80/100)	240	He (30)	MS-EI	2	[$^2H_7$]-Formoterol	Penta-fluoro-propio-nyl; methyl[a]	—	1

[a] Pentafluoropropionyl derivatization of amino groups, and methyl derivatization of the phenolic group.

Extraction — I-1. The sample was incubated at 37°C for 20 min with β-glucuronidase at pH 7. The sample was spiked with 200 ng of the internal standard and mixed with 0.5 g of sodium bicarbonate and the mixture extracted with 4 mℓ of ethyl acetate. The organic layer was back extracted into 3 mℓ of 0.1 *N* HCl. The aqueous layer was made alkaline with 0.8 g of sodium bicarbonate and extracted with 4 mℓ of ethyl acetate. The organic layer was evaporated under reduced pressure. The residue was dissolved in 100 μℓ of 10% pyridine in dichloromethane and 250 μℓ of 25% pentafluoropropionic anhydride in dichloromethane. After 30 min at room temperature, the mixture was evaporated under nitrogen, the residue treated with 4 mℓ of water, 0.8 g of sodium bicarbonate, and extracted with 4 mℓ of ether. The organic layer was evaporated and the residue treated with 100 μℓ of ethereal diazomethane for 5 min. The solvent was removed, the residue dissolved in 50 μℓ of ethyl acetate, and 1 to 2 μℓ of the solution were injected.

REFERENCE

1. **Kamimura, H., Sasaki, H., Higuchi, S., and Shiobara, Y.,** Quantitative determination of the β-adrenoceptor stimulant formoterol in urine by gas chromatography mass spectrometry, *J. Chromatogr.*, 229, 337, 1982.

FORSKOLIN

Gas Chromatography

Specimen (mℓ)	Extraction	Column (m × mm)	Packing (mesh)	Oven temp (°C)	Gas (mℓ/min)	Det.	RT (min)	Internal standard (RT)	Deriv.	Other compounds (RT)	Ref.
Dosage	I-1	1.8 × 4	3% OV-1 GasChrom Q (100/120)	220	N_2 (20)	FID	21.6	Desacetylforskolin (18)	—	—	1

Extraction — I-1. A portion equivalent to one tablet of ground tablets was extracted with chloroform. The organic layer was evaporated. The residue was dissolved in 0.5 mℓ of chloroform and mixed with 0.5 mℓ of the internal standard solution (8 mg/mℓ). Aliquots of 1 μℓ of this solution were injected.

REFERENCE

1. **Inamdar, P. K., Dornauer, H., and de Souza, N. J.**, GLC method for assay of forskolin, a novel positive inotropic and blood pressure-lowering agent, *J. Pharm. Sci.*, 69, 1449, 1980.

FOSFOMYCIN

Gas Chromatography

Specimen (mℓ)	Extraction	Column (m × mm)	Packing (mesh)	Oven temp (°C)	Gas (mℓ/min)	Det.	RT (min)	Internal standard (RT)	Deriv.	Other compounds (RT)	Ref.
Serum, urine (0.25)	I-1	1.5 × 2	3% XE-60 Gas Chrom Q (100/120)	130	He (30)	MS-EI	4	Propenylphosphonic acid (2.5)	Trimethylsilyl	—	1

Extraction — I-1. The sample was treated with 1 mℓ of methanol containing (2 μg/mℓ) of the internal standard. A 100-μℓ aliquot of the supernatant was evaporated under nitrogen. The residue was treated with a 20 μℓ volume of the silylating reagent (bistrimethylsilylacetamide-dichloromethane, 1:1 + 5% trimethylchlorosilane). After 10 min at 60°C, 1 to 2 μℓ of the solution were injected.

REFERENCE

1. **Longo, A., Di Toro, M., Pagani, E., and Carenzi, A.,** Simple selected ion monitoring method for determination of fosfomycin in blood and urine, *J. Chromatogr.*, 224, 257, 1981.

FTORAFUR

Liquid Chromatography

Specimen (mℓ)	Extraction	Column (cm × mm)	Packing (μm)	Elution	Flow (mℓ/min)	Det. (nm)	RT (min)	Internal standard (RT)	Other compounds (RT)	Ref.
Plasma (0.5)	I-1	25 × 6.2	Zorbax SIL (NA)	E-1	1.7	ABS (254)	5.5	—	5-Fluorouracil[a] Uracil	1

[a] Determined by GC-MS.

Extraction — I-1. The sample was adjusted to pH 4 with 5 *N* HCl and extracted twice with 20-mℓ portions of chloroform. The combined chloroform extracts were evaporated at 25°C under nitrogen. The residue was dissolved in 100 μℓ of ethylene dichloride and an aliquot of 20 μℓ was injected.

Elution — E-1. Ethylene dichloride-ethanol (24:1).

REFERENCE

1. **Marunaka, T., Umeno, Y., Yoshida, K., Nagamachi, M., Minami, Y., and Fujii, S.,** High-pressure liquid chromatographic determination of ftorafur[1-(tetrahydro-2-furanyl)-5-fluorouracil] and GLC-mass spectrometric determination of 5-fluorouracil and uracil in biological materials after oral administration of uracil plus ftorafur, *J. Pharm. Sci.*, 69, 1296, 1980.

FURAZOLIDONE

Liquid Chromatography

Specimen (mℓ)	Extraction	Column (cm × mm)	Packing (μm)	Elution	Flow (mℓ/min)	Det. (nm)	RT (min)	Internal standard (RT)	Other compounds (RT)	Ref.
Tissue (50 g)	I-1	25 × 4.6	Hypersyl SAS (NA)[a]	E-1	1.8	ABS (360)	5	—	—	1
Feed	I-2	25 × 4.6	LiChrosorb RP-18 (10)[b]	E-2	1.5	ABS (365)	6.5	—	—	2, 3

FURAZOLIDONE

Liquid Chromatography

Specimen (mℓ)	Extraction	Column (cm × mm)	Packing (μm)	Elution	Flow (mℓ/min)	Det. (nm)	RT (min)	Internal standard (RT)	Other compounds (RT)	Ref.
Tissue	I-3	15 × 4.6	Ultrasphere ODS (5)[c]	E-3	1.5	ABS (365)	5	—	—	4
Plasma (2)	I-4	25 × 4.6	Partisil ODS (10)	E-4	1.1	ABS[d] (362)	4.5	—	—	5

[a] Protected by a 100 × 2.1mm guard column packed with Perisorb RP-8 (30 to 40 μm).
[b] Protected by a 50 × 2mm guard column packed with Perisorb RP-18 (30 to 40 μm).
[c] Protected by a 35 × 4.2mm guard column packed with 30/44um Vydac ODS.
[d] An electrochemical detector at potential -0.75 was also evaluated.

Extraction — I-1. The sample was extracted by macerating it with 200 mℓ of ethyl acetate and 50 g of anhydrous sodium sulphate. After centrifugation, a 100-mℓ volume of the extract was evaporated under vacuum. The residue was dissolved in 2 mℓ of acetonitrile, warmed to 40°C, and immediately cooled and filtered through a G-2 glass filter. Aliquots of 20 μℓ of this solution were injected.

I-2. The sample containing about 0.5 mg of the drug was extracted overnight in a Goldfisch extractor with 50 mℓ of acetone containing 3 mℓ of water. The extract was evaporated on a steam bath, residue dissolved in 5 mℓ of dimethylformamide, and the solution was treated with 5 mℓ of 5% tetraethylammonium bromide in water. After centrifugation the top layer of fat was removed. The solution was then filtered and aliquots of 20 μℓ of the filtrate were injected.

I-3. The frozen tissue sample was homogenized with 100 mℓ of dichloromethane. The organic extract was evaporated at 30°C under vacuum. The residue was dissolved in hexane and back extracted into 0.01 *M* acetic acid. The aqueous layer was washed with hexane and then extracted with dichloromethane. The organic layer was evaporated at 30°C under a stream of nitrogen. The residue was dissolved in 1 mℓ of 30% methyl sulfoxide-0.01 *M* acetic acid for injection.

I-4. The sample was mixed with 100 mg of sodium chloride and extracted with 5 mℓ of ethyl acetate. The organic layer was evaporated under nitrogen. The residue was dissolved in 200 μℓ of the mobile phase for injection.

Elution — E-1. Acetonitrile-water (25:75).
E-2. Acetonitrile-2% acetic acid (20:80).
E-3. Methanol-0.01 *M* acetic acid, pH 5 (30:70).
E-4. Methanol-water (30:70) containing 16.5 m*M* phosphate-citrate buffer, pH 4.0.

REFERENCES

1. **Ernst, G. F. and Van Der Kaaden, A.,** High-performance liquid chromatographic analysis of furazolidone in liver and kidney, *J. Chromatogr.*, 198, 526, 1980.
2. **Smallidge, R. L., Rowe, N. W., Wadgaonkar, N. D., and Stringham, R. W.,** High performance liquid chromatographic determination of furazolidone in feed and feed premixes, *J. Assoc. Off. Anal. Chem.*, 64, 1100, 1981.
3. **Smallidge, R. L.,** Liquid chromatographic method for determination of furazolidone in premixes and complete feeds: collaborative study, *J. Assoc. Off. Anal. Chem.*, 68, 1033, 1985.
4. **Winterlin, W., Hall, G., and Mourer, C.,** Ultra trace determination of furazolidone in turkey tissues by liquid partitioning and high performance liquid chromatography, *J. Assoc. Off. Anal. Chem.*, 64, 1055, 1981.
5. **Veale, H. S. and Harrington, G. W.,** Determination of furazolidone in swine plasma using liquid chromatography, *J. Chromatogr.*, 240, 230, 1982.

FUREGRELATE

Liquid Chromatography

Specimen (mℓ)	Extraction	Column (cm × mm)	Packing (μm)	Elution	Flow (mℓ/min)	Det. (nm)	RT (min)	Internal standard (RT)	Other compounds (RT)	Ref.
Serum, urine (0.1 — 1)	I-1	25 × 4.6	Supelcosil LC-18 (5)[a]	E-1	1.0	ABS (268)	14	5-(3′-Pyridinyl) benzofuran-2-carboxylic acid[b] (12)	—	1

[a] A 50 × 2.1mm guard column packed with Co:Pell ODS (35 μm) was used.
[b] A homologue of furegrelate, used for serum samples.

Extraction — I-1. The sample and 1 mℓ of 5 μg/mℓ of an aqueous solution of the internal standard were applied to a prewashed (3 mℓ acetonitrile, 5 mℓ aqueous 0.008 *M* tetrabutylammonium hydroxide) Sep-Pak C_{18} cartridge, which was then washed with 5 mℓ of water, 2 mℓ of acetonitrile-water (1:9), and finally eluted with 2 mℓ of 30:70 acetonitrile-water. Aliquots of the eluate were injected. Urine samples (50 μℓ) were mixed with 2 mℓ of the mobile phase containing propiophenone as the internal standard, filtered and injected.

Elution — E-1. Acetonitrile-water (30:70) containing 0.008 *M* tetrabutylammonium hydroxide, pH 6.

REFERENCE

1. **Lakings, D. B. and Friis, J. M.,** Liquid chromatographic-ultraviolet methods for furegrelate in serum and urine: Preliminary pharmacokinetic evaluation in the dog, *J. Pharm. Sci.*, 74, 455, 1985.

FUROSEMIDE

Liquid Chromatography

Specimen (mℓ)	Extraction	Column (cm × mm)	Packing (μm)	Elution	Flow (mℓ/min)	Det. (nm)	RT (min)	Internal standard (RT)	Other compounds (RT)	Ref.
Plasma (0.2)	I-1	30 × 4.6	μ-Bondapak-C_{18} (10)[a]	E-1	2.0	Fl (233, 389)	4.5	N-Benzyl-4-chloro-5-sulfamoylanthranilic acid[b] (7.8)	—	1
Plasma (0.5)	I-2	15 × 4.6	LiChrosorb RP-8 (5)[c]	E-2	1.0	Fl (275, 410)	4.2	6-Desmethylnaproxen (6)	d	2
Plasma (0.2)	I-3	25 × 4.6	Brownlee RP-8 (10)[e]	E-3	0.5	ABS (275)	5	*p*-Nitrophenol (16)	—	3
Blood, plasma, urine (2)	I-4	25 × 4.6	Supelcosil LC-18DB (5)	E-4	2.0	Fl[f] (254, 375)	10.2	—	—	4
Plasma, urine (0.5)	I-5	15 × 4.6	Zorbax ODS (5)[g]	E-5	3.0	Fl (235, 389)	3	6-Desmethylnaproxen (6)	—	5
Plasma (0.5)	I-6	50 × 2	Hitachi Gel[h] 3011 (10)[i]	E-6	0.2	Fl (268, 410)	16	Piretanide (24)	—	6
Plasma (0.2)	I-7	30 × 4	μ-Bondapak-C_{18} (10)[j]	E-7	2.0	ABS (254, 280)	5	Phenobarbital (7)	—	7
Plasma (1)	I-8	15 × 4.6	Ultrasphere-ODS (5)	E-8	1.4	Fl (330, 460)	6.5	N-Benzyl-4-chloro-5-sulfamoylanthranilic acid (15)	—	8

Thin-Layer Chromatography

Specimen (mℓ)	Extraction	Plate (Manufacturer)	Layer (mm)	Solvent	Post-separation treatment	Det (nm)	Rf	Internal standard (Rf)	Other compounds (Rf)	Ref.
Plasma (0.1—0.5)	I-9	20 × 20 (Merck)	Silica gel (0.25)	S-1	D: Propylene glycol (45 mℓ) + acetic acid (25 mℓ) + water (130 mℓ)	Fl (Reflectance) (275, k)	0.30	—	4-Chloro-5-sulfamoylanthranolic acid (0.15)	9

[a] A Brownlee (30 × 3.9 mm) guard column packed with a C_{18} 10 μm material was used.
[b] An alternative internal standard (3-butylamino-4-phenoxy-5-sulfamoylbenzoic acid) was also used.
[c] Column temp = 35°C.
[d] Separation of degradation products of furosemide has been investigated.
[e] A 50 × 4.5mm guard column packed with the same material as of the analytical column was used.
[f] An absorbance detector at 254 nm and 280 nm was also used.
[g] Column temp = 45°C.
[h] Styrene-divinylbenzene resin.
[i] Column temp = 30°C.
[j] A 20 × 4mm guard column packed with Corasil C_{18} was used.
[k] A UV exclusion filter was used.

Extraction — I-1. The sample was mixed with 30 μℓ of acetonitrile containing 2.06 μg/mℓ of the internal standard and then an additional 0.4 mℓ of acetonitrile was added. The supernatant was mixed wth 100 μℓ of 0.08 *M* phosphoric acid. An aliquot of 300 μℓ of this solution was injected.
I-2. The sample was mixed with 100 μℓ of a methanolic solution of the internal standard (62.5 μg/mℓ) and 0.5 mℓ of 8.5 *M* acetic acid. The mixture was extracted with 5 mℓ of diethyl-ether-*n*-hexane (65:35). The organic layer was evaporated at 30°C with a nitrogen stream. The residue was dissolved in 0.6 mℓ of methanol, 0.4 mℓ of 0.01 *M* $NaHCO_3$ added and an aliquot of 10 μℓ was injected. The extraction procedure was carried out in subdued light.
I-3. The sample was spiked with 100 ng of the internal standard, pH adjusted to 2 with 4 *N* HCl and then extracted with 1 mℓ of ether. The ether layer was evaporated under a stream of nitrogen. The residue was dissolved in 100 μℓ of methanol and 10-μℓ aliquots were injected.
I-4. The sample was treated with 0.3 mℓ of 4 *N* HCl, 2 mℓ of water, and 2 mℓ of 10% sodium tungstate. The supernatant was extracted twice with 5-mℓ portions of ethyl acetate. The combined extracts were evaporated in the dark. The residue was dissolved in 0.5 mℓ of methanol for injection.

FUROSEMIDE (continued)

I-5. The sample was acidified with 50 μℓ of 8.5 *M* acetic acid and extracted with 5 mℓ of dichloromethane containing (20 ng/mℓ) the internal standard. The extract was evaporated at 30°C under a stream of nitrogen. The residue was reconstituted with 100 μℓ of methanol and an aliquot of 10 μℓ was injected. Reduction of light during extraction procedure was not required.
I-6. The sample was mixed with 100 μℓ of the internal standard solution (60 μg/mℓ in ethanol), 2 mℓ of 1 *M*, pH 1.5, phosphate buffer. The mixture was extracted with 11 mℓ of dichloromethane. The organic layer was back extracted into 2.5 mℓ of 0.001 *M* sodium hydroxide. An aliquot of 2 mℓ of the aqueous layer was mixed with 2 mℓ of phosphate buffer and extracted with 11 mℓ of dichloromethane. An aliquot of 10 mℓ of the organic layer was evaporated at room temperature under vacuum. The dried residue was dissolved in 100 μℓ of ethanol and an aliquot of 20 μℓ of this solution was injected.
I-7. The sample was mixed with 20 μℓ of an aqueous solution of the internal standard (120 mg/ℓ) and filtered through a Millipore sample clarification kit. An aliquot of 20 μℓ of the filtrate was injected.
I-8. The sample was mixed with 10 μℓ of a methanolic solution of the internal standard (2.5 μg/mℓ) and 100 μℓ of 6 *M* HCl. The mixture was extracted with 5 mℓ of anhydrous diethyl ether. A 4-mℓ aliquot of the ether phase was evaporated under a stream of nitrogen. The residue was dissolved in 250 μℓ of 0.02 *M* glycine buffer, pH 11 and a 100-μℓ aliquot of this solution was injected.
I-9. The sample was mixed with 0.2 mℓ of phosphate buffer and extracted twice with 2-mℓ portions of diethyl ether. The combined ether extracts were evaporated at 35°C under a stream of nitrogen. The residue was dissolved in 100 μℓ of methanol and aliquots of 20 μℓ were spotted.

Elution — E-1. Acetonitrile-0.08 *M* phosphoric acid (37.5:62.5).
E-2. Methanol-0.02 *M* phosphate buffer, pH 3 (1:1).
E-3. Methanol-water (25:75).
E-4. Acetonitrile-10% acetic acid (20:80).
E-5. Methanol-0.01 *M* phosphate buffer, pH 3.5 (35:65).
E-6. Ethanol-0.02 *M* $HClO_4$-$NaClO_4$, pH 2 (65:35).
E-7. Acetonitrile-0.02 *M* acetate buffer, pH 5.3 (78:22).
E-8. Methanol-water-acetic acid (40:57:3).

Solvent — S-1. Chloroform-methanol-acetic acid (89:6:5).

REFERENCES

1. **Rapaka, R. S., Roth, J., Viswanathan, Ct., Goehl, T. J., Prasad, V. K., and Cabana, B. E.,** Improved method for the analysis of furosemide in plasma by high-performance liquid chromatography, *J. Chromatogr.*, 227, 463, 1982.
2. **Kerremans, A. L. M., Tan, Y., Van Ginneken, C. A. M., and Gribnau, F. W. J.,** Specimen handling and high-performance liquid chromatographic determination of furosemide, *J. Chromatogr.*, 229, 129, 1982.
3. **Snedden, W., Sharma, J. N., and Fernandez, P. G.,** A sensitive assay method of furosemide in plasma and urine by high-performance liquid chromatography, *Ther. Drug. Monit.*, 4, 381, 1982.
4. **Ray, A. C., Tanksley, T. D., LaRue, D. C., and Reagor, J. C.,** Analytical evaluation of urinary excretion of furosemide in barrows, *Am. J. Vet. Res.*, 45, 1460, 1984.

5. **Lovett, L. J., Nygard, G., Dura, P., and Khalil, S. K. W.,** An improved HPLC method for the determination of furosemide in plasma and urine, *J. Liq. Chromatogr.*, 8, 1611, 1985.
6. **Uchino, K., Isozaki, S., Saitoh, Y.,Nakagawa, F., Tamura, Z., and Tanaka, N.,** Quantitative determination of furosemide in plasma, plasma water, urine and ascites fluid by high-performance liquid chromatography, *J. Chromatogr.*, 308, 241, 1984.
7. **Guermouche, S., Guermouche, M. H., Mansouri, M., and Abed, L.,** Determination of furosemide in rat plasma using HPLC and liquid scintillation, *J. Pharm. Biomed. Anal.*, 3, 453, 1985.
8. **Bauza, M. T., Lesser, C. L., Johnson, J. T.,and Smith, R. V.,** Comparison of extraction and precipitation methods for the HPLC determination of furosemide in plasma and urine, *J. Pharm. Biomed. Anal.*, 3, 459, 1985.
9. **Wesley-Hadzija, B. and Mattocks, A. M.,** Thin-layer chromatographic determination of furosemide and 4-chloro-5-sulfamoyl anthranilic acid in plasma and urine, *J. Chromatogr.*, 229, 425, 1982.

INDEX

A

B

C

D

F

G

H

I

K

L

M

Q

R

S

T

U

V

W

Y

Z

Printed and bound by CPI Group (UK) Ltd, Croydon, CR0 4YY
22/10/2024
01777633-0016